Arid Land Irrigation and Ecological Management

Edited by

S.D. Singh

SCIENTIFIC®
PUBLISHERS

Published by:

SCIENTIFIC PUBLISHERS (INDIA)
5A, New Pali Road, P.O. Box 91
Jodhpur 342 001, India
E-mail: info@scientificpub.com
Website: www.scientificpubonline.com

Print 2026

ISBN: 978-81-72330-65-1
eISBN: 978-93-89061-96-3

Printed in India

Preface

Water limits production in the arid regions, whilst land is far in excess of what it could economically be irrigated. Therefore, the irrigated agriculture in arid areas must operate with the primary goal of maximizing production per unit use of water in the evapotranspiration (ET) process. ET is derived from irrigation, rainfall and water available in soil profile at the time of planting, whilst water losses from evaporation and deep percolation and water in residual storage at the time of harvest are the non-growth related non-ET water uses. The efficient water management is one that ensures the best use of water in the ET fraction of irrigation while keeping non-ET water uses at their minimum. To achieve this goal it is essential to acquaint the students of water management, besides those who are engaged in formulation, planning, implementation and monitoring of irrigation projects, with the new concepts and strategy of managing a limited water supply, most modern method of irrigation (trickle), optimal allocation of water to crops in accordance with the scientific knowledge of yield to water-use relationsships, timely termination of irrigation, choice of crops efficient in water-use and the like.

The new strategy of irrigation relates to deficit irrigation/extensive irrigation approach. Monitoring of deficit irigation requires a precise knowledge of the critical stages of plant growth whilst familiarization with extensive irrigation may bring about considerable increase in the total production apart from distribution of social benefits to as many farmers as is economically possible.

Attainment of optimal crop yields is possible when water use is optimally combined with better crop husbandry practices and nutrient management. A comprehensive knowledge of optimally combining these management practices with an efficient use of limited water supply is the integral part of irrigation management in the arid region. In addition, as the availability of nutrients to roots is directly proportional to soil water content, a chapter on nutrient management in response to irrigation is most relevant.

At times data on crop water requirements are needed, within a limited period, for planning the project operation. A chapter on meteorogical approaches for predicting irrigation needs is a concerted effort towards appraisal of various methods employed from time to time to compute water requirement and suggest strategy for the nineties.

Reviewers have reiterated irrigation scheduling in accord with the root zone soil moisture deficit. This is important in its own right. However, what is more important in arid region is the achievement of the largest yield per unit application of irrigation water. Fulfilment of this objective called for a chapter on the efficient use of water and limitations to increasing water-use efficiency.

A chapter on water for agriculture provides a global picture of water supplies to agriculture, collected from innumerable pieces of information published in as many periodicals and journals as are published in various countries of the world. It also addresses the problem of how water is facing the limits ?, ecological toll of irrigation, water in greenhouse agriculture and strategies for the nineties.

When water sees the signs of abuse, economic and ecological consequences are bound to occur. It may be seen that mounting pressure on water use caused by rapidly increasing population of man and livestock and growing demand of industires manifests itself in pollution, waterlogging, depletion of groundwater supplies, falling water tables culminating into damage to ecological system. Chapters 10 and 11 deal with these aspects of arid land irrigation. It may further be seen that aparat from being scarce, over 60% of the arid areas have the problem of groundwater salinity. A set of different soil, crop water and fertility management together with a track of salt dynamics is needed for combating desertification of the land irrigated with salt affected waters. Chapter 8 in this book would be a guide to those who are engaged in design and installation of most modern system of trickle irrigation.

As important as these scientific observations are the words of acknowledgement with which I shall conclude this preface. Mr Pavan Kumar, Scientific Publisher, Jodhpur, took a keen interest in the preparation and publishing of this book. To him, I owe a special word of thanks. I express appreciation and thanks to Dr. Jean Steiner, Bushland, Texas for his unstinted help in my quest for finding some of the contributors to this book.

January 1993 S.D. SINGH

Contributors

Numbers in parentheses indicate the pages on which the authors contribution begin.

Vincent F. Bralts, Department of Agricultural Engineering, Michigan State University, East Lansing, Michigan 48824 – 1323, USA (263).

M.D. Campbell, Batelle, 7777 PNL, Richland, WA 99352, USA (43).

Raj.K. Gupta, Central Soil Salinity Research Institute, Karnal 132 001, India (159).

R.J. Hanks, Department of Plants, Soils and Biometeorology, Utah State University, Logan, Utah 84322–4840, USA (43).

T.A. Howell, USDA / ARS, Conservation and Production Research Laboratory, P.O. Drawer 10, Bushland, Texas 79012, USA (77).

Suresh Kumar, Central Arid Zone Research Institute, Jodhpur 342 003, India (351).

David E. Miller, IAREC, Rt. 2, Box 2953-A, Prosser, WA 99350–9687, USA (199).

R.W. Miller, Department of Plants, Soils and Biometeorology, Utah State University, Logan, Utah 84322–4820, USA (317).

P.S. Minhas, Central Soil Salinity Research Institute, Karnal 132 001, India (159).

Sandra L. Postel, Worldwatch Institute, 1776 Massachusetts Avenue N.W., Washington, D.C. 20036, USA (1).

Vinod Shankar, Indian Grassland and Fodder Research Institute, Jhansi 284003, India (351).

S.D. Singh, 33 Subhash Nagar, Pal Road, Jodhpur 342 008, India (103).

Y.V. Singh, Central Arid Zone Research Institute, Jodhpur 342003, India (103).

Jean Steiner, USDA / ARS, Conservation and Production Research Laboratory, P.O. Drawer 10, Bushland, Texas, 79012, USA (77).

Claudio O. Stockle, Agricultural Engineering, Washington State University, Pullman, WA 99164-6120, USA (235).

Joseph M. Villar, Department of Meteorology and Soil Science, Politechnic University of Catalunya, Spain (235).

I- Pai Wu, University of Hawai, Honolulu, Hawai, USA (263).
Vladimir S. Zaletaev, Water Problems Institute, Russian Academy of Science, 10, Novaya Basmannaya 107078 Moscow, Russia (401).

Contents

1

Water For Agriculture : Facing The Limits

Sandra L. Postel

I. INTRODUCTION

Among the most ancient of technological activities, irrigation has underpinned the advancement of human societies for several thousand years. Western civilization sprung from the dry soils of Mesopotamia, where survival depended on waters of the Tigris and the Euphrates rivers to give life to land. In what is now Iran, an extensive system of tunnels called kanats channelled water by gravity from mountains to croplands as long as 2500 years ago. Early books of the Bible have referred several times to irrigation. And it was no coincidence that the perennial flows of two great rivers, namely the Nile and the Euphrates of the Biblical region broadly formed the boundaries of the promised land (Framji and Mahajan, 1969).

By enabling farmers to apply water when and where needed, irrigation has turned many of the earth's too hot, dry, and most fertile lands into important crop producing regions. For example, Egypt could grow virtually no crop without water drawn from the Nile or from underground aquifers. The Aral Sea basin and California's Central Valley--the fruit and vegetable baskets of the former Soviet Union and the United States respectively-could barely be cultivated without supplemental water supplies. Without irrigation, yields in the critical crop growing areas of northern

China, northwestern India, and the western U.S. Great Plains will reduce to one-third to one-half.

Although a steady expansion of irrigation has taken place over the centuries, it entered a phase of significant development about 1950. This state of development was possible following the growth of oil as additional source of energy, and a rapid rise in global economy that favoured large capital investments. The two together set conditions favourable for development and use of groundwater resources and construction of some 100 large dams all over the world. During the time period from 1950 until the late seventies irrigated area more than doubled. Today, farming accounts for some 70% of global water use (Leeden, 1975; Williams, 1983; Postel, 1984).

In recent years, however, a confluence of forces has begun to slow irrigation's expansion and to raise questions about agriculture's heavy claim on the world's rivers, streams, and underground aquifers. Rising costs for new water supplies and a host of environmental problems are making new projects harder to justify. In several parts of the world, water demands are fast approaching the limits of resources. Many areas could enter a period of chronic shortages during the nineties, including northern China, virtually all of northern Africa, pockets of India, Mexico, much of the Middle East, and the parts of the western United States. Where scarcities loom, cities and farms are beginning to compete for available water; when supplies tighten, farmers typically lose out.

Moreover, the long-term threat of global warming from the build up of greenhouse gases in the atmosphere casts ominous shadows over future water budgets. As temperatures rise and rainfall patterns shift with the onset of climate change, water supplies will increase in some areas and decrease in others. Whatever the outcome for particular regions, though, adjusting the global irrigation base to the change in water availability will be costly. It seems inevitable that for several decades, irrigation systems will be poorly matched to altered rainfall and redistributed water supplies, jeopardizing agriculture's ability to provide enough food for the world's growing population.

Moving rapidly from a vicious course of life to greater efficiency and equity in the use of water is the surest way to avert shortages and lessen ecological implications of irrigations. Most of the vast quantity of water diverted by and for farmers never benefits a crop, because the efficiency of widely used surface methods of irrigation hardly exceeds 50%. The technologies and knowhow exist to boost that figure substantially. What most needed are policies and incentives that foster efficiency instead of discouraging it (Postal, 1984; Rangeley, 1987).

Greater attention to small-scale irrigation, rainfed farming, and to better integrating water projects with other development goals could do much to minimize adverse effect of water scarcity on food production. Restoration of critical watersheds is needed to stabilize water cycles that have been badly disrupted by deforestation and denuding of the land. Stepped-up research into breeding strains of crops that are more salt and drought tolerants would help prepare for a likely future of increased shortages and hotter, drier climates in some important food- growing regions.

Crucial as they are, though, these measures are but stopgaps. Any hope for balancing the water budgets of many countries with rapidly growing populations hinges as much on slowing birthrates as it does on improving water productivity. In a country such as Egypt, with a population leaping by 1 million every eight months, modernizing irrigation systems is simply not enough. The struggle for a secure water future will not end until societies recognize water's natural limits and begin to bring human numbers and demands into line with them.

II. THE END OF AN ERA

The science of irrigation, grounded in the principles of hydraulics, took root about two centuries ago. In 1800, an estimated 8 million hectares were equipped with watering facilities. A number of large projects constructed during the nineteenth century, especially in India and what is now Pakistan, brought the total area irrigated in 1900 to about 48 million hectares (Framji and Mahajan, 1969).

As world population grew from 1.6 billion to more than 5 billion over the last 90 years, irrigation became a cornerstone of global food security. The higher yields farmers could get with controllable water supplies proved vital to feeding the millions added to our numbers each year, especially as opportunities to cultivate new land dwindled.

Worldwide irrigated area nearly doubled during the first half of this century, rising to 94 million hectares in 1950. An even faster surge since then has brought the total to about 250 million hectares. Today, one-third of the global harvest comes from the 17% of the world's cropland that is irrigated (Table 1.1). Many countries such as China, Egypt, India, Indonesia, Israel, Japan, North and South Korea, Pakistan, and Peru rely on such land for more than half their domestic food production (FAO, 1987; Rangeley, 1989).

Since the late seventies expansion has slowed down considerably. Whereas net irrigated area grew 2-4% per year during the sixties and seventies, the annual rate since 1979 has averaged about 1% - far less than the growth in world population (net irrigated area is the land estimated to have actually been irrigated in a particular year; it is less than the gross area that estimated to be equipped with irrigation facilities). Per capita irrigated area peaked at 47.9 ha per thousand people in 1978, and then fell to 45.2 ha in 1987, a drop of 5.6% (Fig. 1.1). Estimates compiled by the FAO show that only 16 million ha were added to the global irrigation base between 1980 and 1987, an average of 2.3 million ha per year - roughly half the rate of the seventies (FAO, 1988; U.N. DIESA, 1989).

This slow-down occurred for several reasons. Low prices for food commodities, comparatively high energy costs, and general economic conditions during much of the eighties discouraged agricultural investments. The logical presumption is that when crop prices rise, irrigation will pick up speed again. Unfortunately, this optimistic outlook ignores several important constraints.

International agencies reduced sharply the grant over the last decade. Total grant to 23 countries in Asia, North Africa, and the Middle East by the World Bank, the Asian Development Bank, the U.S. Agency for International Development, and the Japanese Overseas Economic Cooperation Fund fell more than 60% in real terms between 1977 and 1987 (Fig. 1.2). Given that large projects often take a decade or more to complete, this funding drop suggests that in a good deal of the Third World, growth in irrigated area will not quicken much in the nineties (Levine et al., 1988; Rosegrant, 1989).

Table 1.1. Gross irrigated area, top 15 countries and world, 1986 (FAO, 1987; Rangeley, 1989)

Country	Gross irrigated area[1] 1000 ha	Share of cropland that is irrigated %
India	55,000	33
China	46,600	48
Soviet union	21,000	9
United States	19,000	10
Pakistan	16,000	77
Indonesia	7,300	34
Iran	5,800	39
Mexico	5,300	21
Spain	3,300	16
Turkey	3,300	12
Egypt	3,200	100
Thailand	3,200	16
Italy	3,000	25
Japan	3,000	63
Romania	3,000	28
Other	52,200	9
World	250,200	17

[1] As used there, gross irrigated area is the area equipped with irrigation facilities; it implies nothing about cropping intensity.

A related cause of the slow-down is that the cost of adding irrigation capacity - building new dams, reservoirs, canals, and distribution channels - has been rising in many countries, making it harder to justify such investments on economic grounds. Engineers naturally exploited the best sites for water development

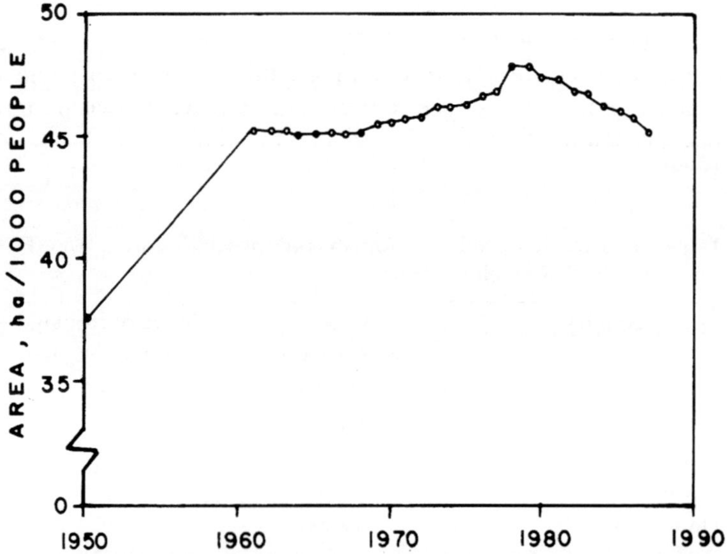

Fig. 1.1. World irrigated area per capita, 1950-87 (FAO, 1988; U.N. DIESA, 1989).

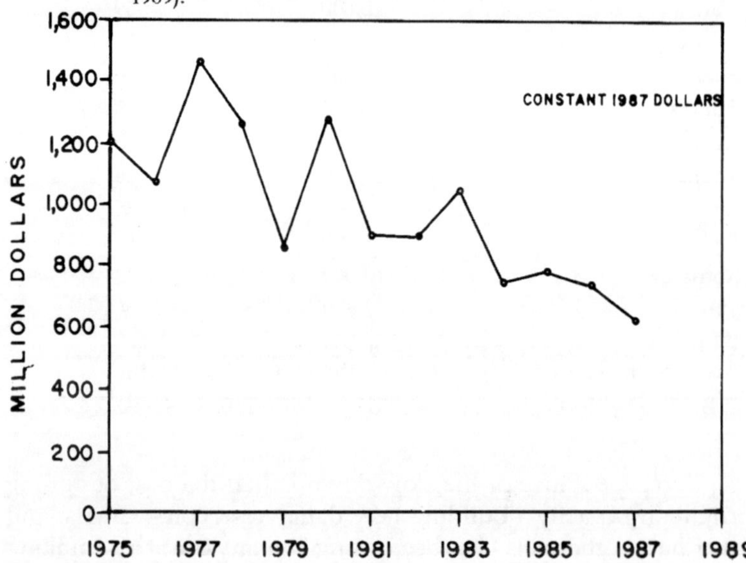

Fig. 1.2. Irrigation lending to 23 developing countries by major donors, 1975-87 (data for figure provided by M. Rosegrant, 1989).

first. Recent projects have been technically more complex, and thus more expensive. In India, for example, the cost in real terms of large canal schemes more than doubled between 1950 and 1980. Today, capital costs for new irrigation capacity ranges from a minimum of $1,500 to a maximum of $10,100 per ha in China, India, Indonesia, Pakistan, the Philippines, Thailand, Brazil, and Mexico (Svendsen, 1988; Gunaratnam, 1989; Repetto, 1986; Ginnsz, 1989; Rangeley, 1989; Yudelman, 1989), depending upon the prevailing rates of labour, materials, machineries, apart from transport facilities, distance of travel, soil types, topography, etc.

Developing new capacity is also most expensive in parts of Africa. Lack of roads and other infrastructure, the relatively small parcels of irrigable land, and the seasonal nature of river flows have driven per hectare costs of many projects to between $10,000 and $20,000, or even higher. Not even double cropping of high value crops can make irrigation systems at the top end of this spectrum economical. Consequently, expansion of irrigated area in Africa has been minimal since 1980, so just a 5% cropland of this continent (excluding South Africa) is irrigated (FAO, 1987; Rangeley, 1989; Scudder, 1989).

The debt burden hanging over many developing countries worsen the situation. Irrigated area in Mexico has actually declined since 1985, because of cuts in capital improvements due to bad economy. Officials in Brazil view accelerated irrigation as critical to their goal of achieving greater food self-sufficiency. But their ambitious targets for the early nineties seem unlikely to be met, given the required investments (Pogrell, 1989; Yudelman, 1989).

As costs rise and public investments decline, a larger share of what new irrigation does come on-line seems likely to support production of high value crops rather than rice, wheat, and other staples, especially if governments cannot afford to subsidize this farming heavily. Should grain prices rise following diminishing global stocks, a possibility during the nineties, investments will pick up again. As the poor suffer most from food shortages, they are unlikely to cultivate crops at the prices needed to inspire such investments. Neither scenario provides relief if irrigation is to enhance food security (Brown, 1988).

For the foreseeable future, therefore, irrigation's contribution will have to come more from improving existing systems than from expanding them to new land. This requires a fundamental shift in thinking : engineers are trained to design and build physical infrastructure, not to operate it effectively from the farmer's point of view. As Indian water analyst Ashok K. Mitra points out, "Irrigation engineers tend to view the output of the canal system as 'water' and not the additional crop which the water should allow". And many government irrigation agencies, intent on maintaining their large budgets, consider expansion to be their ultimate mission, rather than providing a tool to increase food output and raise rural incomes (Mitra, 1986).

There are vast discrepancies as to performance and quality of an irrigation system. As for example, in many large surface-water system less than half the water diverted by and for farmers benefits crops. Much is lost to seepage from unlined farm canals; deep percolation upon over, uneven, or untimely irrigations; evaporation from the soil surface; runoff from irrigation furrows; and use of water by weeds. As a result, efficiencies for large projects in Asia run about 30%. Not all of this water is wasted. Much of it returns to a nearby stream or joins underground supplies, and thus can be used again. In recycling, it picks up salts, fertilizers, and pesticides along the way, and becomes degraded in quality and contaminates water sources downstream (Rangeley, 1989).

Lack of maintenance further reduces the project efficiency. Over time, distribution channels fill with silt, and the likelihood of breach, outlets break, as well as salts build up in the soil that it would enhance. In China, for example, more than 930,000 ha of irrigated farmland have come out of production since 1980, an average loss of some 116,000 ha per year. Similarly, in the former Soviet Union - the world's third largest irrigator - poor performance brought 2.9 million ha out of cultivation in between 1971 and 1985, equal to a quarter of the new area brought under irrigation. World-wide, an estimated 150 million ha - 60% of the world's total irrigated area - need some form of upgrading to remain in good working order. An increasing share of available capital will thus go to rehabilitating existing systems. While this is wise and necessary, it will further diminish funding for new projects (Zhuoyan and Jinan, 1989; USAD/ERS, 1989; Jensen et al., 1989).

Poor management and inequities in the distribution of water are also to blame for irrigation's disappointing performance. Many systems deliver water on a fixed schedule that does not match the needs of crops. In other cases, the supply is so unreliable that farmers cannot risk investing in other yield-raising methods, such as planting better seeds, applying fertilizers, or levelling their land to help ensure efficient watering. As a result, output remains far below its potential. A study in Pakistan, for example, found that the average yields from irrigated fields of wheat, rice, and sugarcane were 60-70% below the upper range of yields some farmers achieved. This "yield gap" can largely be explained by poor management of the water supply and its influence on farmers' other management decisions (Ahmad, cited by Postel, 1989).

Farmers at the tailend of large canals suffer most from irrational distribution of available water supplies. Those who have their farms located at the head of a project area begin irrigating before the main canal and distribution channels reach the end. By cultivating high water requiring crops, such as sugarcane or rice, they stake a claim to more water than was planned for their farms. Often rich farmers with their better political influence continue to maintain heavy withdrawals even when the project is completed. Poor farmers at the end of the line end up without enough water for their crops.

One project in the Indian state of Maharashtra, for example, had sugarcane occupying 12% of the irrigated area when the design called for only 4%. Since the water needed for a hectare of sugarcane can irrigate about 10 ha of wheat, this practice causes much less area actually to be irrigated than was planned. Food production falls short of expectations, and the gap between rich and poor farmers widens (Mitra, 1986).

These factors collectively result in much of the world's existing irrigated area being underused, yielding far below its potential, and, in some cases, failing to enhance food and income security for those who most need it - the rural poor. To be sure, irrigation has greatly expanded world food output, alleviating hunger and improving diets in many areas. But as future gains get harder to come by, an important question becomes, irrigation for whom and for what ?

India, with more than a fifth of the world's irrigated area, perhaps best illustrates the need for a rethinking of ends and means. Analysts credit irrigation for much of the progress Indian agriculture has made over the last three decades. To extend those benefits, the nation seems to be in an all-out race to harness its entire irrigation potential, estimated at 113 million hectares, or roughly double the present area. But much of this existing area is not in use or is performing poorly. Yields of irrigated wheat and rice are only half what Indian agronomists say they could be. Improving water management so as to boost yields closer to their potential on existing irrigated lands could thus produce as much additional food as doubling the irrigated area - at less cost economically and environmentally (Desai, 1988; Dhawan, 1988; Svendsen, 1988).

Former Prime Minister Late Rajiv Gandhi himself bluntly criticized the country's track record in a speech to state irrigation ministers in 1986. Of the 246 large surface-irrigation projects started since 1951, only 65 had been completed, he said, and little benefit had come from projects begun after 1970. "For 16 years we have poured money out. The people have got nothing back, no irrigation, no water, no increase in production, no help in their daily life" (Quoted in Omar Sattaur, 1989).

As the era of easy water development comes to an end, so do the rapid crop production gains from global irrigation expansion. It now seems likely that net irrigated area will increase at an annual rate of only about 2 million ha, or 0.9%, for the near future, with most of the growth occurring in Bangladesh, Brazil, China, India, Nigeria, Turkey, and a few other countries. Since world population is growing at 1.8% per year, a dramatic boost in yields on both existing irrigated and rainfed lands will be needed for food production to keep pace with human numbers (PRB, 1989).

III. ECOLOGICAL TOLL OF IRRIGATION

Irrigation's ecological toll has been discussed in detail in Chapter 10. Since environmental implications of irrigations relate very much to "Water for Agriculture", a brief account of this aspect is also given out here in this chapter.

Each year, some 3,300 cubic kilometers of water - six times the annual flow of the Mississippi - are removed from the earth's rivers, streams, and underground aquifers to water crops. Practiced on such a scale, irrigation has had a profound impact on global water bodies and on the cropland receiving it. Waterlogged and salted lands, declining and contaminated aquifers, shrinking lakes and inland seas, and the destruction of aquatic habitats combine to hang on irrigation a high environmental price. Mounting concern about this damage is making large new water projects increasingly unacceptable. And ironically, the degradation of irrigated land from poor water management is forcing some land to go out of cultivation, offsetting a portion of the gains costly new projects are intended to yield.

By far the most pervasive damage stems from waterlogging and salinization of the soil. Without adequate drainage, seepage from unlined canals and overwatering of fields raises the underlying groundwater. Eventually, the root zone becomes waterlogged, starving plants of oxygen and inhibiting their growth. In dry climates, evaporation of water near the soil surface leads to a steady accumulation of salt that also reduces crop yields and eventually, if the build-up becomes excessive, kills the crops.

Even the best water supplies typically have concentrations of 200-500 parts per million. (For comparison, ocean water has a salinity of about 35,000 ppm; water with less than 1,000 ppm is considered fresh; and the recommended limit for drinking water in the United States is 500 ppm). Applying 100 ha-cm of water per year, a fairly typical irrigation rate, thus adds between 2 and 5 tons of salt to the soil annually. If it is not flushed out, enormous quantities can build-up in a few decades, greatly damaging the land. Aerial views of abandoned irrigated areas in the world's dry regions reveal vast expanses of glistening white salt, land so destroyed it is essentially useless (Kovda, 1983).

No one knows for sure how large an area suffers from salinization. Estimates for the world's top five irrigators clearly suggest, however, that the problem is sufficiently widespread to be reducing world crop output. In India, salinity reduces yields on some 20 million hectares, and an additional 7 million have been abandoned as salty wasteland (Bhumbla and Khare, undated). China has about 7 million hectares of saline and alkaline agricultural land (MWREP, 1987), and Pakistan has 3.2 million (NCA, 1988). In the United States, salt build-up lowers crop yields on 25-30% of the nation's irrigated land (Rhoades, 1989). Some 2.5 million ha are salinized in the former Soviet Union, most of it in the irrigated deserts of central Asia (Mickin, 1989). Assuming the same share of world irrigated area is damaged as the collective share in these top five, salinization is reducing yields on 24% of irrigated land worldwide.

Other areas severely affected include Afghanistan, Turkey, and the Tigris and the Euphrates river basins in Syria and Iraq. In Egygt, where virtually all cropland is irrigated, half is salinized enough to diminish yields. And in Mexico, salinization is estimated to be reducing crop output by the equivalent of 1 million tons of grain per year - enough to feed 5 million people, roughly a quarter of Mexico City (Yudelman, 1989).

In the western United States, excessive irrigation and poor drainage have produced another set of problems. There, scientists have linked alarming discoveries of death, grotesque deformities, and reproductive failure in fish, birds, and other wildlife to agricultural drainage water laced with toxic chemicals. Since 1985, intensive investigations throughout the region have found lethal or potentially hazardous concentrations of selenium at 22 different wildlife sites, including Kesterson National Wildlife Refuge, which has earned an unenviable reputation as the "Three Mile Island of irrigated agriculture". So far, four other sites show a pattern of death and deformity similar to that at Kesterson: the Tulare Basin, also in California's Central Valley; the Stillwater National Wildlife Refuge in Nevada, a haven for the tundra swan, avocet, bald eagle, and more than 160 bird species; the Ouray National Wildlife

Refuge in northeastern Utah; and the Kendrick Project near Casper, Wyoming (Marshall, 1986; Wilderness Society, 1988; Harris, 1989; Peterson, 1989).

Selenium, a natural element, is needed by humans and some other animals in very small amounts, but is highly poisonous at greater concentrations. Irrigation has washed more selenium and other dangerous chemicals out of the soil in several decades than natural rainfall would have done in centuries. Either stored in ponds to evaporate or channelled away from farmland by rivers or canals, this toxic drainage has become a serious hazard to western wildlife - and possibly to humans, as harmful substances move up the food chain (Harris, 1989; NRC, 1989).

Potential solutions to the drainage problem range in cost from \$30 million to more than \$100 million per year for some indefinite period in California alone. With such price tags and the high ecological stakes of continued irrigation, it seems that some of this land will come out of production over the next decade. How much, and in which states, remains to be seen (Harris, 1989; NRC, 1989).

Contamination of land and water by salts and toxic chemicals is only one indicator that some irrigation is unsustainable. In much of the world, falling water tables signal that groundwater withdrawals exceed the rate of replenishment. As water levels decline, overpumping eventually can make irrigation too costly to continue, and can even drain some aquifers dry. In either case, land is forced out of irrigation temporarily or permanently.

In the United States, more than 4 million hectares - roughly a fifth of the nation's irrigated area - are watered by pumping in excess of recharge. By the early eighties, the depletion was already particularly severe in Texas, California, Kansas, and Nebraska, four important food-producing states (Dickason, 1988, 1989). In Texas, water tables were falling at least 15 cm per year beneath 1.54 million hectares, 72% of the state's total irrigated area. During the late seventies and early eighties, falling groundwater levels combined with high energy costs and low commodity prices to pull much irrigated land out of production. The total area in Texas fell 30% between 1974, its peak, and 1987 (Tuck, 1989).

Current year-to-year fluctuations in the U.S. irrigated area reflect crop prices and government farm policies more than water availability and cost. Indeed, the 1989 total was likely the highest since 1982, with much of the recent growth occurring in the Southeast. Nonetheless, overpumping cannot continue indefinitely. The 4 million hectares currently watered unsustainably will eventually come out of irrigated production unless farmers reduce pumping to no more than the rate of aquifer recharge (Anon., 1989).

No other country has systematically assessed the extent of excessive groundwater use. But the situation is serious elsewhere as well, including China and India, two of the three other major food producers. Groundwater levels are falling up to a metre per year in parts of northern China, and heavy pumping in portions of the southern Indian state of Tamil Nadu reportedly dropped water tables as much as 25-30 meters in a decade. In the western state of Gujarat, overpumping by irrigators in the coastal districts has caused saltwater to invade the aquifer, contaminating village drinking supplies (Widstrand, 1980; Chengappa, 1986; Nickum and Dixon, 1989).

Irrigation schemes in some extremely dry regions of the world rely on water drawn from fossil aquifers - ancient water supplies, often deep underground, that get very little recharge. Withdrawals from such sources constitute water mining, one-time extractions from a nonrenewable reserve. While it is reasonable to put this water to productive use, much of it is being exhausted rapidly and wastefully (Margat and Saad, 1984).

The Ogallala Aquifer, for example, which stretches from southern South Dakota to northwestern Texas, has been heavily depleted in its southern portions. Overpumping in the Texas High Plains has diminished supplies there by about a fourth in just a few decades - with much of the water going to surplus crops. In its efforts to achieve food self-sufficiency, Libya plans to irrigate some 200,000-240,000 hectares with fossil waters lying deep in the deserts. Dubbed "The Great Man-made River Project", this scheme aims to pump water from deep wells and transfer it north through underground pipes to the Mediterranean coast with exorbitant cost (Gutentag et al., 1984; Sieb, 1985).

Signs of irrigation surpassing ecological limits are evident among the world's rivers, lakes, and streams, as well. By far the most notable is the shrinking Aral Sea in the central Asian republics of the former Soviet Union. Fully 95% of the Soviet cotton harvest is grown in this region, as well as a third of the country's fruits, a quarter of its vegetables, and 40% of its rice (Micklin, 1989).

Because of a dry climate, 90% of the croplands in these republics is irrigated. As production expanded during the sixties and seventies, increasing amounts of water were diverted from the region's two major rivers, the Amu Dar'ya and Syr Dar'ya - the only sources of replenishment for the Aral Sea other than meager rainfall. By 1980, flows in these rivers' lower stretches had been reduced to a trickle, and the sea had contracted markedly (Volftsun et al., 1988; Micklin, 1989) as seen from Fig. 1.3.

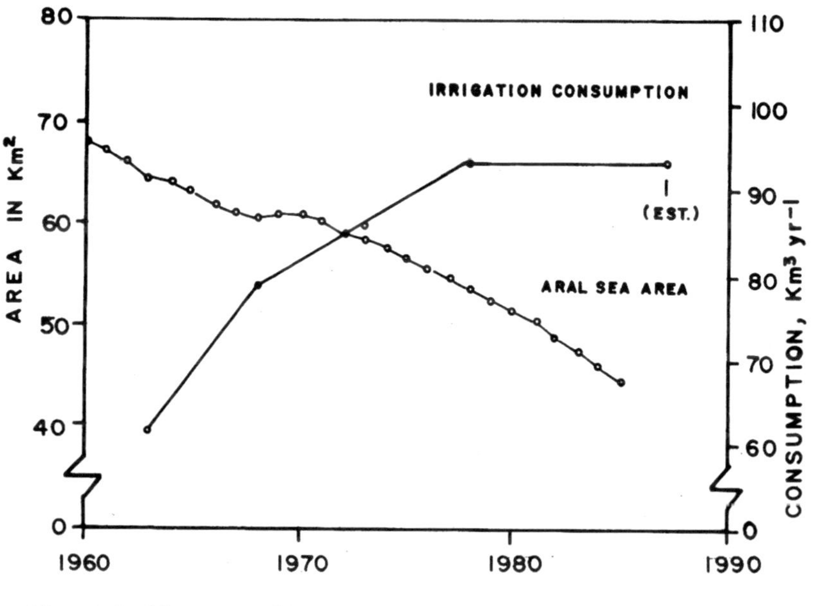

Fig. 1.3. Aral Sea area and irrigation consumption in Aral Sea basin, 1960-87 (Volftsun et al., 1988; Micklin, 1989).

The Aral's surface area has shrunk by more than 40% since 1960, volume has dropped by two-thirds, and salinity levels have risen threefold. All native fish species have disappeared, devastating the region's fishing industry. Winds pick up salt from the dry seabed and annually dump 43 million tons of it on surrounding cropland, damaging harvests. If no corrective measures are taken, by the end of this decade the Aral - once the world's fourth largest freshwater lake - "could consist of a main body in the south with a salinity well above the ocean and several small brine lakes in the north (Micklin, 1989).

Saving the Aral, albeit in some diminished form, was a high priority in the former Soviet Union. A long-standing plan to divert water from north-flowing Siberian rivers southwest into central Asia has been shelved, at least for the time being, because of concerns over the scheme's environmental effects and its exorbitant cost (Micklin, 1989).

Planners believe that curtailing the use of water for irrigation in the Aral Sea basin will be sufficient to preserve the sea. In a September 1988 decree, the then Communist Party Central Committee laid out a strategy aimed at raising average flows into the sea threefold over average levels during the eighties by the year 2005, largely by making irrigation systems more efficient. The plan outlay is high, and whether the needed investments materialize remains to be seen. It may well turn out that preventing further ecological havoc will require taking some of the then Soviet central Asia's prized farmland out of production (Perera, 1988; Micklin, 1989).

All this visible damage from large-scale irrigation has spawned strong opposition to new dams and diversion projects, even in the Third World, where water development remains a high priority. In February 1989, 10,000 villagers demonstrated against the Sardar Sarovar Dam, a centerpiece of India's Narmada Valley Development Project, which encompasses 30 large dams, 135 medium-sized ones, and 3,000 small ones. Sardar Sarovar alone would flood nearly 40,000 hectares and displace up to 100,000 people. Government officials are pushing ahead, in part with a World Bank loan, even though the dam's ecological impact has not been adequately investigated. Nor, apparently, have studies been completed on how to manage the watershed to prevent erosion from quickly silting up the planned reservoir and canals (Sattaur, 1989; Udall, 1989).

In Africa, similarly, planners often overestimate the benefits of large projects and underestimate social and environmental costs. Development anthropologist Thayer Scudder points out that large irrigation schemes have tended to benefit a small minority there while often destroying local production systems, such as flood-water farming and fisheries, that are vital to the poorer majority. Following completion of the Kainji Dam in northern Nigeria, fish catches and dry-season harvests from traditional floodplain agriculture dropped by more than half. A similar pattern was noted when the Bakolori Dam came on-line. And after the nearby Taiga Dam was completed, fish and floodplain harvests reportedly dropped even more than during the disastrous drought in the early seventies (Scudder, 1989).

In the United States, where the era of dam-building has largely ended and where water in many river basins is already overallocated, concerns center around the loss of free-flowing rivers, the destruction of fisheries from stream-flow depletion, and damage to riverine and other wildlife habitat. Efforts to restore and protect natural ecosystems could involve shifting water away from agriculture, by far the biggest consumer in the West.

Heightened awareness about these environmental effects could also upset the renewal of federal irrigation contracts, hundreds of which come due in the next couple of decades. The U.S. Environmental Protection Agency, backed by the President's Council on Environmental Quality, contends that extending at least some of these contracts requires the preparation of an Environmental Impact Statement under the National Environmental Policy Act. A dozen environmental and fishing groups have taken the federal Bureau of Reclamation to court, challenging its authority to renew contracts without analyzing irrigation's environmental damage. Should thorough impact studies be required, farmers would almost certainly be forced to reduce harm to wildlife, damage to the land, and toxic contamination generally - steps that, if too costly, could pull more land out of irrigation (NRDC, 1988; U.S. Water News, May and August, 1989).

IV. SCARCITY, COMPETITION, AND CONFLICT

Regional shortages cropping up around the world form another growing threat to agriculture's claim on water resources (Table 1.2).

Table 1.2. Water scarcity, selected countries and regions (Source: Worldwatch Institute).

Country/ Region	Observation
Africa, North and East	Ten countries are likely to experience severe water stress by 2000; Egypt, already near its limits, could lose vital supplies from the Nile as upper-basin countries develop the river's headwaters.
China	Fifty cities face acute shortages; water tables beneath Beijing are dropping 1-2 m per year; farmers in the Beijing region could lose 30-40% of their supplies to domestic and industrial uses.
India	Tens of thousands of villages throughout India now face shortages; plans to divert water from Brahmaputra River have heightened Bangladesh's fear of shortages; large portions of New Delhi have water only a few hours a day.
Mexico	Groundwater pumping in parts of the valley containing Mexico City exceeds recharge by 40%, causing land to subside; few options exist to import more freshwater.
Middle East	With Israel, Jordan, and the West Bank expected to be using all renewable sources by 1995, shortages are imminent; Syria could lose vital supplies when Turkey's massive Ataturk Dam comes on-line in 1992.
Soviet Union (former)	Depletion of river flows has caused volume of Aral Sea to drop by two-thirds since 1960; irrigation plans have been scaled back; high unemployment and deteriorating conditions have caused tens of thousands to leave the area.
United States	One-fifth of total irrigated area is watered by excessive pumping of groundwater; roughly half of western rivers are overappropriated; to augment supplies, cities are buying farmers' water rights.

Scarcities loom in a number of naturally water deficient areas where irrigation is critical to farming, often accounting for 80% or more of a region's total consumption. Many of these areas also are experiencing rapid population growth and urbanization - setting the stage for heated competition between cities and farms, and imposing constraints on agricultural development.

In much of northern and eastern Africa, expanding human numbers are on a collision course with scarce water resources. Swedish hydrologist Malin Falkenmark has found that societies typically experience "water stress" when annual renewable supplies approach 2,000 m^3 per person. By the end of the nineties, six out of seven East African countries (Burundi, Ethiopia, Kenya, Rwanda, Somalia, and Tanzania) (Falkenmark, 1989) and all five North African countries bordering the Mediterranean Sea (Algeria, Egypt, Libya, Morocco, and Tunisia) will exceed this measure of water scarcity. Six of these countries will have less than 1,000 m^3 per person. All of North Africa except Morocco already imports half or more of its grain. Under conditions of such severe water scarcity, boosting food self-sufficiency and meeting the domestic needs of populations growing on average at about 3% a year could well be impossible.

Egypt, where rain is sparse, may be the region's most extreme case. Its 55 million people depend almost entirely on the waters of the Nile, none of which originates within the desert nation's borders. Virtually all its cropland requires irrigation. Because of its burgeoning population, drinking-water and food needs are rising rapidly: water demands will likely exceed reliable supplies within a decade or so. Moreover, Ethiopia, which controls the headwaters of 80% of the Nile's flow,has plans to divert more upper-basin water for itself. Supplies downstream could thus diminish, making Egypt's situation dire (Starr and Stoll, 1987, 1988).

Worse, the Nile's flow can drop markedly during periods of low rainfall. During the summer of 1988, after several years of drought in the watershed, the Nile fell to its lowest level in a century. Fortunately, good rains since then have restored supplies. But the reprieve may not last long: the Nile's runoff pattern exhibits periods of low flow early in each century. Diminished supplies at a time of escalating demands could lead to a painful period of chronic water shortages (AID, 1987, 1988; Starr, 1989).

Just across the Suez Canal, the situation looks equally grim. Israel, Jordan, and Syria get most of their water from the Jordan River basin. Israel already uses 95% of the renewable supplies available to it, largely because of its sixfold expansion of irrigated area since 1948. The World Bank predicts that if current consumption patterns continue, water demands in Israel, Jordan, and the occupied West Bank will exceed all renewable supplies within six years. Both Israel and Jordan may soon face the politically difficult choice of shifting water away from farmers to supply growing domestic and industrial needs (Starr and Stoll, 1987, 1988; Cowell, 1989).

Competition over water is heightening in China as well, where dozens of cities already face acute shortages. The predicament is especially severe in and around Beijing, in the important industrial city of Tianjin, and in other portions of the North China Plain, a vast expanse of flat, fertile farmland that yields a quarter of the nation's grain. Water tables beneath the capital have been dropping 1-2 m per year; already, a third of its wells have gone dry (Nickum, 1987; Crook, 1988; Levied, 1989; Hong, 1989).

Most supplies on the North China Plain have already been tapped, yet demands continue to grow rapidly. At current rates of use, planners project Beijing's water needs to increase by 50% over the next decade. Demands in Tianjin, where water tables also have been falling precipitously, are expected to more than double. A long-awaited diversion of the Chang Jiang (Yangtze River) from central China to the North Plain would ease the crunch, but only slightly: even this grandiose scheme would meet only a fraction of Beijing's anticipated needs in the year 2000, still leaving a gap. Moreover, even if support for the project continues, it is probably a decade or more away from completion, and so will do nothing to alleviate shortages during the nineties.

Shifting water from farm to city use may be the only way to balance the region's water budget. A management study for Beijing suggests that farmers in the vicinity could lose 30-40% of their current supply within the next decade. In dry years, some already lose out: when the levels of Beijing's two major reservoirs fell to record lows in 1985, supplies to farmers not growing vegetables were cut off (Nickum, 1987; Nickum and Dixon, 1989).

India's water problems stem largely from gross mismanagement and degradation of its land. Deforestation and the denuding of critical watersheds have upset the natural water cycle, causing more rainfall to run off in damaging floods and less to percolate into the ground to recharge aquifers. According to Jayanto Bandyopadhyay of the Research Foundation for Science and Ecology in Dehra dun, this loss of recharge largely explains why high-rainfall areas now petition for drought relief. Tens of thousands of villages across the subcontinent are living with shortages, and their numbers are growing. Says Bandyopadhyay, "Water is only a renewable resource if we respect the ecological processes that maintain and give stability to the water cycle. This, India has signally failed to do (Bandyopadhyay, 1988, 1989).

In pockets of the western United States, a region plagued by numerous water deficits, supplies are being siphoned away from agriculture by thirsty cities willing to pay a premium to ensure water for their future growth. Where allocation systems and laws establish clear property rights to water, markets can operate to transfer supplies between willing buyers and sellers for an agreed-upon price. Marketing in the American West was quite common around the turn of the century, but the pace of transactions slowed when the government began building and subsidizing large water projects. As that era draws to a close, many western cities are finding that buying water from another user is an attractive alternative to developing costly new supplies.

If a farmer can earn more by selling water to a nearby city than by spraying it on cotton, alfalfa, or wheat, shifting that water from farm to city use is economically beneficial. If it prevents the city from damming another river to increase supplies, the transfer can also benefit the environment. Sale prices have varied greatly, with perpetual water rights having been sold for less than $16 per ha-cm in the Salt Lake City area, but for as high as $243 to $486 per ha-cm in the rapidly urbanizing Colorado Front Range (1,233 m^3 or an acre-foot is enough to supply a typical four-person U.S. household for about two years) (Zeilig, 1988).

Many large trades of this kind obviously could threaten crop production. But not all farm-to-city water transfers require shifting rights permanently out of agriculture. For example, the Metropolitan Water District of Southern California (MWD) - water wholesaler for roughly half the state's 28 million residents - has agreed to finance conservation projects in the neighbouring Imperial Irrigation District in exchange for the estimated 1.23 million ha-cm of water per year the investments will save. The annual cost per ha-cm conserved is estimated at $10, far lower than the MWD's best new-supply option. Enough water is being traded this way to meet the needs of 800,000 Californians, yet no cropland is being taken out of production and no irrigation water rights are actually being sold (Water Market Update, December 1988).

At the other extreme are transactions in Arizona, where Tucson, Phoenix, and other rapidly expanding cities have taken to "water ranching". State law makes it difficult to buy rights to water independent of the land. As of April 1988, some 232,800 hectares of agricultural land had been purchased as water farms. Some of it will stay in production until the cities actually need the water, but all of it will likely be retired eventually. In Pima County, where Tucson is located, irrigated agriculture is expected to disappear by 2020 (Checchio, 1988; Woodard et al., 1988).

The extent to which marketing leads to reductions in irrigated area will depend on how desperate for water cities become, and on how farmers respond. But whether marketing is encouraged or not, agriculture will likely lose supplies wherever demands are nearing water's natural limits. Almost everywhere, this resource's value in crop production is far less than in other activities. Planners in China, for example, calculate that a given amount of water used in industry generates more than 60 times the economic value of that same amount used in agriculture.

Growing concern about excessive stream-flow depletion could also enhance competition for farmers' water. In the United States, the doctrine of "prior appropriation" governing western water allocations traditionally assigned rights only to water withdrawn from a river or stream for some "beneficial use". Historically, diversions for agriculture, industry, and domestic purposes met this criterion, but no rights were granted for water left "instream" to preserve fish and wildlife habitat, recreational uses, and other ecological and esthetic values. In recent years, all the western states except New Mexico have taken steps to protect instream water values, including legal recognition of these functions as a beneficial use, and, in some cases, authorization for state agencies to buy water rights from private holders in order to boost stream-flow levels (Huffman, 1983; Metzger and Haverkamp, 1984; Jackson et al., 1989; U.S. Water News, May 1989).

Another powerful lever is the "public trust doctrine", which asserts that governments hold certain rights in trust for the people and can take steps to protect those rights from private interests. This doctrine has been invoked, for example, to protect northern California's Mono Lake, a beautiful water body that has been shrinking because of diversions from its tributaries to supply southern California with more water. If applied more broadly, this doctrine could have sweeping effects, since even existing water rights could be revoked to prevent violation of the public trust (Johnson, 1988).

IV. WATER IN THE GREENHOUSE WORLD

Beyond the unsettling prospects of rising irrigation costs, worsening environmental damage, and increasing water scarcity is climate change from the build-up of carbon dioxide and other greenhouse gases in the atmosphere. Despite its many uncertainties, global warming guarantees that the future will not be a simple extrapolation of the past. Changes in water availability, among the more profound transformations in store, add an entirely new dimension to the constraints irrigated agriculture faces during the next several decades.

Like a one-way filter, greenhouse gases allow the sun's energy to pass through the atmosphere but trap the heat waves radiated back from the earth. As a result of the emission of these gases from fossil fuel burning and other human activities, climatologists project that the earth's lower atmosphere will warm to the point where the hydrological cycle - the transfer of water between the sea, air, and land - will accelerate. Both precipitation and evaporation, which balance each other in the planetary water budget, are expected to increase by 7-15% when the heat-trapping effect of these gases doubles over preindustrial levels. Global averages, however, say little about what particular regions will experience. Temperatures will rise almost everywhere, but climate models predict a greater increase in the temperate and polar regions than in the tropics. Rainfall will increase in some areas, decrease in others (Schneider, 1989).

Runoff, the primary measure of a region's renewable surface water supply is a function of the amount of water gained through precipitation and lost through deep drainage, evaporation, interception, and transpiration by plants. Global warming will alter not only temperature and precipitation, but winds, humidity, and cloud cover - all influences on evapotranspiration and, thus, on runoff. Scientists have also found that plants grown in carbon-dioxide-enriched air partially close their stomata, through which they exchange water vapour and other gases with the atmosphere. This enables them to use water more efficiently. Studies have shown that doubling carbon dioxide levels can reduce transpiration by a third to half; across an entire watershed, this could augment runoff substantially. Scientists do not know, however, whether water efficiency would actually increase like this in natural environments, given the many other factors at work (Moss and Lins, 1989).

Much of the limited research done to explore global warming's impact on regional water supplies has focused on the western United States, a food-producing area of great importance that relies heavily on irrigation. A large share of the West's runoff derives from the melting of high-mountain snowpacks, which are in effect vast reservoirs that release their supplies gradually through the spring and summer. A good deal of western agriculture depends on this meltwater becoming available during the growing season. Moreover, engineers designed the numerous reservoirs in the West to meet hydropower, flood protection, and water supply needs assuming that historical runoff patterns would continue. Changes in these patterns would greatly complicate water management, and cost billions of dollars - either in increased flood damage, reduced hydropower, and more severe summer shortages, or in new reservoirs and other measures to avoid those consequences (Gleick, 1987, 1988).

The anticipated higher temperatures from the greenhouse effect will dramatically alter the timing of runoff in key parts of the West (Gleick, 1987; Schaake). It has also been found that a temperature increase of 2 degrees Celsius would reduce summer runoff in California's Sacramento River basin by 22%, a 4-degree increase, well within the predicted range, could lower summer runoff by 62% (Gleick, 1987).

Schaake analyzed changes in runoff for the Animas River at Durango, Colorado. He found that a 2-degree increase, with no change in precipitation, would affect total annual runoff very little. Seasonal runoff patterns, however, would change markedly as a result of reduced winter snowpack and faster melting of the snow that does fall. Schaake's model shows that average runoff for January through March would increase by 85%, while in the critical months of July through September it would drop by 40% (Table 1.3). Such drastic shifts would raise the risk of winter flooding, require costly trade-offs between hydropower and water supply needs, and lead to scarcities during the hot, dry summer, when water demands for crops and urban areas are greatest.

Table 1.3 : Changes in seasonal runoff from a 2-degree Celsius temperature increase, Animas River at Durango, Colorado (J.C.Schaake)

Season	Runoff in current climate 1000 ha-cm	Runoff in warmer climate 1000 ha-cm	Change %
January-March	493	912	+85
April-June	4081	4192	+3
July-Sept.	1529	925	–40
Oct.–Dec.	604	740	+22
Annual avg.	6708	6769	+1

To assess the possible impact for the entire West, Stockton and Boggess (1983) examined the effects on seven river basins of a 2-degree-Celsius increase and a 10% decrease in precipitation - within the range of possible futures, according to several leading climate models. They found that supplies in each of the seven basins dropped by 40-76%. In four of them - the Upper and Lower Colorado, the Missouri, and the Rio Grande - severe water deficits would likely arise. Since increasing scarcity has already shifted water out of agriculture, large reductions in available supplies would undoubtedly hasten that trend. Indeed, balancing the water budgets of these seven regions could require taking as many as 4.6 million hectares out of irrigation, nearly a quarter of the U.S. total.

Even as forces act to shift water away from farms, a greater share of cropland may need the protection from low soil-moisture offered by irrigation. Peterson and Keller have estimated that a rise of 3 degrees Celsius would push irrigation water needs up by 15%. If, in addition, rainfall increased by 10%, irrigation needs would rise only 7%. But the combination of the same temperature increase and a 10% drop in rainfall would up the need for irrigation water by 26%.

With too little water available to expand irrigation in the West, cropland needing additional water for profitable production but unable to get it would likely come out of cultivation altogether. Peterson and Keller estimate that cultivated area in the western states could shrink by 24, 15, and 31% under the three different scenarios.

A look at the effects of a real-life water shortage in the West confirms these computer predictions that climate change's impact on water resources will have major consequences for agriculture. During 1976 and 1977, California experienced its severest two-year drought since 1906. According to Gleick (1988), runoff in the state's river basins dropped to half of normal levels in 1976 and to less than a quarter in 1977. As a result, hydropower's contribution to total state electricity production fell by nearly two-thirds. Increased fossil fuel burning made up for these losses, at a cost to the state of $500 million. Groundwater levels declined by as much as 6 m in some location and added an estimated $25 million to pumping costs in the drought's second year (Gleick, 1988).

Gleick estimates that the water shortage could have reduced supplies to farmers in the San Joaquin Valley, the state's prime agricultural region, by 60%. Fortunately, state officials actively encouraged water exchanges during the drought, and enough was transferred to farmers to hold their supply cutbacks to 11%. A large portion of the water shifted to crop production came from the Metropolitan Water District (MWD) of Southern California. MWD stepped up imports from the distant Colorado River and gave San Joaquin Valley farmers a share of its allotment of state water. But this key fallback would likely disappear in the future. Colorado River water is now virtually fully appropriated, and the climatic changes that may diminish California's runoff would probably reduce those of the Colorado as well (Gleick, 1988).

With supplies as tight as they are in the West, any reduction in runoff will almost certainly cause land to come out of irrigation. A region can muddle through a two-year shortfall, but coping with long-term supply reductions of the magnitude global warming could bring would require more lasting and drastic adjustments. Putting the above findings together would suggest that 10-30% of existing irrigated lands in the western United States could be pulled out of production because of climatic changes by the year 2030.

Global warming's impacts on water and agriculture have not been studied as closely for other locations in the world. Yet for several reasons, climate change seems likely to act as an additional drag on expansion of the irrigation base over the next several decades. To be sure, some land will become suitable for irrigation, but even more will likely have to be retired.

It is probably conservative to assume that water scarcity induced by climate change will force irrigation to cease on 5% of the world's existing irrigated cropland. That would remove 13 million hectares from the global base. At today's rate of expansion, it would take half a decade just to recoup those losses. And building the dams, canals, and other infrastructure to regain that large an area would require supplemental investments on the order of $26 billion to $52 billion (Table 1.4).

In addition, some rainfed area will need irrigation to replace soil moisture lost to the atmosphere from the increased evapotranspiration that higher temperatures will bring. Peterson and Keller looked at the eastern United States, a region with little permanent irrigation now, and concluded that this increased need for new irrigation could be substantial. If 5% of the world's rainfed cropland requires irrigation to stay profitably productive, even after farmers switch crops and make other adjustments, some 60 million more hectares of irrigated area would be necessary. That would raise investments to adapt to a new climate by $120-240 billion, bringing the global total to roughly $150-300 billion (Gleick, 1988).

Whether such investments would materialize is far from certain. As noted earlier, lending for irrigation has been declining over the last decade. Perhaps more important, new investments will be hostage to uncertainties about just how the climate is changing in a given locale. Regional projections of soil moisture, evapotranspiration, runoff, and the other factors affecting water availability are not accurate enough to be the basis of important investment decisions. And it could be decades before climate models acquire that degree of precision.

On a brighter note, some existing irrigated areas could benefit from increased rainfall that boosts regional water supplies. Models suggest such a possibility for India, for example, where irrigated crop production is currently constrained by water scarcity. Additional runoff could allow some irrigated lands to get optimal water amounts instead of the current limited quantities, helping raise crop yields. But if the greater availability of water comes in the form of intensified monsoons, as models also suggest could happen, it could be more of a curse than a blessing. Much of it would run off in catastrophic floods rather than augmenting soil moisture and stable supplies (Hansen, 1989).

On balance, it appears that as climate changes over the next 20-30 years, irrigation is unlikely to be able to provide the degree of crop protection and yield enhancement the world now counts on. For some time, a mismatch between the cropland needing more water and that getting it could dampen irrigation's contribution to the global food supply.

Table 1.4 Illustrative irrigation adjustments from global climate change (Worldwatch Institute)

Adjustment	Area affected (million ha)	Estimated investment cost[1] (billion dollars)
Shifting 5% of world's existing irrigated area because of changes in water availability	13	26-52
Expanding irrigation to 5% of existing rainfed cropland to compensate for increased evapotranspiration	60	120-240
Total	73	146-292

[1] Assumes range of \$2,000-4,000 h^{-1}; costs in some areas could be substantially higher. Future costs were not discounted.

V. STRATEGIES FOR THE NINETIES

Securing water to sustainably meet the world's growing food needs will not be easy. The slow-down in irrigation growth, mounting environmental damage from irrigation and from water projects generally, worsening regional scarcities, and the prospects of climate change all combine to severely constrain water use for crop production. Together these trends point to the need for a more creative and diverse approach to watering crops - one that focuses on raising water efficiency, on integrating irrigation more fully with basic development goals, and on improving water's productivity in rainfed farming.

Much of the pervasive overuse and mismanagement of water in agriculture stems from the near-universal failure to price it properly. Irrigation systems are often built, operated, and maintained by public agencies that charge next to nothing for these services. Farmers' fees in Pakistan, for example, cover only about

13% of the government's costs. Indeed, in most of the Third World, government revenues from irrigation average no more than 10-20% of the full cost of delivering the water. Such undercharging deprives agencies of the funds needed to maintain canals and other irrigation works adequately (Repetto, 1986).

Pricing reforms are equally needed in the United States. The federal Bureau of Reclamation supplies water to a quarter of the West's irrigated land - more than 4 million hectares - under long-term contracts (typically 40 years) at greatly subsidized prices. Farmers benefiting from the huge Central Valley Project in California, for example, have repaid only 5% of the project's costs over the last 40 years: $50 million out of $931 million. Largely because of this underpricing, one-third of the bureau's water irrigates hay, pasture, and other low-value forage crops that are fed to livestock, even though so many other higher-value activities need water (LeVeen and King, 1985; Moore and McGuckin, 1988).

This free ride in large part explains why so few western farmers invest in efficiency improvements. As noted earlier, hundreds of federal irrigation contracts will be coming up for renewal during the nineties. If the bureau seizes this opportunity to establish contracts that foster efficiency rather than discourage it, water stresses in the American West could ease measurably.

When prices reflect water's scarcity, when the costs of obtaining water go up, or when governments regulate its use, farmers begin to use the resource more prudently. Irrigators in Israel, parts of the former Soviet Union, and Texas have shown that they can boost their efficiencies dramatically by adopting modern technologies and better management practices.

Large sprinklers can be made more efficient, for instance, by vertical drop tubes attached to the sprinkler arm. Called low-energy precision application (LEPA), these systems deliver water closer to the ground and in large droplets, cutting evaporation losses. Farmers can also conserve by using lasers as a surveying aid to help level fields so that water gets distributed evenly, recycling used irrigation water, installing water economizing drip systems, and monitoring soil moisture so they irrigate only when necessary. The average amount of water applied per hectare by Texan farmers dropped 28% between 1974 and 1987 because they adopted LEPA and other efficiency measures. Similarly, upgrading irrigation systems has reduced per hectare withdrawals in the former Soviet Union by 30% since 1970 (Postel, 1985).

In many developing countries, improving the performance of canal systems is critical to boosting irrigated crop yields closer to their potential. Involving farmers in the layout, operation, and management of the systems and providing flexibility in the scheduling of water deliveries are effective steps towards this end. Studies in the Philippines have shown, for example, that when farmers actively participate, canals and other infrastructure function better, a greater proportion of the project area gets irrigated, and rice yields increase more (Reyes and Jopillo, 1986).

Variously called "irrigation associations" or "water user associations", organized groups of farmers who share a common water source can be instrumental in involving them in projects and in collecting fees. The Philippines National Irrigation Agency now receives no governmental funding for operating and maintaining irrigation systems: it depends on revenues from the irrigation associations. This has made the agency more responsive to farmers' needs and more apt to solicit their views. Similar approaches are followed in China and South Korea, where canal systems tend to perform quite well (Repetto, 1986; Wade, 1988).

Redesigning projects so that farmers get water when they want it rather than according to some fixed schedule is another key to increasing irrigation's productivity. In many cases, this requires bridging an unfortunate gap between irrigation ministries, who often view their mission as merely supplying water, and farmers, who can only make optimum use of water if they have some control over it. A reliable and flexible supply enables farmers to invest in fertilizer and other yield-raising agricultural inputs and to diversify cropping patterns, helping boost production and income (Keen, 1988).

Drawing on groundwater in conjunction with large surface-water systems is one way of doing this. In many areas, seepage from canals and watering of fields creates a large reservoir underground. If not drawn out or drained away, this can lead to waterlogging and salinization, as has happened in many large irrigation schemes. Encouraging private development of wells to use this untapped source, however, can help manage this problem while at the same time augmenting farmers' dependable supplies. Especially if the groundwater is drawn during the dry season, when surface waters are short, this dual approach can make water available year-round, allowing farmers to plant two or three crops a year. In Punjab, Andhra Pradesh, and other parts of India, for example, irrigators with groundwater wells produce 5-6 tons of grain per hectare, compared with 2.5-3 tons with canal- delivered surface water (Dhawan, 1988).

Greater attention to small-scale irrigation and techniques to make better use of rainwater can do much to increase food production at the village level - and to eliminate the need for expensive large dams with their attendant social and environmental costs. Smaller projects are decentralized, and focus on improving the productivity of water where it falls, rather than transporting it great distances. In contrast to the top-down character of large projects, they can be developed at the grassroots, inspiring local self-reliance. As local initiatives, they also tend to be more responsive to farmers' needs as they perceive them.

Simple techniques aimed at increasing soil moisture in the root zones of crops can markedly boost yields and make production more reliable. For instance, farmers can build check dams of earth and stone to capture runoff from hillsides, and then channel this water to their fields. Similar forms of "water harvesting" allowed some ancient farming cultures to thrive where annual rainfall averaged only 100 millimetres. Vetiver grass (*Vetiveria zizanioides*), known in its native India as khus, shows promise for increasing sustainable cropping on sloping land. When planted along the contours of a hillside, vetiver creates a vegetative barrier that slows runoff, allowing rainwater to seep into the soil. Sediment trapped behind it forms a terrace that further conserves soil and moisture for crops. Yields have increased by half over those in similar areas not using this technique (Postel, 1985; World Bank, 1987; Greenfield, 1988).

Although such down-to-earth practices rarely get the visibility and fanfare afforded a large, new dam, they nonetheless can produce the needed results. In Karnataka, India, for example, a watershed management effort involving some 600,000 hectares - nearly a third of the cultivated area - is based on low-cost techniques that farmers can use to increase soil moisture in their fields. Cropping intensity reportedly doubled in the project areas (Deccan Herald, 6 January 1989).

An analysis by the Environmental Defense Fund in Washington, D.C., shows that a variety of proven small-scale techniques collectively constitute a workable alternative to the irrigation component of the huge Sardar Sarovar Dam. Even the most expensive small-scale methods - which include small reservoirs to store rainfall, percolation tanks to replenish groundwater, and check dams and "microcatchments" to increase rainwater productivity - are estimated to cost less than half as much per hectare as irrigation with water from Sardar Sarovar. Besides being cheaper, these alternatives would allow water development benefits to be distributed more equitably, reduce construction time, afford farmers more control over their supply, and promote local employment.

Besides directly funding more small-scale projects, development organizations can help foster private initiatives by giving small-landholders access to credit and by bolstering local industries that supply needed parts and equipment. In Bangladesh, the progressive Grameen Bank is acquiring control of 100 tubewells and then offering loans to groups of five farmers who wish to buy one (World Water, May 1989). The International Fund for Agricultural Development is among the global organizations now giving greater emphasis to small, farmer-managed systems. Its experience has been that they are "more economical, easier to manage, and better targeted at the poor" (Hunger Notes, May 1989).

Tackling head-on the question, Irrigation for whom and for what ? is particularly critical in Africa. For much of the continent, irrigation will never work the miracles it did in Asia. Water and irrigable land are too scarce, and the costs of developing water too high. Thus far, irrigation has tended to expand production of higher - valued crops grown for export or urban consumption rather than food staples (FAO, 1987).

Projects that supplement rather than replace traditional agricultural practices may be the most promising ways to boost water productivity in Africa over the next few decades. African cultivation strategies tend to aim more toward minimizing risks of crop failure in bad years than maximizing yields in good ones. For many farmers, complete dependence on irrigation would increase risk, since spare parts and power to run the systems may not always be available or reliable. Moreover, outside of Egypt and the Sudan, large-scale irrigation is foreign to most African cultures and has the potential to alienate farmers rather than benefit them (Hutchinson, 1989).

An important test of integrating modern water management with traditional practices is now under way in the Senegal River valley. Farmers there practice "flood-recession agriculture", an indigenous cultivation method used by millions that involves planting crops on river flood-plains after seasonal flood waters recede. By engineering controlled floods, the reliability and productivity of this method can be improved - indeed, some say yields can be doubled or tripled - without incurring much of the ecological damage, exorbitant cost, and disruption of local traditions that full-scale irrigation often brings. At least until irrigation facilities are in place, the Manantali Dam is being operated for optimal recession cropping of sorghum. The first controlled flood was released in September 1988, and plans are to continue them through 1993. If successful and sufficiently high-yielding, this modified version of traditional floodplain agriculture could become a permanent part of the dam's operations (Scudder, 1989).

Restoring deforested watersheds urgently needs greater international assistance. Especially in tropical regions, a large share of potentially useful water runs off in damaging floods, causing far more harm than good. Reforesting mountainous catchments will help slow runoff, enhance the infiltration of rainwater into the soil, and thereby increase downstream groundwater supplies for the dry season. Revegetating watersheds is especially important if, as some climate models predict, tropical monsoons begin striking with even greater ferocity. An important step would be for the Asian Development Bank, the World Bank, or another international body to work with the countries sharing the Himalayan watershed on a plan to restore this badly degraded region (Postel and Heise, 1988).

As freshwater becomes increasingly scarce, and as cities bid more supplies away from farmers, the use of treated urban wastewater for irrigation is likely to become commonplace. This returns valuable nutrients to the land and helps keep them out of rivers and streams, where they become troublesome pollutants. With proper treatment, and with care in how and where reclaimed wastewater is applied, the practice can be very beneficial.

In California's Monterey County, a decade-long study concluded that irrigating vegetable crops with wastewater that had received advanced (tertiary) treatment proved just as safe as using fresh drinking-quality supplies - and cost only a fifth as much as a new freshwater source. Israel, with virtually no new fresh sources to tap, is now reusing 35% of its municipal wastewater, generally for irrigation. Some 15,000 hectares, most of it planted to cotton, are now irrigated with the reclaimed water. Current plans are to reuse 80% of the nation's total wastewater flow by the end of the decade (Water Newsletter, 30 April 1987 ; Shuval, 1987).

A relatively inexpensive way governments can buy insurance against future water constraints, especially in light of climate change, is to increase funding for international agricultural research centers working to develop new strains of crops. Plants that are more salt-tolerant, drought-resistant, and water-efficient could play a vital part in securing adequate food supplies. Research suggests that wheat, for one, is a good crop for breeding-in greater tolerance to salt. This could allow it to perform better under soil and water salinity where strains common today would not thrive (Charnock, 1988; Forster, 1988).

Salicornia, a succulent, can be irrigated with seawater and shows promise as a substitute for thirsty fodder crops in watershort regions. Its yield of oil seed compares well with soybeans, and it can contribute up to 10% of a fodder mix for cattle, sheep, and other livestock. Developing new strains that are sufficiently high-yielding and profitable to grow, however, takes time. Greater support for such efforts now could pay back handsomely in the decades ahead (Charnock, 1988; U.S. Water News, September 1988).

No quick fix is going to solve agriculture's water problems any time soon. Transforming crop production into a water-thrifty but still highly productive enterprise is a monumental task. Added up, the varied ways of using irrigation water more efficiently and of increasing water productivity on rainfed lands can go a long way toward preventing water scarcity from undermining food security. But these diverse technologies and measures will only spread widely if pricing policies, markets, regulations, and international development agencies promote them.

Ultimately, emerging scarcities and other water constraints call upon societies to slow the growth in human numbers and to adjust their activities to water's natural limits. In Africa, for instance, some 240 million people live in countries where per capita water availability is fast approaching, or is already below, a reasonable human-support level. The prospect of these populations doubling within a generation undoubtedly means rising food imports, declining living standards, and more hunger and malnutrition. Using scarce water to grow thirsty cotton crops in the desert, as in former Soviet central Asia, or to irrigate hay for cattle to eat, as in the water-short American West, may simply not be possible for much longer.

When something so fundamental as water becomes scarce, only fundamental responses will suffice. The consequences of overstepping water's limits are inescapable; for much of the world, time to avert them is rapidly running out.

REFERENCES

Ahmad, M. *In* Jordan (ed.) Water and water policy (cited by Postel, 1989).

AID. 1987. Irrigation briefing paper, Caro, April 1987.

AID. 1988. Agricultural briefing paper, Caro, December 1988.

Anonymous. 1989. Agricultural outlook, May 1989.

Bandyopadhyay, J. 1988. The Ecologist, 18(2).

Bandyopadhyay, J. 1989. Ambio, 18(5).

Bhumbla, D.R. and Khare, A. undated. Society for promotion of wasteland development, New Delhi.

Brown, L.R. 1988. Worldwatch Paper 85, Worldwatch Institute, Washington, D.C.

Charnock, A. 1988. New Scientist, 3 December 1988.

Checchio, E. 1988. Water Farming : The promise and problems of water transfers in Arizona, University of Arizona, Tucson.

Cowell, A. 1989. New York Times, 16 April 1989.

Crook, F.W. 1988. Agricultural statistics of the People's Republic of China, 1949-86 (Washington, D.C., USDA, ERS).

Desai, C.G. 1988. Proceeding planning workshop on policy related issues in Indian irrigation, Ootakmund, Tamil Nadu, India, April 26-28, 1988.

Dhawan, B.D. 1988. Economic and Political Weekly, 7 May 1988.

Dickason, C. 1988. J.Soil and Water Conservation, May-June 1988.

Dickason, C. 1989. USDA, ERS, private communication, 19 October 1989.

Falkenmark, M. 1989. Ambio, 18(2).

FAO. 1987. Consultation on irrigation in Africa, Washington, D.C., June 1989.

FAO. 1987. Production Yearbook 1987.

FAO. 1988. Production Yearbook 1988.

Forster, B. 1988. New Scientist, 3 December 1988.

Framji, K.K. and I.K. Mahajan. 1969. Irrigation and drainage in the world : a global review, Caxton Press Private Limited, New Delhi.

Ginnsz, Jean-Louis. 1989. Brazil agriculture operations division, World Bank, Washington, D.C.

Gleick, P.H. 1987. Climatic change, Vol. 10.

Gleick, P.H. 1988. *In* M.H. Glantz (ed.) Societal response to regional climate change : Forecasting by analogy, Westview Press, Boulder, Colo.

Greenfield, J.C. 1988. Seminar on the vegetative system presented at the World Bank, Washington, D.C., 4 August 1988.

Gunaratnam, D. 1989. China agriculture operations division, World Bank, Washington, D.C.

Gutentag, E.D. et al. 1984. U.S. geological survey paper 1400-B, U.S. Government printing office, Washington, D.C.

Hansen, J.E. 1989. Modeling greenhouse climate effects, Testimony before Subcommittee on Science, Technology, and Space, Committee on Commerce, Science and Transportation, U.S. Senate, 8 May 1989.

Harris, T. 1989. Saccromento Bee, 16 July 1989.

Hong, Li. 1989. China Daily, 17 October 1989.

Huffman, J. 1983. *In* T.L. Anderson (ed.) Water rights : Scarce resource allocation, bureaucracy and the environment, Ballinger Publishing Co., Cambridge, Mass.

Hutchinson, C.F. 1989. Resources, Spring 1989.

Jackson, W.L. et al. 1989. J. Soil Water Conservation, March/April 1989.

Jensen, M.E. et al. 1989. *In* B.A. Stewart and D.R. Nielson (eds.) Irrigation of agricultural crops, Am. Soc. Agron. Monograph No. 30, Medison, WI.

Johnson, R.W. 1988. U.S. Water News, July 1988.

Keen, M. 1988. Ceres, May-June 1988.

Kovda, V.A. 1983. Ambio, 12(2).

Leeden, Frits van der. 1975. Water resources of the world, water information centre, Inc., Port Washington, NY.

Leied, F. 1988. China Daily, 18 May 1989.

LeVeen, E.P. and L.B. King. 1985. Turning off the tap on federal water subsidies, vol.1, NRDC and California Legal Assistance Foundation, San Francisco.

Levine, G. et al. 1988. Irrigation in Asia and the Near East in the 1990s : Problems and prospects, prepared for the irrigation support project for Asia and the Near East at the request of the Bueaux for Asia and the Near East U.S. AID.

Margat, J. and K.F. Saad. 1984. Nature and Resources, April-June 1984.

Marshall, E. 1986. Science, 10 January 1986.

Metzger, P.C. and J.A. Haverkamp. 1984. Stream flow protection : Adaptation to intensifying demands, Conservation Foundation, Washington, D.C., June 1984.

Mickin, P.P. 1989. Western Michigan University, Kalamazoo, Michigan, private communication, 13 October 1989.

Micklin, P.P. 1989. Final report to the National Council for Soviet and East European Research, Washington, D.C., February 1989.

Mitra, A.K. 1986. Economic and Political Weekly, 26 April 1986.

Moore, M.R. and McGuckin, C.A. 1988. Agricultural Resources : cropland, water, and conservation situation and outlook report, Washington, D.C., September 1988.

Moss, M.E. and H.F. Lins. 1989. U.S. Geological Survey Circular No. 1030.

MWREP. 1987. Ministry of Water Resources and Electric Power, Irrigation and Drainage in China, Beijing, 1987.

NCA. 1988. Report of the National Commission on Agriculture, Islamabad.

Nickum, J.E. 1987. Beijing's rural water use, prepared for East- West Centre North China Project, Honolulu, March 1987.

Nickum, J.E. and J. Dixon. 1989. *In* C.E. Morrison and R.E. Dernberger, Focus : China in the Reform Era, Asia-Pacific Report 1989.

NRC. 1989. National Resource Council, irrigation-induced water quality problems : what can be learned from the San Joaquin valley experience ? National Academic Press, Washington, D.C.

NRDC. 1988. Natural Resources Defence Council, environmentalists and fishing groups announce lawsuit to block federal water contracts, press release, 21 December 1988.

Perera, J. 1988. New Scientist, 26 November 1988.

Peterson, C. 1989. Washington Post, 19 March 1989.

Peterson, D.F. and A.A. Keller. *In* AAAS, climatic variability, climatic change, and the planning and management of U.S. Water Resources, John Wiley & Sons, New York.]

Pogrell, H. van. 1989. Latin American Agriculture Operations Division, World Bank, Washington, D.C.

Postel, S.L. 1984. Worldwatch Paper 62, Worldwatch Institute, Washington, D.C.

Postel, S.L. 1985. Worldwatch Paper 67, Worldwatch Institute, Washington, D.C.

Postel, S.L. and L. Heise. 1988. Worldwatch Paper 83, Worldwatch Institute, Washington, D.C.

PRB. 1989. World Population Data Sheet, Washington, D.C.

Rangeley, W.R. 1987. *In* W.R. Jordon (ed.) Water and water policy in world food supplies. Texas A & M University Press, College Station, Texas.

Rangeley, W.R. 1989. Private communication.

Rangeley, W.R. 1989. Irrigation and Drainage in the world.

Repetto, R. 1986. Skimming the water : Rent-seeking and the performance of public irrigation systems, World Resources Institute, Washington, D.C.

Reyes, Romana P. de las and Sylvia Ma. G. Jopillo. 1986. An evaluation of the Philippine participatory communal irrigation program, Institute of Philippine Culture, Ateneo de Manina University, Quezon City.

Rhodes, J. 1989. U.S. Salinity Laboratory, Riverside, Calif., private communication, 1 September 1989.

Rosegrant, M. 1989. International Food Policy Research Institute, Washington, D.C.

Sattaur, O. 1989. New Scientist, 27 May 1989.

Schaake, J.C. *In* AAAS climatic variability, climate change, and the planning and management of U.S. water resources, John Wiley & Sons, New York, in press.

Schneider, S.H. 1989. Global warming : are we entering the greenhouse century ? Sierra Club Books, San Francisco.

Scudder, T. 1989. Conservation versus development : River Basin Projects in Africa, Environment, March 1989.

Seib, G.F. 1985. Wall Street J., 3 October 1985.

Shuval, H.L. 1987. Ambio, 16(4).

Starr, J.R. and D.C. Stoll. 1987. U.S. foreign policy on water resources in the Middle East, Centre for Strategic & International Studies, Washington, D.C.

Starr, J.R. and D.C. Stoll (eds.) 1988. The politics of scarcity : water in the Middle East, Westview Press, Boulder, Colo.

Starr, J.R. 1989. Global water scarcity : The Middle East as a case study, Testimony before Subcommittee on Natural Resources, ARE, CSST, U.S. House of Representatives, 28 February 1989.

Stockton, C.W. and W.R. Boggess. 1983. *In* NRC, Changing climate, National Academy Press, Washington, D.C.

Svendson, M. 1988. Sources of future growth in Indian irrigated agriculture, presented to the planning workshop on policy related issues in Indian irrigation, Ootacamund, Tamil Nadu, India, 26-28 April 1988.

Tuck, C. 1989. Texas water Development Board, Austin, Texas, private communication, 14 September 1989.

Udall, L. 1989. Testimony before the Subcommittee on Natural Resources, Agricultural Research and Environment (ARE) Committee on Science, Space and Technology (CSST), U.S. House of Representatives, 24 October 1989.

UN. 1989. Department of International and Social Affairs, World Population Prospects, NY.

USDA / ERS. 1989. USSR : Agriculture and Trade Report, Washington, D.C.

Volftsun, I.B. et al. 1988. Vodnyye Resursy No. 3.

Wade, R. 1988. World Development, 16 (4).

Widstrand, C. (ed.) 1980. Water conflicts and research priorities, Pergamon Press, Elmsford, NY.

Wilderness Society. 1988. Ten most endangered national wildlife refuges, Washington, D.C., October 1988.

Williams, P. 1983. Daming the world, not man apart, October 1983.

Woodard, G.C. et al. 1988. The water transfer process in Arizona : analysis of impacts and legislative options, college of Business and Public Administration, University of Arizona, Tucson, April 1988.

World Bank. 1987. Vetiver grass (*Vetiveria zizanioides*) : A method of vegetative soil and moisture conservation, New Delhi.

Yudelman, M. 1989. Sustainable and equitable development.

Yudelman, M. 1989. *In* H. Jeffrey Leonard et al., environment and the poor : development strategies for a common agenda, Transaction Books, New Brunswick, N.J., 1989.

Zeilig, Nancy. 1988. Interview with S.J. Shupe by Nancy Zeilig, face to face-water marketing : an overview, J. Am. Water Works Association, March 1988.

Zhuoyan, Li and Gao Jinan. 1989. China Daily, 10 March 1989.

2

Field Water Balance For Irrigation Planning

R. John Hanks and Melvin D. Campbell

I. INTRODUCTION

The interception, transformation, and storage of solar energy to photosynthesis (and thus crop growth) at optimum efficiency is the ultimate objective of agriculture. Where water is the limiting factor in optimum crop production, irrigation can facilitate photosynthesis very efficiently. However, irrigation is the one of most energy intensive cultural practices in agriculture and thus warrants careful and thoughtful management to optimize energy use.

Optimizing energy integration by agricultural crops involves well-planned investment of resources with the specific intent of gathering only the most efficiently available return on the investment. Irrigation represents a major resource investment that must be justified by commensurate returns in crop yield and quality. Irrigation is almost never practiced to achieve maximum production. Instead, the minimum investment possible is almost always pitted against the highest likely economic returns-not maximum possible crop yields. Thus, if the purpose of irrigation is to provide an environment to promote plant growth, then the purpose of irrigation planning is to optimize crop energy integration. This perspective sets the stage for optimum irrigation management.

Every wise farmer that irrigates to enhance crop production considers the resource tradeoffs: How much water is available? How much water will be needed and when? How large will the delivery system be? How much will it cost? When will irrigation be required? Who will do the work? If water is expensive, the farmer pays careful attention to how much water he uses and avoids unnecessary waste. If the water is cheap, little care is given to the amount of water used, except to avoid other losses such as nutrient leaching or soil waterlogging.

"When and how much to irrigate" is much more an idiom from irrigation design engineers or irrigation management consultants than it is from farmers. Many farmers, when asked if they would like an irrigation management service, have said to us, "I can tell when to irrigate while I'm driving' down the road eighty miles an hour; why should I pay for irrigation management?"

Many farmers have demonstrated good water management skills if their fields have water available on demand. Astute farmers in areas of relatively invariant weather, such as Arizona or California, have developed an uncanny sense that allows them to irrigate with the consistency of measurement-based models. What the farmers cannot do without careful measurements and models, however, is plan ahead well enough to achieve distribution system efficiency. Consulting services can make a favourable impact on crop production efficiency by providing a field-measurement based irrigation schedule where many fields depend on a common water distribution system. Such a service is especially effective in optimizing costs where water or power is costly or where weather is highly variable.

The purpose of irrigation is to provide an environment to promote plant growth. This purpose may also be constrained by the need to use a limited amount of water to stretch water supplies, reduce drainage, pollution, etc. To accomplish this it is necessary to consider the field water balance. The relation of field water balance to irrigation planning is fundamental because the field water balance determines the soil water content or status which in turn determines the plant growth. Thus irrigation needs to be managed to obtain the optimum growth considering the other effects of irrigation that may result.

Irrigation water is often of low quality. Use of "high technology" herbicides, fungicides, as well as fertilizers may have caused contamination of the soil and groundwater to an "unknown" degree. Thus future irrigation planning may be forced to consider negative effects caused by contamination as well as positive effects of increased plant growth.

II. BASIC PRINCIPLES

The basic relation between irrigation and the field water balance is the hydrological cycle, which in one dimensional form can be written:

$$I = Es + T - P - Dep + Ru + Dr \qquad [2.1]$$

where I is irrigation, Es is soil evaporation, T is plant transpiration, P is precipitation or rainfall, Dep is soil water depletion, Ru is runoff, and Dr is drainage below the root zone. As can be seen from Eq.[2.1] the amount of irrigation required for best plant growth is dependent on many factors. Soil factors influence Es, T, Dep, Ru, and Dr; plant factors influence Es, T, Dep, Ru, and Dr; climatic factors influence P, Es, T, Dep, Ru, and Dr; and irrigation and management factors influence Es, T, Dep, Ru, and Dr. Thus, considerable interaction exists among the factors.

The field water cycle is essentially Eq.[2.1] applied to a particular field for a particular period. It is useful to expand Dep (depletion) to be more specific as follows :

$$Dep = SWB - SWE \qquad [2.2]$$

where SWB is the sum of the soil water at the beginning of a time period and SWE is the sum of the soil water at the end of a time period. The stored soil water must be measured to include the entire root zone of the crop to conform to Eq. [2.1]. In practice SWB and SWE are measured by the following :

$$SWB = \Sigma\ (\theta\ vzb*\Delta z) \qquad [2.3]$$

where θvzb is the volumetric water content at the beginning of a time period in depth increment Δz. Thus the measurement of Dep requires measurement of water content by depth increments at the beginning (θvzb) and end (θvze) of the time period in question.

As an example, assume measurements were made in 30 cm increments as shown in Table 2.1. For the example, Dep = 3.6 cm. If for the same period I was 5 cm, R was 0.5 cm, and Ru = Dr = 0 then Es + T (ET) would be 9.1 cm.

Table 2.1. Example of data needed to measure SWB and SWE.

delz cm	θvzb	θvzb*Δz cm	θvze	θvze*Δz cm
0–30	0.24	7.2	0.18	5.4
30–60	0.22	6.6	0.16	4.8
60–90	0.12	3.6	0.12	3.6
		SWB = 17.4 cm		SWE = 13.8 cm

It is useful to consider each of the components of the field water cycle because the evaluation of each component is not simple and there is usually some uncertainty about their values.

Soil evaporation, Es, is primarily dependent upon soil water status and climate at or near the soil surface. If the soil is wet, Es will be dependent on the climatic environment, will be in the "constant rate stage" and will be the potential soil evaporation "Esp" (Hanks and Ashcroft, 1980). If the soil has dried to the point that the soil surface is air dry, Es will be less than Esp and will be in the "falling rate stage". The value of Esp depends on the crop because canopy properties, such as shading, will influence the microclimate. The value of Esp might be measured approximately by a free water evaporimeter within the canopy or by measuring net radiation. The value of Es depends on irrigation or rain because soil water content will increase thereby. Rainfall may have an additional effect because it will usually be accompanied by cloudy weather that would influence Es and or Esp. A crop has an additional effect on Es, other than influencing Esp, because crop growth will be associated with transpiration, T, which could lower the soil water content in the soil surface layer and thus lower Es. A further complication is the difficulty of separating the Es and T processes when using the usual methods of measuring soil water content and thus deducing Dep. Thus more complicated procedures are called for as will be discussed later.

Measurements of soil evaporation have been made using microlysimeters by Kidman et al. (1990). Figure 2.1 shows data for seasonal soil evaporation for sweet corn (*Zea Mays 1.*) at a 76 cm row spacing. There was a rapid decrease of Es/Esp at the beginning of the season followed by a moderated decrease later in the season. The rapid decrease early in the season was caused by high climatic demands (high Esp) with no plant shading where the soil water flow to the surface was less than Esp. The moderated decrease in Es/Esp later in the season was caused by lower values of Esp resulting from plant shading so the soil water flow to the surface equalled Esp for a longer period of time. Figure 2.1 also shows the relation of Esp/Ep and Tp/Ep for the same crop for reference where Ep is the sum of Esp and Tp (potential transpiration). It can be seen that the highest values of measured Es/Ep define the upper

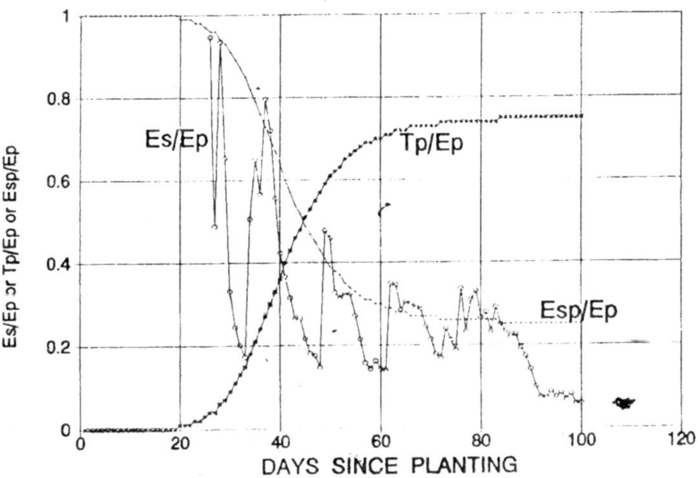

Fig 2.1. Measured Es/Ep vs. time for sweet corn in 76 cm rows in 1989 at Logan, UT as compared to Esp/Ep and Tp/Ep.

limits of the Esp/Ep curve. The curve of Tp/Ep vs. time was calculated as 1 - Esp/Ep.

All of these complications make estimation of Es in the field somewhat uncertain. However, there are some simplifications that are possible and that allowed reasonable estimates to be made under field conditions. One simplification is related to the observations made that the soil evaporation process involves primarily the surface 20 to 30 cm of soil (Hanks, 1985). Another simplification made is

$$Es = f(\theta vsur) \quad [2.4]$$

where θvsur is function relating Es to the soil water content of the surface layer (θvsur) where climatic demand is very high so that Es is determined by the water supplying power of the soil. In practice the value computed by Eq. 2.4 is compared with Esp and the smallest of the two is taken as Es. For the data shown in Fig. 2.1 the following equation was found to hold

$$Es = [S/(ST-s2)]^2 \quad [2.5]$$

where s is actual soil water in the top layer, ST is the potential soil water in the surface layer between field capacity and air dry, and s2 is a calibration constant (average soil water lost when Es = Esp).

Transpiration is a more complicated process than Es because it involves biological as well as physical processes. Water is withdrawn from the root zone in a complicated manner. Not only does the root zone change as the plant grows but the extent and amount of root activity changes with depth and time. Plant growth processes also change with time. The lowest water content that the plants can extract from the soil because of T, the wilting point, is considerably higher than the lowest water content caused by Es (air dry). In most situations there is a much larger volume of water available in the soil for T than for Es because roots can extract water to a much greater depth (100-300 cm). This means that after a soil is thoroughly wet, T is larger and lasts much longer time than does Es. Transpiration is also dependent on climatic conditions if soil water is not limiting. This "climatic determined" value is defined as Tp herein.

Transpiration is also dependent on many other factors. From the point of view of the plant the important factor is the soil water content in the plant root zone. Thus It Is dependent on any factor that influences soil water. Since T also influences soil water content a feedback relation is involved just like the Es processes.

As will be discussed in the section on modeling to estimate T simplifications are often used. We estimate T from an estimate of the soil water content in specific root zone as follows (Hanks, 1974):

$$T = Tp \text{ if } RSWCz > swfac \qquad [2.6]$$

$$T = Tp^{*}RSWCz/swfac \text{ if } RSWCz > swfac$$

where RSWCz is the relative soil water content of the wettest soil layer at depth "z" (the wettest layer) and swfac is a coefficient (like 0.5 etc.). There are generally several layers involved so a choice has to be made of which layer to withdraw the water from or how to withdraw water from several layers at a time.

Precipitation (rainfall) occurs in an uncontrolled manner so irrigation planning has to adjust to it. General climatic data gives information about what has happened historically so planning needs to adjust to what can be expected or has happened recently. An uncertainty involved with the effectiveness of precipitation is the determination how much of it is stored in the soil that implies a knowledge of how much runoff or drainage has resulted. Models, discussed below, may help with this estimation. One problem not always acknowledged, is an uncertainty about how much rainfall actually occurred. Measurements may have been made at locations different than the desired locations. This points to the need for some on-site measurements to assist model simulations or other estimations.

Irrigation has some of the same uncertainty as rainfall but has a possibility of management about how much irrigation is applied and when it occurs. Historically, irrigation has been applied in sufficient amounts to assure good plant growth. This has undoubtedly resulted in excess leaching that may need to be minimized in the future. Different irrigation methods may also have their own built-in uncertainties. For example, gravity irrigation may be associated with more water infiltrated in the field where water has been applied (at the top of the field) than on the average or at the bottom of the field. Thus the distribution of applied water is a problem. Again field checking of soil water status is useful to determine what is actually happening. Use of models can help to put ranges on possibilities of what has happened or can happen.

Drainage, Dr, is probably the most uncertain water balance component because it is so difficult to measure what is happening within the soil at the bottom of the root zone. Flow is very slow and may be highly variable and soil properties also may be highly variable. Use of models, discussed below, is probably the best approach to estimate Dr based on water balance concerns. This is contrast to a flux measurement approach (measuring the components of the Darcy's law) that requires continuous measurements whereas the water balance approach requires only periodic measurements. Another difficulty in estimating Dr, even with the water balance approach, is that errors in estimations of ET would make large errors in measurements of Dr since Dr is estimated as a difference of all other water balance components.

Runoff, Ru, is also a difficulty in many instances. There are very few measurements made of Ru in irrigation (or other circumstances) so additional uncertainty is introduced. However, there are many conditions where Ru is zero and can be predicted as such. If Ru is not zero the amount is uncertain. Again this uncertainty can be minimized if some field measurements are made to put a range on the extent of Ru. Models can also help to put limits on Ru.

Figure 2.2 shows the measured cumulative amounts of Es, T, and ET for one of the plots of the 1989 corn study made by Kidman et al. (1990). This short season crop with widely spaced rows resulted in 15 cm of Es at the end of the season, 23 cm of T, and 38 cm of ET. Irrigation frequency was every two weeks. Over the season irrigation was 29 cm so 9 cm of water was depleted from the soil profile during the season.

The main purpose of irrigation is to provide plant water needs to improve crop production, and all components of equation [2.1] must be related to crop growth or yield. One of the simplest relations (Hanks, 1983) is

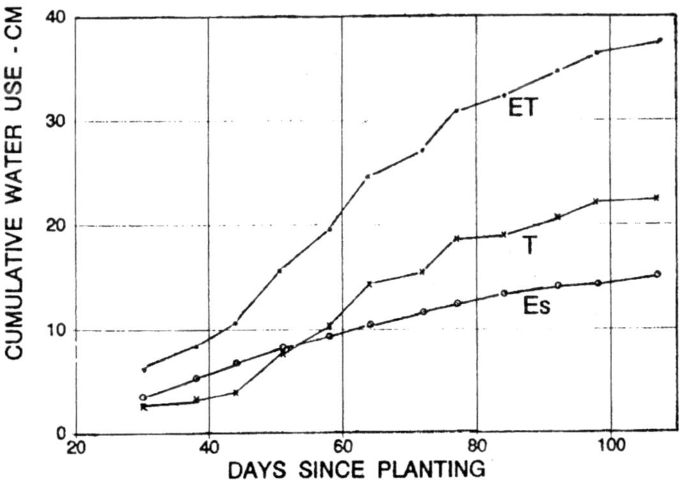

Fig. 2.2. Cumulative ET, Es, and T for sweet corn at Logan, UT for 1989 (Kidman et al., 1990).

$$Y/Yp = T/Tp \qquad [2.7]$$

where Y is yield, Yp is potential yield where water is never limiting or when T = Tp. Thus if water is supplied by I or P or coming from soil water storage (Dep) T can be made to equal Tp. In areas where rain is very small, both before and during the season, most of the water used must come from I. To take advantage of soil water storage, irrigation may be applied before planting the crop because water may be more readily available then.

If large amounts of water are applied at any one time, Ru may be significant and water may not be efficiently used by the crop. Even if Ru is minimized by applying water slowly, water may be lost as Dr if the soil water storage capacity is not enough to retain the water applied. If Dr occurs there may be leaching of nutrients or contaminants that may be undesirable. In arid regions, where irrigation water may be saline and rainfall is insufficient for natural leaching, it may be desirable to systematically leach accumulated salts. Thus, irrigation is often managed to include some leaching.

There are very few methods available to measure Dr with any reliability. The authors' opinion is that computation based on equation [2.1] can often give good estimates of Dr since most of the other components of the water balance equation can be estimated with reasonable accuracy. It is also apparent that few users have any appreciation for the factors that influence Dr.

Another consideration in irrigation planning is the time interval between measurements. If the time interval is too large then it will be impossible to differentiate between the various components of equation [2.1] under some situations.

The importance of soil water storage is not always appreciated in irrigation planning. Table 2.2 shows components of the water balance equation for three locations, in western USA, for "normal" and "limited" irrigation situations. At Davis, with almost no irrigation or precipitation, there was 409 mm of ET (Es + T) coming mostly from soil water storage. Where ET was increased under the irrigated treatment, the proportion of the ET coming from water storage (Dep) decreased dramatically. At Logan and Ft. Collins, where there was some rainfall and irrigation for the "dry" treatment, significant water withdrawal for ET from soil water storage still occurred. For the normal water levels at Logan, ET coming from soil water storage, irrigations, and precipitation was optimized to the maximum such that Dr was reduced to 14 mm.

III. EXAMPLE OF A COMPUTERIZED IRRIGATION SCHEDULING MODEL

A large number of computerized irrigation planning models have been developed by many researchers for this purpose. The model PLANTGRO (Hanks, 1974; Retta and Hanks, 1980) will be discussed in this chapter. In model PLANTGRO the water balance of the soil is computed daily as a function of boundary conditions (I, P, Esp and Tp) and initial conditions (soil water content). Soil, crop, and climate properties all influence the result. Crop yield is estimated at the end of the season as a function of relative T.

Table 2.3 illustrates the input-output data of the PLANTGRO model, for corn (maize). This model is based on the estimation of

Table 2.2. Components of the hydrological cycle for dry and irrigated treatments (mm) at three locations for maize (corn) as related to yield. Stewart et al. (1977). Data in mm.

Location	Treat	I	P	ET	Dep	Dr	Ru	Yield $t\ ha^{-1}$
Davis,	Dry	3	4	409	402	0	0	15.4
Calif., USA	Wet	406	4	611	243	42	0	22.0
Ft. Collins	Dry	11	186	365	233	65	0	10.7
CO	Wet	346	186	526	104	110	0	15.8
Logan, UT	Dry	64	81	366	222	1	0	8.8
	Wet	334	81	543	142	14	0	15.6

soil water content only with no information needed on soil water hydraulic properties. Even for a simple model, there are still considerable data requirements of soil, crop, climate, and management parameters. PLANTGRO has somewhat limited use because it is very simple and it ignores many important characteristics of both soils and plants.

Some of the soils information needed is the soil water content at the beginning of the season (BGSM array for five layers of thickness defined in the THK array), and the available water holding capacity (WHC array expressed as volumetric water fraction for each depth). Also needed is the amount of water that could be lost by evaporation in the top layer between wilting and air dry (AD = – 20 m in Table 2.3)

The crop factors include, the time of maximum root growth (RM = 68 days in Table 2.3), the length of the growing season (DA = 112 days), and the relative water content below which T < Tp (AF = 0.50). The available water holding capacity is a combined soil and plant factor that involves root depth and potential extraction with depth. Note that the WHC array has decreasingly lower values as depth increases reflecting less ability of the crop to extract water from the soil at deep depths.

A combined crop-climate condition is indicated by the partitioning of potential evaporation from a wet surface (often measured as pan evaporation Ep in Table 2.3) into potential soil evaporation (Esp) and potential transpiration (Tpot in Table 2.3). Potential transpiration is dependent on crop cover, crop type, planting date etc. and is defined as

$$Tp = Ep - Esp. \qquad 2.8$$

Another crop factor is the coefficient used to indicate the effect of crop stress at different growth stages (E array). Note that Ep and Esp are given as average values over 10-day intervals in Table 2.3. Climatic variables include, besides Ep, the days rain occurs, the amount of rain, and the evaporation pan coefficient (VA).

Management variables include the days of irrigation and the amount of irrigation applied (I). However, management also dic-

Table 2.3. Input and output data for PLANTGRO for an example of maize (corn) at Logan, Utah for 1986. Data in mm.

Input Data

DA = 112 AD = −20 AF = .5 RM = 68 PP = 100 VA = 1 CY = 1

E ARRAY 0 .25 .25 .25 0

THK ARRAY 300 300 300 300 300

WHC ARRAY 0.09 0.09 0.08 0.07 0.05

BGSW ARRAY = 27 25 21 16 12

DAY	RAIN	DAY	RAIN	DAY	RAIN	DAY	RAIN	DAY	RAIN
6	1	8	10	11	3	24	1	30	3
35	10	36	11	37	18	39	2	57	5
58	5	68	2	78	20	79	4	105	3

DAY	EP	ESP	DAY	EP	ESP	DAY	EP	ESP	DAY	EP	ESP
1	6.6	6.6	11	4.0	4.0	21	2.7	2.7	31	4.6	3.1
41	4.2	2.0	51	6.5	1.7	61	8.7	2.0	71	7.0	0.7
81	7.4	0.7	91	7.0	0.8	101	7.6	2.4	111	8.3	4.0
121	8.0	4.0									

DAY	IRR	DAY	IRR	DAY	IRR	DAY	IRR	DAY	IRR
53	14	54	21	64	38	65	17	75	33
82	10	86	25	89	29	92	35	96	35

Table contd.....

Output Data

DAY	S–STOR	A–STOR	IRRIG	RAIN	DRAIN	TACT	TPOT	EVSOIL
10	70	114	0	11	0	0	0	41
20	58	114	0	14	0	0	0	57
30	53	114	0	18	0	1	3	64
40	64	114	0	58	0	16	18	79
50	36	114	0	58	0	38	40	86
60	25	114	35	68	0	88	88	96
70	13	114	90	70	0	142	155	107
80	19	114	123	94	0	188	218	111
90	20	114	187	94	0	245	285	117
100	23	114	257	94	0	307	347	122
110	–02	114	257	98	0	328	399	130
112	–03	114	257	98	0	328	408	132

YIELD WAS 80.4%

tates the crop grown, the date of planting and, possibly, the field of soil used.

The output data shown in Table 2.3 are for 10-day intervals. The column 'S-STOR' is the total amount of available soil water storage for the day (on day 20 = 58 mm). Column 'A-STOR' is the potential amount stored. Columns IRRIG, RAIN and DRAIN, are cumulative amounts of irrigation, rain, and drainage below the root zone, respectively. The columns TACT, TPOT, and EVSOIL are cumulative amounts of T, Tp and Esp respectively. The computed dry matter yield is determined from the end-of-season ratio of T/Tp using Eq (2.6)

With this model and the many simplifying assumption associated with it, considerable calibration is required. For example, the data for the WHC array is determined from field measurements using the soil and crop in question. The partitioning of Esp and Tp also comes from calibration. Measured data of "crop coefficients" (of Wright, 1982), are used to estimate the transpiration coefficient, Tc, (Hanks 1985) :

$$Tc = Tp/Ep \qquad [2.9]$$

Figure 2.3 shows the transpiration coefficient, Tc, as well as the soil evaporation coefficient, Esp/Ep, and the crop coefficient (Kc = (Es + Tp) / Ep). The value of Kc varies considerably from day to day because Es varies so much from day to day. The smoothed value of Kc also changes considerably. This illustrates a basic problem using the crop coefficient method to estimate ET. The value of Kc is highly dependent on Es. Thus use of this approach requires adjustments to the "standard" Kc be made for a specific site especially for rainfall or irrigation events. In contrast Tc is a smooth curve.

Unfortunately it is very difficult to measure Tc directly. Relatively easy methods of estimating Tc are available as discussed by Hanks (1985).

The model described above has many limitations so it should be used only by those who realized the implications of the limitations. Some important assumptions not discussed above are :

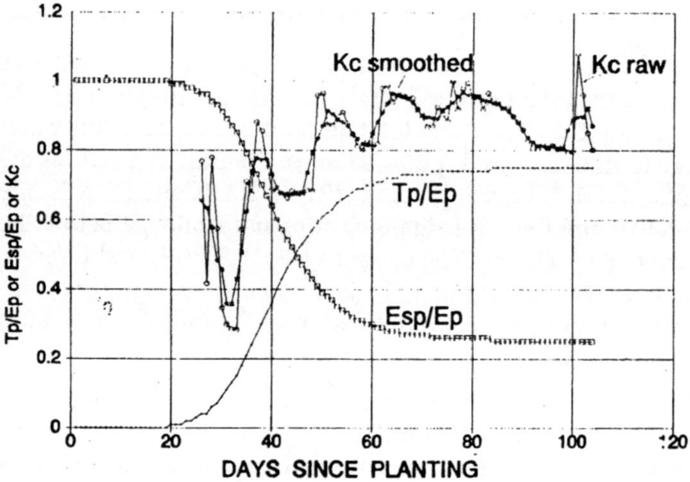

Fig. 2.3 Comparison of measured and smooth Kc for sweet corn to Tc (Tp/Ep) and Es coefficient (Esp/Ep).

1. Water flow is only in the downward direction, is maximum at field capacity, and occurs instantaneously.
2. Potential evaporation, Ep, is known and knowledge is available to estimate Tp and Esp (both as a function of time).
3. Soil evaporation, Esp, occurs from the top soil layer only.
4. Water uptake by plant roots is calculated as a function of T and the relative available water content in the soil layer having the highest relative water content where there are significant roots.

An "irrigation scheduling" version of this model has been made into a game that demonstrates how some soil, plant, and climatic factors are interrelated (Hanks and Rasmussen, 1984).

Table 2.4 shows a comparison of the simulated results of PLANTGRO from variation of the input data. When 60 mm more irrigation was applied on day 82, than shown in Table 2.3, ET was increased from 460 to 503 mm, T was increased from 328 to 372 mm and T/Tp was increased from 80.4 to 91.1% and there was no drainage. Where 120 mm of water was applied on day 82, Et and T were increased very slightly because most of the additional water was lost as drainage.

Where 120 mm of additional irrigation was applied on day 64, not day 82, T increased so T/Tp increased from 92.2 to 94.4% even though there was an increase in Dr from 46 to 54 mm. If 120 mm of irrigation was applied on day 92, Dr was higher than for the same irrigation increase on day 64 or 82 and T/Tp was lower. This demonstrates the importance of timing of irrigation.

Table 2.4 also shows the effect of changing the beginning soil water and water holding capacity. Where the beginning water content (total) was 57 mm instead of the 101 mm shown in Table 2.3, ET was decreased from 460 to 416 mm, T was decreased from 328 to 298 mm, and T/Tp was decreased from 80.4 to 73%. If the soil water holding capacity was decreased by half with the decreased beginning water content, the results were essentially the same. Thus, changes in soil properties or input conditions may or may not influence the end season results.

When there was decreased beginning soil water content and decreased soil water holding capacity an irrigation of 120 mm on day 82 had less of an effect on ET, T, and T/Tp than it did for the higher beginning water content and water holding capacity (T increased from 299 to 318 compared to from 328 to 376 mm). This result was undoubtedly caused by increased Dr. (99 compared to 46 mm) with the low beginning water content and water holding capacity.

Many other simulations could be run using a model like PLANTGRO to illustrate the complex interaction of the many variables. The model demonstrates why such "if" questions as "what would be the benefit if I irrigated with 120 mm of water on day 64?" cannot be answered directly. A specific answer is possible only in all other factors (inputs) are known. For example, in Table 2.4, if an irrigation was applied on day 64, the answer would be an increased relative yield from 80 to 94%.

Table 2.4 Influence of variations in the input variables (mm) of Table 2.3 on output results.

Input variation	ET	Dr	I	T	T/Tp
From Table 2.3	460	0	257	328	80.4%
60 mm more I added on day 82	503	0	317	372	91.1
120 mm more I added on day 82	508	46	377	376	92.2
120 mm more I added on day 64	515	54	377	383	94.4
120 mm more I added on day 92	498	56	377	366	89.8
Beginning water content decreased from 101 to 57 mm	416	0	257	298	73.0
Same as above but soil water holding capacity decreased by 1/2	417	0	257	299	73.3
Same as above but 120 mm I added on day 82	436	99	377	318	78.0

However, we must be aware of what we are doing in the context of the overall system and be especially concerned with the assumptions made. Since all models are imitations of reality with many assumptions and simplifications, measurements should be integrated with models to assure some validity to the simulations. Those who measure should be aware of the model they are applying and those who model should 'keep one foot in the field".

IV. SOME PRACTICAL FIELD IRRIGATION SCHEDULING METHODS

Practical irrigation planning tools or methods that can be used in the field are as follows:

1. Calculation of soil water budgets using soils, crop, weather, and irrigation management information. This can be done simply by using hand calculations (checkbook method that is well adapted for computer spreadsheets) or by the use of models as discussed above.
2. Monitor soil water status by use of instruments or sampling techniques such as 'feel", gravimetric, gypsum blocks, tensiometers, and neutron probes. These methods have the advantage of on-site checking but may not be well adapted to telling the amount of water to apply, determine drainage or leaching, etc. They can best be used in combination with other methods to complete the estimation of the water balance. These methods can be used to provide feedback data for the models at specified time periods during the growing season.
3. Observation and measurement of plant indicators, such as when the crop shows visible evidence of stress by colour change or leaf wilt or by use of canopy temperature measurements. A useful technique has the advantage of being able to measure on-site plant water stress but has the disadvantage of allowing some yield reduction to occur before

measurable symptoms detected. If used alone this method does not give information about the amount of water to apply so it is best used in combination with some other method, such as models.

While irrigators have used these and other means to schedule irrigations, a considerable amount of assistance is required in training users to successfully apply these tools and methods to their situation. A continuing time commitment is also required to maintain current information to insure efficient use of irrigation water.

Many field procedures for irrigation scheduling have been developed. We well now discuss a method that has been used successfully in many field locations.

V. A PERSPECTIVE ON IRRIGATION MANAGEMENT

1.Water Balance Accounting

Water balance is the accounting of water gain, storage, and loss. During the growing season, water income is normally from rain or irrigation. Water loss is runoff, evaporation, transpiration, and drainage. Water storage is actually a change in storage (Dep in Eq. 2.1). Irrigation management balances these components to optimize crop production. Only the area of influence and degree of control vary.

Soil water storage is more easily understood if we refer to it as change in storage or depletion. ET is often expressed in terms of potential evapotranspiration (PET) and the crop coefficient (Kc) (Doorenbas and Pruitt, 1977; Jensen, 1974). Thus, Eq. 2.1 becomes

$$I = Ru + Kc \bullet PET + Dep + Dr - P \qquad [2.10]$$

Where optimum crop production is the main objective of water management, cultural practices are usually designed to minimize runoff and drainage, thus leaving.

$$I = Kc \bullet PET + Dep - P \qquad [2.11]$$

The amount of irrigation required for an interval is equal to Kc • PET minus rain plus the allowable depletion. This is a common condition in arid, irrigated areas where periodic irrigations refill the soil reservoir to maximum storage capacity. This assumes that ET is sufficient enough that water is not limiting T (T = Tp). It is almost impossible to do this on a practical scale without causing some drainage.

2. Irrigation Scheduling Based on Depletion

Irrigation scheduling is commercially available in many parts of the world. Since about 1975, scheduling based on soil moisture measurements by a neutron probe has been commercially available. Accurate, efficient, individual field irrigations have been possible since that time, with substantial reductions in total water consumption. If the water distribution and irrigation systems can deliver water rapidly and efficiently whenever needed, then irrigation scheduling is a simple matter of measuring soil moisture periodically and irrigating when the previously selected storage refill point is reached (Gear et al., 1977). This method may be compared to a thermostat used for heating. When the temperature reaches a low set point heat is applied until the temperature rises to a high set point at which time heat is turned off. When the soil water content reaches a low set point irrigation is applied to bring the soil water profile back to "field capacity" or to fill the profile back to it's maximum capacity. This method has the advantage of simplicity and it is quite robust (many errors are self correcting). To illustrate this, simulations were done with PLANTGRO model which had been modified to determine the time and amount of irrigation as shown in Table 2.5.

The simulations show that the yield is not very sensitive to Depf (the fraction of depletion allowed before irrigation is called for) nor to Irrigf (the fraction of the depletion that is filled by irrigation). This illustrates a very important attribute of this method. The soil water status is actually measured so any estimations that are in error at one measurement time may be corrected the next time the measurement is made and illustrates the advantage of field measurements (keeping one foot in the field).

This method is commonly used where there is little concern for the amount of water used or the cost of the water. The data of Table 2.5 show that there is considerable difference in the amount of

Table 2.5 Simulations of results of using soil water depletion to determine irrigation amount and time using data for sweet corn grown at Logan, UT in1990. Water amounts are in mm.

Depf*	Irrigf**	No.	Amount	Dr	Es	T	T/Tp
0.3	1.0	17	470	0	272	257	99.9%
0.4	1.0	13	447	0	246	257	99.9
0.5	1.0	10	419	0	244	257	99.9
0.6	1.0	8	392	0	222	257	99.9
0.7	1.0	7	388	0	190	257	99.8
0.8	1.0	5	309	0	175	253	98.5
0.9	1.0	5	335	0	170	247	96.1
1.0	1.0	4	285	0	164	221	86.1
0.6	0.7	12	402	0	242	257	99.9
0.6	0.8	10	369	0	216	256	99.7
0.6	0.9	9	387	0	236	257	99.9
0.6	1.0	8	392	0	222	257	99.9
0.6	1.1	7	383	0	195	257	99.9
0.6	1.2	7	409	14	190	257	99.9
0.6	1.3	7	437	44	189	257	99.9
0.6	1.4	7	470	77	189	257	99.9
Irrigate every 10 days		9	392	58	252	213	82.9

* Depf is the fraction of available water depletion allowed before irrigation.

** Irrigf is the fraction of depletion applied for each irrigation.

water used depending on the criteria. If Depf is low then the number of irrigations is higher because smaller amounts of water are applied each irrigation. Most of the extra irrigation added with higher irrigation frequency is lost to increased soil evaporation. For example, a Depf of 0.3 required 17 irrigations and 470 mm compared to 7 irrigations and 388 mm for Depf of 0.7. The difference of 82 mm is the same as the difference of soil evaporation. Soil evaporation is essentially wasted water since it does not contribute to plant growth. Thus the most efficient irrigation regime will minimize soil evaporation.

If Depf is held constant at 0.6 the data of Table 2.5 show that Irrigf can be changed from 0.7 to 1.4 without any noticeable decrease in yield. However as Irrigf increases, irrigation amount decreases from 0.7 to 1.1 and then increases caused by water loss as drainage. This illustrates one problem with this method in that drainage is not considered. Since it is desired to fill up as much of the depletion as possible errors in the measurement of Dep and irrigation applied could cause some drainage. Since it is very difficult to measure drainage it may occur and not be recognized. Note that drainage could occur as a result of one irrigation or rainfall event though monthly or seasonal water balance calculations may not predict it. Nevertheless using this method is a vast improvement over not using any method at all, which is the common practice in much of the world.

There are, however, problems with measuring soil water conditions as discussed above. One problem is that the measurements are made at one point in a field and can only made at specified times (weekly). Thus many measurements have to be made to account for different crops and soil spatial variability. Also estimates have to be made of what will happen between measurements. Field workers usually plot data on a graph and extrapolate the data with time to tell the next time to irrigate. This works well after experience is gained. Still a lot of work and effort is required to make this method useful and the manual measurements are tedious.

It is interesting to compare the results of irrigation according to depletion to the commonly used equal time method (last line of

Table 2.5). Equal water application resulted in a relative yield of 83% with 392 mm of irrigation compared to 100% for the same irrigation using the depletion method. The result was caused by less water being available for T because it was lost as drainage, 58 mm,or soil evaporation, 252 mm.

Another variation of the field measurement scheme is to use tensiometers or gypsum blocks instead of actual soil moisture measurement. This has done by several service groups but has the disadvantage of not knowing the amount of water depletion and limited range of tensiometers. There is an advantage, however, in that these instruments are more easily connected to an electronic control system so that irrigation can be turned on and off under instrument control in an automatic system. Thus the analogy to the thermostat is closer. Commercial instruments are available for this purpose especially for lawn irrigation.

We have done some research on this method and have gained insight into the characteristic of such a system. An experiment was conducted in 1988 (Or and Hanks, 1991) to evaluate this system. One problem encountered is determining a suitable set point to turn the irrigation on and off. Our study used four regimes as shown in Table 2.6. This system as such does not lend itself to control of drainage. Drainage was very high where the set point was set high. A set point of 15-5 kPa (irrigation on at 15 kPa matric suction and off at kPa matric suction) required 3110 mm of irrigation of which 2360 mm drained through the soil. This set point is around field capacity. The best yields were for a set point of 60-22 that corresponded to Depf (Table 2.5) of about 0.5. The two wettest treatments had decreased yields because of excess leaching of nutrients and/or aeration problems and the driest treatment had insufficient water. The sensors were located at 450 mm deep in the soil that might have caused too much water application before turning off. This is,. of course, an extreme situation but shows the negative effects of excess drainage on yield. Conditions could certainly be improved if the amount of water applied was limited or determined more efficiently.

Table 2.6. Results of water applied, number of irrigations, drainage and yield for several set points for an automatic irrigation system.

Set point	Irrigation	No. irrig.	Dr	Rel. yield
15–5 kPa	3110 mm	64	2360 mm	43%
35–15	1570	16	890	64
60–22	860	10	215	100
120–30	615	4	75	87

These treatments shown in Table 2.6 were all located near the middle of a point source sprinkler that also provided information on dry matter yield and irrigation for a wide range of water applications both in amount and frequency as shown in Fig. 2.4. Regardless of the set point these treatments produced a yield-irrigation relation that fits all of the treatments reasonably well. There is a fairly wide range of irrigation where yield was near maximum before decreases occurred due to drainage.

Where the delivery systems cannot serve all fields at once, advance planning must be practiced to avoid unacceptable losses or excessive costs. Suitable computer models of PET and Kc become important tools in water balance management. Thus, estimates of PET and Kc play a major role where advance planning is required for efficient water delivery.

However, on many large farms, water pumping and delivery are still unnecessarily costly because of a lack of coordinated water delivery. The highest costs result from peak power demand charges at certain times and from excessive working pressures in the water distribution system when advance planning and coordinated delivery are not practiced. Also, drainage losses and fertilizer leaching are the consequence of inefficient irrigation management.

With coordinated water delivery and irrigation scheduling to optimize crop production costs, peak-demand power costs from pumping irrigation water can often be reduced more than 30%. Whereas simple graphic scheduling based on field measurements may suffice for individual fields (Gear et al., 1977), a more complex

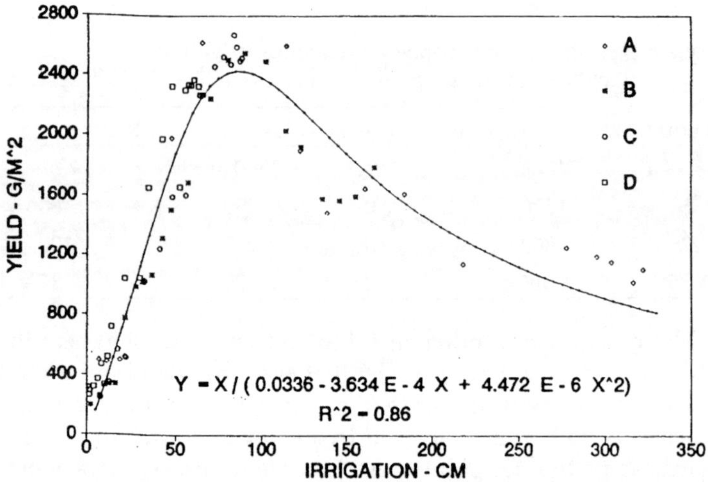

Fig. 2.4. Dry matter yield-irrigation water production function for corn.

scheduling system is needed for large irrigation systems. The main computer function is to coordinate water delivery to distribute the water pumping load uniformly over time and to avoid unnecessary peak power demand costs. It thus becomes necessary to predict ET.

3. Irrigation Scheduling Process Using ET Predictions and Soil Water Measurement

Despite efforts to the contrary, reliable and efficient water management requires periodic soil moisture measurements, regardless of the quality of predictive models. Even a perfect mathematical model of water use requires a starting point. With measured water balance as the numerator and modeled water balance as the denominator, a ratio called a field-factor (ff) is calculated (Campbell and Berrett, 1990). The field-factor is used to produce reliable projections of water balance for each field and to reveal individual field variations from the average use for each crop type. A method for selecting appropriate crop coefficients from a crop curve was also developed. The field-factors were then used to evaluate and refine the crop coefficients. Crop coefficient (Kc) are the ratio of actual water use divided by a reference, such as evaporation from a Class A pan (Wright, 1981). Crop coefficients are essential when computer models of potential evapotranspiration are used to plan and optimize water delivery to many fields.

When computers and field measurements are used to model history, crop growth and water use can then be predicted, and advance planning becomes possible. This kind of predictive modeling is useful in scheduling water delivery for optimum efficiency. The crop coefficient becomes the essential link between the past and the modeled future. Reliability depends on the quality of water use measurements and on the repeatability of the reference (Wright, 1982).

The daily water use coefficient for a crop, taken over a growing season, is a crop curve. Some people divide the crop curve into two parts (Jensen, 1974). The first part provides the crop coefficients from planting to full cover, with the x-axis designated in per cent of time to full cover. The second part provides the crop coefficient from full cover to harvest or dormancy, with the x-axis designated in days after full cover.

Problems with the crop coefficients arise from normal but non-average influences on the crop, such as early or late planting, variety, abnormally cold or not weather, pesticide effects, plant population density, and cultivation. Some mechanism is needed to account for each non-average influence. The field-factor serves this purpose. The field-factor allows us to compute a reliable water balance from which irrigations can be scheduled correctly, regardless of model shortcomings. We will describe some of the most useful aspects of the field-factor and show how to use it automatically to adjust crop coefficients.

A nearby Class A evaporation pan with long-term evaporation record may serve as a good reference standard for long-term PET. A polynomial fit to the long-term evaporation pan data may be used to project long-term PET for irrigation scheduling. Neutron probes are efficient tools for accurate measurements of soil water content and crop water use. The crop coefficient can be found as the ratio of measured crop water use to pan evaporation during a season. The Kc-time curves can be fitted to (Jensen, 1974) separate polynomials to each part of the crop curve. A long-term weather record can be used in the modified version of the Penman equation to estimate the long-term pan record. Once fit, current weather data and weather forecasts may be used in the modified version of the Penman equation (Jensen et al., 1989) to model Class A pan evaporation and include short-term weather effects. We suggest modifying the linear wind function commonly used with the Penman equation to a log function to make the modeled PET match pan evaporation. Projected PET should be modified to account for recent and forecast weather, with convergence on the long-term pan loss in two weeks.

While irrigation scheduling may begin with two-part crop curves found in published literature, use of these curves can cause significant errors, particularly during early and late-season irrigations when peak power demand costs are the highest. Consequently, we recommend using a field-factor to obtain new crop coefficients unique to each field. In the process, each field will give its own unique Kc because of the cultural practices to which the crop is subjected.

As a unique crop coefficient is fit to each field, it will become evident that irrigation scheduling could begin with any crop curve because the field-factor would modify the curve to fit the field. Also, the average two-part crop coefficients from published literature can have a polynomial fit to the front half and its integral used to adjust for different planting dates and weather conditions. The back half of the curves may likewise be adjusted to irrigate for pre-harvest conditioning and scheduled harvest dates.

Solar Thermal Unit (STU) accumulation (Caprio, 1971) was used to deduce for each field the correct coefficient from the general crop curve.

$$STU = C_t\,(T - 31) \qquad [2.12]$$

where C_t, is the reciprocal of average temperature during the hottest month, and T is the average temperature of the day in question. The summation of equation [2.12] represents PET, as shown in equation [2.13].

$$PET = STU \times 10^{-5} \qquad [2.13]$$

The Sum (STU) may be used to obtain the potential water use between planting and full cover by the best-yielding fields of each type crop. Then, the following expression may be used to obtain the correct crop coefficient

$$Kc = Kco\,(J - P) \,/\, (C - P) \qquad [2.14]$$

where Kc and Kco are the corrected and original coefficients, respectively, and (J - P) / (C - P) is the ratio of two integrals. The denominator is the integral of the original crop curve polynomial, expressed as solar-thermal units, integrated from the lower limit P to the upper limit C, where P and C are average planting and cover dates. The numerator is the same polynomial integrated from the lower limit P to the upper limit J at which the integral equals the current Sum (STU). The value J is the current per cent of full cover. The crop curve at the J value is Kc.

The integral of the denominator for potatoes and corn grown in Southcentral Washington is about 1,200,000 STU, or 305 mm of pan evaporation. Using these models, one can correct for influences of cold and hot weather and early and late planting on the crop coefficient.

Finally, after projected crop water use has been estimated as closely as possible, remaining errors may be detected and removed by adjusting each field model by a unique field-factor computed as

$$ff = \frac{\{[(HMcur - HMprev)/Hp] - (NPcur - NPprev) + Rain\}}{Kc * PET} \qquad [2.14]$$

where HMcur and HMprev are current and previous hour metre readings, NPcur and NPprev are current and previous neutron probe readings, Hp is the irrigation system hours per unit water delivery, and Kc • PET is the modeled ET. A perfect ratio would yield 1.00, which is obtained for scheduling by dividing both sides of equation 2.14 by the field-factor. A value less than 1 indicates that actual use was less than predicted or that modeled use or delivery was greater than expected. A value greater than 1 indicates that use was greater than predicted or that delivery or modeled use was less than expected. Deviations must always be held suspect until their cause is proven. A fixed sequence of possible causes of error should always be checked until the causes of deviation is known. The average of computed field-factors should be allowed to vary between 0.8 and 1.1, with an additional 10% allowed for any individual field. Causes of greater deviations must be found and corrected or otherwise accommodated in the models. The field- factors may then be used to generate a field-specific irrigation schedule.

The irrigation scheduling scenario outlined here has significantly refined irrigation planning, improved water and nutrient use efficiency on irritated crops, and reduced loss to groundwater, as evidenced by disappearance of seeps. This type of irrigation scheduling helps smooth fluctuations and reduce peak power demand by providing better advance planning. Simultaneously, higher crop yields, reduced electric power consumption, and decreased water application have resulted from use of this type of irrigation scheduling. Some of the unexpected side benefits include the ability to detect incorrect nozzles, poor efficiency practices, and other aberrations in equipment and practice, such as high-per cent run-time setting and faulty hour-metres on centre pivots.

Table 2.7 shows an actual field-factor summary from an irrigation schedule in which field 12 had the wrong size irrigation nozzles. The associated schedule is shown in Table 2.8, confirming the extra water application resulting from the large nozzles and showing by the "H" that field 12 has been over watered for more than a week. In this case, we identified the problem from the Field-Factor Report and confirmed it by field inspection of the centre pivot system.

Thus, periodic measurements of soil moisture coupled with automatic adjustment of crop coefficients to fit in individual field conditions can provide a useful tool for accurate irrigation scheduling while optimizing use of water and power.

Table 2.7 Field-factor summary sheet showing unusually small field factors for field 12. Investigation revealed oversize sprinkler nozzles on field 12.

		Field-factor Repot (Measured/Modeled)				
Fld	CR	8/01	7/29	7/25	7/22	7/18
10	P	1.00	0.88	0.82	0.77	0.86
12	P	0.76	0.53	0.54	0.49	0.51
14	P	1.02	0.85	0.83	0.85	0.94
58	P	1.14	0.84	0.90	0.91	0.89
60	P	1.18	0.87	0.90	0.95	0.97
61	P	1.05	0.89	0.92	0.87	0.95
Avg		1.03	0.81	0.82	0.81	0.85
SD		0.15	0.14	0.14	0.17	0.17

Table 2.8 Irrigation schedule showing Field 12 over-irrigated and that soil moisture had been above optimum longer than one week as indicated by the "H" at the left of the field number.

Irrigation Schedule (by priority)

	Measurement				... mm to apply by midnight ...					
	FLD	Time	Metre	FC%	Fri	Sat	Sun	Mon	Tue	Wed
	10	9	5607	89	3.3	5.3	5.8	5.1	5.3	5.3
	58	10	6710	89	4.3	5.8	6.4	5.6	5.8	5.8
	14	12	5683	94	1.8	5.3	5.8	5.1	5.3	5.6
	60	12	1228	100	0.3	5.8	6.4	5.6	5.8	5.8
	61	12	1185	100	0.0	5.3	6.1	5.3	5.6	5.6
H	12	11	4466	100	0.0	1.3	5.3	4.6	4.8	4.8

REFERENCES

Caprio, J.M. 1971. Circ. 251; NOAA : 71052401.

Clawson, K.L. and B.L. Blad. 1982. Agson. J. 74 : 311-316.

Doorenbos, J. and W.O. Pruitt, 1977. Crop water requirements (revised). FAO Irrigation and Drainage Paper 24, United Nations, Rome, Italy.

Gear, R.D., A.S. Dransfield and M.D. Campbell. 1977. J. of Irrig. and Drain. Div., ASCE. 103(IR3) : 291-298.

Hanks, R.J. 1974. Agson. J. 66 : 660-664.

Hanks, R.J. 1983. Ch. 9 in Amer Soc. Agron. Limitations to efficient water use in crop production. Madison.

Hanks, R.J. 1985. Proc. Nat. Sym. Advances in Evapotranspiration. ASAE 114-121.

Kidman, R.L. 1990. Ph. D. thesis, Utah state university, Logan, Utah.

Jensen, M.E. (ed.). 1974. Consumptive use of water and irrigation water requirements. ASAE, New York.

Jensen, M.E., R.D. Burman, and R.G. Allen (eds.). 1989. Evapotranspiration and irrigation water requirements. A manual prepared for Irrig. and Drain. Div., ASCE, New York.

Or, Dani and R. J. Hanks. 1992. Irrig Sci (in press).

Retta, A. and R.J. Hanks. 1980. Utah Agr. Exp. Stn. Res. Rep. 46.

Stewart, J.I.,R.E. Danielson, R.J. Hanks, E.B. Jackson, R.M. Hagan, W. O. Pruitt, W.T. Franklin, and J.P. Riley. 1977. Optimizing crop production through control of water and salinity levels in the soil. Utah Water Lab. PRWG 151-1. Logan.

Wright,J.L. 1982. Proc. of ASCE, J. Irrig. and Drain Div. 108(IR1) : 57-74,

Wright, J.L. 1981. Irrigation scheduling for water and energy conservation in the 80's. ASAE, 18-26.

3

Meteorological Approaches For Predicting Irrigation Needs

J.L. Steiner and T.A. Howell

I. DEFINING IRRIGATION REQUIREMENTS

The goal of irrigation is to supply water to a crop so that its water requirements are met and it can produce an acceptable yield. However, the amount of water that must be applied as irrigation is often less than the crop water requirement, because part of the crop water requirement can be met by precipitation, stored soil water, or sometimes by upward flow from a shallow water table (Abdulmumin, 1989; Ghali and Svehlik, 1988; Steiner et al., 1985; Tripathi et al., 1987). The water requirements for various crops, grown in different times of the year, under different types of management vary considerably. Growing season heat unit accumulation, frost-free periods, and maximum and minimum temperatures, determine which crops are suited to production in various times of the year. Rainfall and soil water storage patterns determine the extent to which crop water needs must be met by irrigation to avoid the risk of water stress at various times of the year.

In initiating an irrigation project or in evaluating the efficiency of existing projects, there are many tools available to analyze the agroclimatic potentials and limitations for crop production. An excellent review of characterization and classification of agricultural environments is given in Bunting (1987). Agroclimatic analyses for Africa (FAO, 1978a), southwest Asia (FAO, 1978b), South and Central America (FAO, 1981) and southwest Asia (FAO, 1980) give a good overview of agroclimatic characteristics of these regions, organized around the concept of agro-ecological zones. While these analyses were not conducted with irrigation requirements in mind, they provide a clear guideline to identifying the need for irrigation for crop production in various regions.

To have a successful, sustainable irrigation system, more water must be applied than the net irrigation need because some of the water will be evaporated from the soil and/or from plant surfaces at rates greater than unstressed plant transpiration rates. Drainage and salinity management must also be considered in any irrigation system. Additionally, irrigation system losses during conveyances and application have to be considered. Meteorological conditions drive the evapotranspiration (ET) process, which is a major factor in the water balance. Thus a clear understanding of the effects of meteorological factors on ET is important in order to correctly assess a regional or field scale water balance and to properly manage irrigation systems or projects. A knowledge of many non-meterological factors is required to understand drainage requirement and irrigation efficiency, so this chapter will emphasize evaporation and crop water use and the reader is referred to other chapters in this book for more detailed discussion of drainage requirements and irrigation efficiency. Other reviews describing ET processes and estimation of ET which cover aspects not considered in this chapter (e.g., historical aspects, equation forms for many prediction methods, microclimatic aspects of ET, radiation interception) are given by Brutsaert (1982), Hatfield (1990), and Monteith (1973).

Both over-and under-irrigation can reduce the effectiveness and sustainability of irrigation. Historically, more attention has been given to the problems of under-irrigating than to over-irrigating. The effects of under-irrigating are generally exhibited on a field level often within the immediate growing season while many of the problems associated with over-irrigating are exhibited on a regional basis over a longer time period. Two major hazards of under-irrigating are yield loss, which reduces gross economic returns to the system, and salinization of the soil. Salinization exhibits itself over a longer time period than does yield loss; therefore, it has often been ignored until serious degradation of a soil has already occurred.

Over-irrigation can likewise have negative economic and ecological impacts. It increases costs through drainage of water below the root zone with associated leaching of nutrients and chemicals. In situations where these inputs are undervalued or costs are subsidized, the likelihood of over-irrigation increases. Yield reduction may be associated with over-irrigation if leached nutrients and chemicals are not replaced or if waterlogging of a field inhibits crop growth. Waterlogging may be a particular problem during crop establishment (Kanwar, 1988; Kanwar et al., 1988), and is obviously more of a hazard on some soils and with some crops than others. Water applied at rates greater than the soil infiltration rate or in quantities beyond the water storage capacity are lost to runoff or subsurface flow and are associated with ecological degradation, usually off-site. The magnitude of the water quality and environmental problems associated with agricultural use of water resources has gained increased visibility in recent years (Chen and Druliner, 1987; Higgins et al., 1988; Mapp, 1987; National Research Council, 1989) and will likely force improved water management in the future.

II. THE WATER BALANCE

Irrigation is the most common and direct form of manipulating a water balance to improve the conditions for crop production. The water balance for a one dimensional system is defined as

$$SW_t = SW_i + Pr + I - R - D - E - T \quad [3.1]$$

where SW_i and SW_t are soil water content at the beginning and end of a period, respectively, and Pr is precipitation, I is irrigation, R is surface runoff, D is drainage or flux across the lower boundary of the soil profile, E is evaporation from the soil surface, and T is transpiration, all over the same time period. The E and T terms are often jointly referred to as ET because of the difficulty of measuring the components separately. Water balances have historically been applied in two predominant ways in irrigation science : (1) use of soil water measurements to determine ET or (2) use of estimated or measured ET to estimate soil water content over time. When equation [3.1] is rearranged to define ET in terms of the other components of the water balance, ΔSW represents the measured change in soil water content (SW_i - SW_t) over the period. Water balance approaches for determining ET have been used to develop, test, and calibrate numerous meteorological methods of estimating ET.

The accuracy of the water balance method for determining ET depends on the accuracy with which the components can be measured or estimated. Drainage across the root zone cannot be measured easily under field conditions so ET estimates should be made only when the flux across the lower boundary of the soil profile can be determined (such as with the use of lysimeters) or assumed to be negligible. Precipitation, irrigation, and surface runoff can be measured if the defined area is reasonably small and uniform. A water balance for a watershed may be inaccurate because effective precipitation and irrigation will not be uniform across the area and is very difficult to measure. The accuracy in the measurement of the temporal change in soil water content depends on the method of measurement, and the magnitude of the change in soil water content.

Several methods of measuring SW are discussed in detail by Cary (1981), Schmugge et al. (1980), and Stafford (1985). A widely used method of determining soil water content is by gravimetric sampling. There are inherently large errors involved in gravimetric sampling due to the spatial variability of soils and soil water content in the field and the small volume of soil sampled. Therefore, gravimetric sampling should not be used to estimate ET on a short term basis (daily or weekly) when ΔSW would be small, relative to the other components. Neutron attenuation provides a more accurate measurement of SW because repeated measurements are made over time at the same sampling location and a larger volume of soil is sampled. However, neutron probes are expensive and require trained (and, in the U.S.A., licensed) operators. They are most practical when one group or organization is responsible for soil water measurement on many fields in a region.

Only weighing lysimeters can be used to determine changes in soil water content accurately enough under field conditions to determine ET on a daily or shorter time-scale. Therefore, much of the work leading to the development of ET prediction equations and improvement in our understanding of the physical principles involved in the ET processes has been done using lysimeters. Design and management factors which affect the precision, accuracy, and representativeness of lysimeters is discussed by Aboukhaled et al. (1982), Marek et al. (1988), Pruitt and Lourence (1985), and Tanner (1967).

III. ESTIMATING ET

Relationships between weather and evaporation processes have been analyzed throughout the development of civilization, as excellently reviewed by Brutsaert (1982). The application of ET prediction for irrigation design and management has roots in the work of Penman (1948), Thornthwaite (1948), Blaney and Criddle (1950), and others. The most predominant approach to calculating ET for irrigation management involves (1) calculation of a "potential" ET (ET_p) that depends only on climatic factors, and is independent of surface conditions, (2) relating "actual" ET of a specific crop at a given period of growth to ET_p using empirically derived "crop coefficients", and (3) using the estimated actual ET in the water balance to determine specific irrigation needs based on irrigation efficiency, soil water holding characteristics, and management objectives (Kanemasu et al., 1983). Jensen (1974), Jensen et al. (1990), Doorenbos and Pruitt (1977), and Doneen and Westcot (1984) present comprehensive and practical overviews of estimating ET for irrigation water management applications. Characteristics specific to sprinkler (Rolland, 1982), drip or trickle (Vermeiren and Jobling, 1984), or other types of irrigation that affect management options and system efficiency should be considered in quantifying the irrigation requirement.

A. Climatic Factors Related to ET

Climatic factors that affect ET include temperature, solar or net radiation, precipitation, vapour pressure deficit, wind, and barometric pressure. These data should be collected at a site that is maintained in a standard condition which is generally over short grass away from buildings and trees. If data are collected in non-standard conditions, they must be empirically adjusted to a standard condition before application of equations for predicting ET. Jensen et al. (1990) discuss errors in ET prediction that might be associated with climatic data collected from nonstandard sites. The sensitivity of an ET model to the climatic input variables was analyzed by Camillo and Gurney (1984) who found the prediction was sensitive to errors in radiation and relatively insensitive to errors in wind over both bare soil and wheat surfaces, but was much more sensitive to errors in vapour pressure and temperature measurement over a wheat crop than over a bare soil. Heermann (1988), in a discussion of priorities in ET research for irrigation, identified the uncertainty of limits of transferability of meteorological data for ET prediction as a high priority research need. Harcum and Loftis (1987a, 1987b) conducted a statistical analysis of regional weather networks required for ET prediction for irrigation or hydrologic studies. Analysis of model sensitivity to input parameters will require increased attention as more realistic water balance models are implemented and as physically-based models are substituted for empirical models, where local calibration of equations often corrects for systematic biases in input data.

Temperature and precipitation are the most widely measured weather variables in weather networks, because of the relatively low expense and ease of maintenance of instrumentation. Many of the climatic variables can be extrapolated from weather stations on a regional basis, but precipitation is so spatially variable that on-site measurement is required for accurate water balance work and field-level irrigation management. However, long-term precipitation records from standard sites are invaluable for agroclimatic analyses and irrigation system design. Because thermometers are sensitive to radiation errors if exposed to direct sunlight, weather station temperature measurements are often made in white, ventilated shelters with the sensors mounted about

1.5 m above the surface. Temperature is a major controller of crop growth and phenological development in addition to the ET processes.

Solar radiation is less commonly measured, and much of the data available to determine solar radiation are observations of cloud cover or per cent of possible sunshine hours. The actual driver of ET is net radiation but it is seldom measured outside of the research environment. For most applications, net radiation is calculated as a function of solar radiation, temperature, vapour pressure, and albedo. Radiation is a major factor influencing crop growth as well as ET.

Wind is widely measured, but the siting of the measurement stations is highly variable. A high percentage of wind measurement sites were established for aviation purposes. Measurement of wind is highly sensitive to the surrounding conditions and is often affected by buildings or vegetation. Because the windspeed increases logarithmically with height within the planetary boundary layer, it is essential that the height of measurement be known.

Dewpoint or wet bulb temperature data for determination of vapour pressure deficit are very difficult to collect on a routine basis in a weather network. Relative humidity is often collected instead, but unless air temperature at the time of the relative humidity measurement is known, it can be difficult to accurately interpret the numbers. Vapour pressure deficit is inversely related to water-use efficiency (Tanner and Sinclair, 1983).

Barometric pressure is generally estimated from altitude for application to ET equations and this estimate is simple and adequate for ET prediction. The influence of barometric pressure on ET prediction is sometimes overlooked by practitioners. Use of mean sea level barometric pressure can lead to slight overestimation of ET at high altitudes because of the effect of pressure on the psychrometric constant. Altitude also interacts with the thermal environment as well as the radiative environment.

B. Estimating Potential ET from Meteorological Data

1. Empirical ET equations : Many researchers have attempted to describe mathematical relationships between ET and various climatological or meteorological factors. Temperature methods were developed by Thornthwaite (1948) and Blaney and Criddle (1950) to estimate seasonal ET from mean temperature data. Data for these methods are readily available. Thornthwaite's method has been widely used for climatological classification but has not been widely applied for irrigation planning because it is generally not accurate when applied to dry, advective climates. Doorenbos and Pruitt (1977) described the Blaney-Criddle method and gave tabulated coefficients to estimate latitude effects on monthly radiation and to relate regional windspeed, humidity, and cloudiness conditions to ET. This method is appropriate to estimate monthly or seasonal ET and, therefore, is most commonly used for irrigation system design considerations. It is not applicable for in-season irrigation management decisions. Numerous other limitations to this method are discussed in Doorenbos and Pruitt (1977).

Radiation methods have been proposed by Hargreaves and Samani (1982), Makkink (1957), Jensen and Haise (1963), and Turc (1961), where radiation data are required in addition to temperature. These approaches were developed for in-season irrigation management, and thus are applicable to shorter time-scales than monthly or seasonal. A radiation method presented by Doorenbos and Pruitt (1977)requires observed sunshine hours or measured radiation and temperature with empirical coefficients based on general humidity and wind conditions. The Jensen and Haise (1963) model uses daily temperature and solar radiation as well as long- term temperature data and altitude. It has been widely used in the western U.S.A. for irrigation scheduling programmes.

Christiansen (1968) and Christiansen and Hargreaves (1969) developed regression equations relating ET to pan evaporation or to radiation data and temperature, wind, humidity, and sunlight functions. Evaporation pans have been widely used, and, if properly sited, calibrated to a specific crop and region, and carefully maintained, can provide estimates of crop water use requirements. However, the data cannot be extrapolated beyond the immediate region and application for crops grown in non-typical seasons is questionable (e.g., equations developed for a rainy season crop should not be applied to a post-rainy season crop).

The success of empirical methods using limited weather data relies upon the correlation among climatic factors. Radiation is highly correlated with ET_p since radiation supplies much of the energy required for the vapourization of water. Temperature methods rely upon the correlation of temperature to radiation and vapour pressure deficit. Errors can arise because seasonal storage and release of heat from the soil produces annual temperature and radiation cycles that are out of phase.

2. Aerodynamic methods : The equation developed by Penman (1948) predicts potential ET (ET_P) as the sum of a radiative component and an aerodynamic, water vapour pressure component. This model combines energy balance and aerodynamic principles and is often termed the combination equation. It was developed to predict ET_P (mmd^{-1} defined as evaporation from a short, well-watered green crop

$$ET_P = [\Delta/(\Delta + \gamma)]\ [(Rn + G)/L] + [\gamma/(\Delta + \gamma)\, f(u)\, (VPD)] \qquad [3.2]$$

where Δ is the slope of the saturated vapour pressure- temperature curve (kPa/C), γ is the psychrometric constant (kPa/C), L is the latent heat of vapourization (MJ/kg), Rn is net radiation (MJ $m^{-2}d^{-1}$), G is soil heat flux (positive toward the soil surface, MJ $m^{-2}d^{-1}$), f(u) is a wind function [mm/(d kPa)], and VPD is vapour pressure deficit (kPa). In this equation, decreased atmospheric pressure with increased altitude decreases the psychrometric constant and the contribution of the advective component to ET. Penman described the f(u) as an empirical, linear function of the form

$$f(u) = a + b\,U \qquad [3.3]$$

where U is average daily windspeed at 2 m (ms^{-1}) and a and b are regression coefficients. Penman (1948) reported values of 2.63 and 1.38 for a and b, respectively, for a grass surface. Doorenbos and Pruitt (1977) evaluated data from many lysimeter sites worldwide and derived a linear wind function with a = 2.70 and b = 2.33 for a grass surface, which they recommended if a locally calibrated wind function is not available. Many other functions have been locally fit. Phene et al. (1986) reported that their data were relatively insensitive to different a and b coefficients, but reported that ET was highly sensitive to mowing the grass on and around the lysimeter that indicates a possible problem with the assumption of Penman (1948) that evaporation would be insensitive to the surface vegetative cover as long as there was a short, green, fully-transpiring canopy. Tanner and Fuchs (1968) pointed out that plant surface can affect soil heat flux, albedo, emissivity, and roughness lengths and, therefore, surface properties should affect ET. The Penman (1948) equation has been used successfully in many forms to predict ET. The combination equation was recommended by Doorenbos and Pruitt (1977) as the method of choice for situations where adequate climatic data are available.

Many attempts have been made to develop a less empirical form of the wind function for the combination equation. van Bavel (1966) developed an aerodynamic form of the resistance term, similar to earlier work of Businger (1956) and others, which incorporates the standard wind profile theory assuming neutral atmospheric conditions, and similarity of the transfer coefficients of sensible heat and vapour. Thom (1972) developed a meteorological based model, again with the aerodynamic resistance term based on the standard log profile for wind, which accounts for separate sinks for sensible heat and vapour. Monteith (1965) developed an aerodynamically-based form of the combination equation that incorporates a canopy or surface resistance as follows

$$ET_P = [\Delta/(\Delta + \gamma^*)]\,[(Rn + G)/L] + [\gamma/(\Delta + \gamma^*)]\ Ea \qquad [3.4]$$

with Ea (mm d^{-1}) given as

$$Ea = [(\rho\ Cp\ VPD)/(L\,\gamma\,r_{av})]\,(8.64E\text{–}2) \qquad [3.5]$$

where ρ is air density (kg m^{-1}), Cp is specific heat of dry air at constant pressure (J/kg/C) and r_{av} is the aerodynamic resistance (s/m).

The γ^* is defined as

$$\gamma^* = \gamma\,(1 + r_c/r_{av}) \qquad [3.6]$$

where r_c is the canopy resistance (s/m). The r_{av} can be calculated from the log wind profile as

$$r_{av} = \ln[(Z\text{-}d)/Z_{om}] \ln[(Z\text{-}d)/Z_{ov}]/k^2 Uz \qquad [3.7]$$

where U_Z is windspeed (m s^{-1}) measured at height Z, d is the zero-plane displacement height, and Z_{om} and Z_{ov} are the surface roughness lengths for momentum and vapour, respectively and k is von Karman's constant (0.41).

Under conditions with a flat, uniform, vegetative surface such as grass, alfalfa (*Medicago sativa* L.), or a full canopy cover row crop, the d, Z_{om}, and Z_{ov} terms are often estimated as a function of vegetation height (Monteith, 1973). With complex surfaces, surface properties may need to be accounted for (Abtew et al., 1989; Azevedo and Verma, 1986; Shaw and Pereira, 1982). For non-stressed, uniform, full cover vegetative surfaces, r_c can be estimated as a function of radiation and leaf area index. For other conditions such as partial canopy cover, sloped fields, stressed vegetation, or rapidly changing partly cloudy conditions, lack of understanding of the behaviour of r_{av} and r_c might limit application of the resistance forms of the combination equation. Idso (1983) proposed a method of calculating r_c by measuring the canopy temperature and solving the energy balance, which might provide an estimate of r_c over a wider range of conditions. Allen (1986) and Allen et al. (1989) have derived and tested ET_p equations using the Penman-Monteith form for general application to irrigation scheduling.

The combination method does not account for interactions between the plant and soil surfaces. When small, isolated plants are surrounded by hot, dry soil surfaces, the microclimatic conditions at the leaf surface are vastly different from ambient weather conditions. Shuttleworth (1976) described a multi layer model with the lowest level at the soil surface and the highest level above the canopy. The data required to apply the model are not routinely available, but his work provides a theoretical framework to consider cases where flux at the soil surface affects fluxes within and from the canopy.

Tanner and Fuchs (1968) discussed the limitations of the Penman- Monteith equation, which assumes no saturation deficit at the surface, and proposed that inclusion of a surface temperature term eliminates the requirement for this assumption and makes the form applicable to actual as well as potential ET prediction. For conditions where a complex surface is involved (such as with a sparse canopy, a drying soil surface, or a residue or mulch layer that partially covers the soil), additional resistance terms must be estimated to estimate the "average" surface resistance. Examples of this are given by Shuttleworth and Wallace (1985) for a sparse canopy, Staple (1974) for a drying soil, and Snane et al. (1984) for a surface with crop, residue, and soil components.

Many attempts have been made to simplify the Penman equation, particularly for conditions where wind and vapour pressure measurements may not be available. Priestley and Taylor (1972) hypothesized that a strong correlation would exist between the radiative and advective components of the Penman equation under conditions of near-saturation (the potential ET condition requires non-water limiting conditions, by definition) and that the radiative component would be dominant. They described potential ET as

$$ET_P = \alpha \ [\Delta/(\Delta + \gamma)] \ [(Rn + G)/L] \qquad [3.8]$$

and proposed a value of 1.26 for α, based on empirical analysis. The Priestley-Taylor model was developed for large-area prediction, but because of its simplicity it has been adopted for many short-term, local applications.

Jury and Tanner (1975) proposed that α would vary as a function of VPD and proposed a modification that would improve the applicability of the Priestley-Taylor model to regions with variable, advective climate. In order that the data requirement to implement the model not be increased, Jury and Tanner (1975) proposed that dewpoint temperature could be estimated as the daily minimum temperature, but this method would lead to underestimation of the VPD and ET in semiarid or arid climates where minimum temperature is generally considerably warmer than the dewpoint. Kanemasu et al. (1976) proposed an advective modification of α based on maximum air temperature.

Steiner et al. (1989a) tested several forms of ET prediction equations against well-watered, full-cover grain sorghum ET measured on weighing lysimeters and found that the most accurate daily ET predictions were given by the Penman-Monteith form of the combination equation. The Priestley-Taylor model worked quite well with either VPD - or temperature-based advective modifications of α, but the original Priestley-Taylor model was not suited for ET prediction in the semiarid, variable environment at Bushland, Texas, in the U.S. southern Great Plains. Forms of the Penman equation with empirical, linear wind functions over-predicted sorghum ET considerably in this windy environment. Application of any wind function, including the Penman-Monteith form, worked best if daily mean 2 m windspeed was capped at about 3 m s^{-1} (Howell et al., 1989b), which is lower than most daily mean windspeeds in the region.

C. Estimating Actual ET

1. Crop cover effects : Estimating actual ET from a ET_P generally involves gross simplifications of very complex processes. When the plant is not stressed, but the surface is not completely covered by vegetation, actual ET is limited by the amount of crop cover and the condition of the surface that is not covered by vegetation. There is some understanding of the dynamics of evaporation from a bare soil, but little quantitative description of evaporation from residue covered or mulched soils. Prediction of ET from stressed crops may have been de-emphasized because the purpose of irrigation was to prevent stress. however, with increased emphasis on optimization of irrigation and yield, rather than maximization of yield, and on off-site effects of irrigation, the possibility of allowing mild plant stress cannot be discounted and prediction of ET as water becomes limiting is more important.

a. *Crop coefficients.* The most common method of estimating actual ET from ET_p, particularly for irrigation applications, involves empirical crop coefficient (K_c) curves as described by Wright (1985). A crop coefficient curve basically gives the time course of the ratio of actual ET to ET_p for a crop at a location. The K_c is a basal coefficient which assumes that the soil is not wet but that the soil profile is moist enough that some evaporation occurs. Doorenbos and Pruitt (1977) describe how to develop a K_c curve from generalized tables based on crop, sowing date, rate of crop development, length of crop growth stages, and prevailing climatic conditions.

Because irrigation frequency differs tremendously with different irrigation systems and crops, it is necessary to consider the effect of soil surface wetness, particularly with frequent irrigation. Doorenbos and Pruitt (1977) describe the basal K_c for a bare soil as a function of irrigation and rainfall frequency. Following rain or irrigation, the K_c must be adjusted for wet soil surface conditions as described by Wright (1982) who also gives K_c curves for several crops in the northwestern part of the U.S.A. Sammis et al. (1985) and Hattendorf et al. (1988) presented crop coefficients with the time scale represented by cumulative heat units, which they propose would make the coefficients more transferrable across sites and seasons. de Jager and van Zyl (1988) present an analysis of crop coefficients which accounts for many of the possible surface conditions such as partial canopy cover, soil surface wetness, and plant stress.

Howell et al. (1989b) developed crop coefficient curves for grain sorghum [*Sorghum bicolor* (L.) Moench] at Bushland, Texas, and showed that the crop coefficient curves vary depending on the ET_p equation that is chosen. In their analysis, the peak K_c was 0.78 and 0.99 when referenced to alfalfa and grass ET_p, respectively, calculated by the Penman-Monteith form as described by Allen et al. (1989). When referenced to a Priestley-Taylor form with a VPD-based advective modification as described by Steiner et al. (1989a), the K_c peaked at about 0.9 and had a basically different shape than when referenced to the Penman- Monteith equation forms. Feddes (1987) and Hussien and El Daw (1989) also discuss the importance of the reference equation and surface in the application of crop coefficients.

b. *Leaf area index functions*. An alternative method of calculating actual ET partitions ET_p to soil and plant components based on the crop leaf area index (LAI), which is the ratio of green leaf area to soil surface area (Ritchie, 1972). Potential ET for a full-cover, non-stressed condition is usually calculated by one of the equations described above. The partitioning of ET_p is based on Beer's law for transmission of radiation through a medium. For a crop canopy

$$R_t/R_i = EXP(-k'LAI) \qquad [3.9]$$

where R_t is transmitted radiation, R_i is incoming radiation, and k is the extinction coefficient. Soil evaporation (E_s) is assumed to equal $ET_p(R_t/R_i)$ for a wet soil surface and to decline as a function of the square root of days since precipitation for a dry soil surface. The potential transpiration (T_p) is assumed to equal ET_p (1-Rt/Ri). Transpiration (T) may be less than the T_p if the soil in the root zone cannot provide adequate water to meet the plant needs.

In order to apply this model, the LAI over time must be measured or predicted. Because leaf area index is a labour intensive measurement, estimation is often preferred over measurement, particularly for operational applications. There are many crop growth models which predict LAI development, generally as a function of heat unit accumulation over time. Since it may not be practical to run a crop growth model for all crops or fields of interest, there is interest in obtaining estimates of LAI or crop cover from remotely sensed data (Ottle et al., 1989; Bausch and Neale, 1987).

Prediction or measurement of LAI for a specific growing season is more flexible than the assumption of a constant crop coefficient curve (K_c) because it allows for different rates of development to be considered dependent on season, soil, climate, and management. Use of crop growth models also allows analysis of the probability distribution, related to climatic variability, to yield and water use under different management scenarios and provides a powerful tool to predict irrigation requirements, to design irrigation systems, and identify improved irrigation practices for a region (Howell et al., 1989a).

2. Water limited conditions : The effects of limited water on transpiration and on plant productivity are often considered to be similar, since transpiration and plant growth are generally considered to be linearly related. The work of Denmead and Shaw (1962) showed that the relative transpiration rate (T/T_p) of corn (*Zea mays* L.) was a function of both soil water content and evaporative demand. Their work was conducted with restricted root zones for each plant (each plant had about one-tenth the soil volume compared to a field with a 1.5 m deep soil profile planted at 7 plants m^{-1}) and, therefore, the availability of water was restricted by volume as well as by soil water potential and flow rate to the root. For crops with a more normal root volume, it has been difficult to establish an effect ET_p on the T/T_p ratio until the soil water content is relatively low, which is consistent with the theoretical paper of Gardner (1960). Water balance models reported by Jones and Kiniry (1986), Ritchie (1985), Rosenthal et al. (1989), have assumed that transpiration would proceed at the potential rate until about 50 to 65% of the plant available water was depleted. These models have been applied successfully under a wide range of conditions.

IV. NEW CONCEPTS AND STRATEGIES

A. Potential ET

There have been few theoretical advances in the concept of potential ET since the work of Monteith (1965). However, there have been tremendous advances in our ability to apply the best theoretical approaches to ever more complex problems. Allen (1986) and Allen et al. (1989) discuss some of the simplifying assumptions required to apply the Penman-Monteith resistance form of the combination equation to calculate ET_p on an operational basis for irrigation scheduling applications. Howell et al. (1989b) have shown that these simplified Penman-Monteith forms can be applied to predict ET for well-watered grain sorghum throughout the growing season.

B. Actual ET

As we increase our quantitative description of the processes and factors that control ET, there are many benefits to abandoning the traditional approach of calculating ET_p and then relating ET_p to actual ET_a through use of crop coefficients. Models to directly estimate ET_a generally use an equation similar to the Penman-Monteith form described above and calculate the surface resistance as a function of the surface condition. Jagtap and Jones (1989) calculated a surface resistance for bare soil and for full crop canopy conditions and interpolated between the values as a linear function of LAI to estimate a surface resistance of a developing canopy. Lascano et al. (1987) developed and tested a model that includes numerous surface resistance components, including an aerodynamic resistance based on logarithmic wind profile theory, sensible heat resistances for soil and canopy as a function of leaf area index, stomatal resistance as a function of light, leaf area index, and leaf water potential, and leaf water potential based on soil water potential and root water uptake. Steiner et al. (1989b) reported excellent prediction of daily ET from planting to maturity of grain sorghum using the model described by Lascano et al. (1987) and van Bavel et al. (1984).

The direct calculation of actual ET, rather than calculating ET as an empirical function of ET_p, will require further theoretical development of explanations of soil water flow processes in field situations where flow in macropores may dominate over flow in the soil matrix (White, 1988). Most of theoretical soil physics is based on the assumption of a uniform matrix. A problem very much in need of better theoretical description is the role of roots in water flow in the soil-plant-atmosphere continuum. If roots are currently accounted for at all, they are treated as a sink term empirically related to soil and plant factors (Clothier, 1988). Even more limiting than our lack of understanding of the physical aspects of root distribution and activity is our lack of understanding of the biology of roots under variable field conditions and the role of roots as sensors of the environment and controllers of biological activities through hormone production (Clothier, 1988).

Shuttleworth and Wallace (1985) and Chen (1984) describe multilayer models of vapour and heat flux for application to situations where the single-layer model of Penman is not appropriate. As computing capability becomes more powerful and as our understanding of the surface resistance and transfer terms in the aerodynamically based equations continues to increase, more complex models can be embedded in the management models which currently rely on many simplifications and empiricisms. However, the input requirements for the more complex models are substantial and multi-layer models will be used for research applications, rather than operationally, for some time to come.

Morton (1983) described the theory of complementarity in which actual ET and potential ET are inversely related and, when summed, equal twice the areal wet ET. The original work was intended for hydrological applications over a large area and, therefore, might be most suited to evaluation of irrigation needs during the planning stages. Morton (1985) discussed the possible application of complementarity theory to field level ET estimation, but this application has yet to be thoroughly tested. Application of complementarity theory to ET prediction requires that meteorological data be collected over the crop surface of interest, rather than over a standard surface in a weather station.

In most applications, soil heat flux has been ignored in making daily ET predictions. This introduces a small error and is reasonable during periods of full canopy cover. When analyses are extended to partial cover crops or to bare soil, omission of soil heat flux can introduce considerable error. Fuchs and Hadas (1972) reported that daily soil heat flux represented up to 35% of net radiation for a bare soil, and Idso et al. (1975) reported that the ratio was 30-50% and varied as a function of soil water content. Soil heat flux cannot be ignored for ET estimation on less than a daily time step, which might be required to evaluate aspects of sprinkler irrigation efficiency. Clotheir et al. (1986) reported that mid-day soil heat flux was 10% of net radiation even under a dense, full-cover alfalfa canopy.

C. Dynamic Soil Water Availability

A major limitation of water balance models commonly used for ET prediction and irrigation scheduling applications is the concept of a static soil water availability dependent only on soil characteristics. In reality, water availability is a function of soil, plant, and climatic factors (Hillel, 1990) but, because our understanding of plant water relations and of root growth, function, and dynamics are so limited under field conditions, the role of the plant is often ignored or greatly simplified. Clothier (1989) identified study of root function as one of the major underdeveloped research areas essential to improved irrigation management. Ahuja and Nielsen (1990) review the changing concepts of soil water availability which can greatly contribute to improved irrigation scheduling and management models. Plant water stress estimations must be made based on the soil water potentials and root densities within different layers of the root zone at a given time. This area of research will be very important in developing deficit or limited irrigation strategies.

D. Irrigation Water Requirements

With the cost of new irrigation development in the range of \$5,000 to \$6,000 U.S. dollars ha^{-1} or even as high as \$10,000 ha^{-1} in small projects in areas without a history of irrigation (Rangley, 1987) and with a very high percentage of the existing irrigation projects in need of renovation (Jensen, 1990), it is essential that the best possible management practices be implemented. In the past, the lack of available climatic data often justified the use of empirical approaches to prediction of crop water requirements. While these approaches have had an important role in irrigation planning and management, they are basically inflexible in their application and often inaccurate when transferred to a new area. Inexpensive, automated weather stations are now available that are suited to operation in remote areas and that can provide standardized data with lower maintenance costs than manually operated stations. The output from automated weather stations must be monitored in real-time by knowledgeble staff to ensure high quality weather records, as discussed by Oldeman (1987), but monitoring of several sites can be done from a centralized location with today's technologies (Hubbard, 1989).

Often, some of the most advanced technologies have been rejected as unsuited to developing countries, but it is essential that the best management approaches and scientific methods and models available be used for new irrigation districts and to improve existing irrigation projects (Jensen, 1990). The goal should be to increase the flexibility of on-farm management decision making, leading to a demand driven irrigation system, rather than a centralized supply driven system. Because all of the models currently available for operational prediction of ET and irrigation water requirements involve some empiricism and/or simplifying assumptions that may not be appropriate for all situations, it is highly desirable, and perhaps necessary, that each major irrigation region or project has a research programme that adapts the available technologies to the specific conditions and needs of the region. These researchers should use the best available technologies such as geographical information systems (Brinkman and Stein, 1987; Dyke, 1987), crop growth models (Jones and O'Toole, 1987; Nix, 1987), and standardized data bases (for example, see discussion of

International Benchmark Sites Network for Agrotechnology Transfer (IBSNAT) project in Brinkman and Stein; 1987 and Jones and O'Toole, 1987) to enable transfer or extrapolation of research results from other regions to their irrigation region and from their experimental plots to farm level recommendations.

In order to seriously consider ecological impacts of irrigation, drainage and subsurface flow must be treated more realistically than they have been in the past. A valid consideration of the hydrologic balance as affected by irrigation will have to consider all components of the balance. This will require use of sound physically-based models instead of the simplified models which have been used when maximizing yield has been treated as the primary or only objective of irrigation. Variability over space and time has a major influence on irrigation efficiency and on ecological impacts of irrigation and will have to be considered in integrated system management models. Irrigation system design and management will increasingly have to balance the often conflicting needs of maintaining high yields without causing environmental degradation, either on-site or off-site.

REFERENCES

Abdulmumin, Salisu. 1989. J. Irrig. and Drainage Eng. 115:938- 953.

Aboukhaled, A., A. Alfaro, and M. Smith. 1982. "Lysimeters." FAO irrigation and drainage paper No. 39. Food and Aric. Org. of the United Nations, Rome. 68 p.

Abtew, Wossenu, J.M. Gregory, and J. Borrelli. 1989. Trans. ASAE 32:521-527.

Ahuja, L.R. and D.R. Nielsen. 1990. *In* B.A. Stewart and D.R. Nielsen (eds.) Irrigation of agricultural crops. Agron. Monogr. No. 30, p. 143-190, Am. Soc. Agron., Madison, Wisconsin, U.S.A.

Allen, R.G. 1986. J. Irrig. and Drainage Eng. 112:348-368.

Allen, R.G., M.E. Jensen, J.L. Wright, and R.D. Burman. 1989. Agron. J 81:650-662.

Azevedo, P.V. and S.B. Verma. 1986. Agric. and Forest Meteor. 38:193-204.

Bausch, W.C. and C.M.U. Neale. 1987. Trans. ASAE 30:703-709.

Blaney, H.F. and W.D. Criddle. 1950. "Determining water requirements in irrigated areas from climatology and irrigation data." USDA, Soil Conserv. Serv. SCS-TP 96. 44 p.

Brinkman, R. and A. Stein. 1987. *In* A.H. Bunting (ed.) Agricultural environments. characterization, classification and mapping, p. 119-126, C.A.B. International, Wallingford, Oxon, U.K.

Brutsaert, W. 1982. Evaporation into the atmosphere. Theory, history, and applications. D. Reidel Publ. Co, Dordrecht, Holland. 299 p.

Bunting, A.H. (ed.), 1987. Agricultural environments. characterization, classification and mapping. C.A.B. International, Wallingford, Oxon, U.K.

Businger, J.A. 1956. Neth. J. Agric. Sci. 4:77-80.

Camillo, P.J. and R.J. Gurney. 1984. Water Resour. Res. 20:105- 112.

Cary, J.W. 1981. *In* Irrigation scheduling for water and energy conservation in the 80's. Publ. 23-81, p. 81-90. ASAE, St. Joseph, Michigan, U.S.A.

Chen, J. 1984. Boundary-Layer Meteor. 28:213-225.

Chen, H. and A.D. Druliner. 1987. Nonpoint-source agricultural chemicals in ground water in Nebraska - Preliminary results for six areas of the High Plains Aquifer. Water Resources Investigation Report 86-4338. U.S. Geological Survey, Denver, Colorado, U.S.A. 68 p.

Christiansen, J.E. 1968. J. Irr. and Drain. Div., Amer. Soc. Civ. Eng. 94:243-265.

Christiansen, J.E. and G.H. Hargreaves. 1969. Trans. Intern. Comm. on Irrig. and Drain., Vol. III, 23.569-23.596.

Clothier, B.E. 1988. *In* W.L. Steffan and O.T. Denmead (eds.). Flow and transport in the natural environment : Advances and applications. p. 86-94. Springer-Verlag, Berlin.

Clothier, B.E. 1989. J. Irrig. and Drainage Eng. 115:421-448.

Clothier, B.E., K.L. Clawson, P.J. Pinter, M.S. Moran, R.J. Reginato, and R.D. Jackson. 1986. Agric. and Forest Meteor. 37:319-329.

de Jager, J.M. and W.H. van Zyl. 1988. Water SA 15(2):103-110.

Denmead, O.T. and R.H. Shaw. 1962. Agron. J. 45:385-390.

Doneen, L.D. and D.W. Westcot. 1984. Irrigation practice and water management. FAO Irrig. and Drainage Paper 1, Rev. 1. Food and Agric. Org. of the United Nations, Rome.

Doorenbos, J. and W.O. Pruitt. 1977. Crop water requirements. FAO Irrig. and Drain. Paper No. 24. Food and Agric. Org. of the United Nations, Rome.

Dyke, P.T. 1987. *In* A.H. Bunting (ed.) Agricultural environments. characterization, classification and mapping. p. 127-136. C.A.B. International, Wallingford, Oxon, U.K.

FAO 1978a. Report on the agro-ecological zones project. Vol. 1, Methodology and results for Africa. "Food and Agri. of the United Nations, Rome. World Soil Resources Report No. 48/1.

FAO. 1978b. Report on the agro-ecological zones project. Vol. 2. Results for southwest Asia. Food and Agric. Org. of the United Nations, Rome. World Soil Resources Report No. 48/2.

FAO. 1980. Report on the agro-ecological zones project. Vol. 4. Results for southeast Asia. Food and Agric. Org. of the United Nations, Rome. World Soil Resources Report No. 48/4.

FAO. 1981. "Report on the agro-ecological zones project. Vol. 3. Methodology and results for South and Central America." Food and Agric. Org. of the United Nations. Rome. World Soil Resources Report No. 48/3.

Feddes, R.A. 1987. *In* Evaporation and weather : Technical meeting 44. p. 33-45. The Hague, The Netherlands.

Fuchs, M. and A. Hadas. 1972. Boundary-Layer Meteor. 3:191-200.

Gardner, W.R. 1960. Soil Sci. 89:63-73.

Ghali, G.A. and Z.J. Svehlik. 1988. Agric. Water Management 14:307-315.

Harcum, J.B. and J.C. Loftis. 1987a. Trans. ASAE 30:1673-1678.

Harcum, J.B. and J.C. Loftis. 1987b. Trans. ASAE 30:129-136.

Hargreaves, G.H. and Z.A. Samani. 1982. J. Irrig. Drain. Div., Am. Soc. Civ. Engin. 108(1R3)223-230.

Hatfield, J.L. 1990. *In* B.A. Stewart and D.R. Nielsen (eds.) "Irrigation of agricultural crops." Agron. Monogr. No. 30, p. 435-474. Amer. Soc. Agron., Madison, Wisconsin, U.S.A.

Hattendorf, M.J., M.S. Redelfs, B. Amos, L.R. Stone, and R.E. Gwin, Jr. 1988. Agron. J. 80:80-85.

Heermann, D.F. 1988. Trans. ASAE 31:497-502.

Higgins, G.M., P.J. Dieleman, and C.L. Abernathy. 1988. J. Hydrol. Sci. 33:43-59.

Hillel, D.I. 1990. *In* B.A. Stewart and D.R. Nielsen (eds.) Irrigation of agricultural crops. Agron. Monogr. No. 30, p. 5-30. Am. Soc. Agron., Madison, Wisconsin, U.S.A.

Howell, T.A., K.S. Copeland, A.D. Schneider, and D.A. Dusek. 1989a. Trans. ASAE 32:147-160.

Howell, T.A., J.L. Steiner, and A.D. Schneider. 1989b. ASAE Paper No. 89-2668.

Hubbard, K.G. 1987. *In* A. Weiss (ed.) Proc. climate and agriculture : Systems approach to decision making, p. 53-61. Am. Meteor. Soc. Boston, MA.

Hussein, A.S.A. and A.K. El. Daw. 1989. J. of Irrig. and Drainage Eng. 115:1018-1033.

Idso, S.B. 1983. Agr. Meteor. 29:213-217.

Idso, S.B., J.K. Aase, and R.D. Jackson. 1975. Boundary-Layer Meteor. 9:113-122.

Jagtap, S.S. and J.W. Jones. 1989. Trans. ASAE 32:1342-1350.

Jensen, M.E. (ed.). 1974. Consumptive use of water and irrigation requirements. Am. Soc. Civ. Eng., New York. 215 p.

Jensen, M.E. 1990. *In* B.A. Stewart and D.R. Nielsen (eds.) Irrigation of agricultural crops. Agron. Monogr. No. 30, p. 31- 68. Amer. Soc. Agron., Madison, Wisconsin, U.S.A.

Jensen, M.E., R.D. Burman, and R.G. Allen. (eds.) 1990. Evapotranspiration and irrigation water requirements. ASCE, New York, NY. (in press).

Jensen, M.E. and H.R. Haise. 1963. J. Irr. and Drain. Div., Amer. Soc. Civ. Eng. 89(IR4):15-41.

Jones, C.A. and J.R. Kiniry (eds.). 1986. CERES-Maize, A simulation model of maize growth and development. Texas A&M University Press, College Station, Texas. 194 p.

Jones, C.A. and J.C. O'Toole. 1987. *In* A.H. Bunting (ed.) Agricultural environments. characterization, classification and mapping, p. 199-210. C.A.B. International, Wallingford, Oxon, U.K.

Jury, W.B. and C.B. Tanner. 1975. Agron. J. 67:840-842.

Kanemasu, E.T., J.L. Steiner, A.W. Biere, F.D. Worman, and J.F. Stone. 1983. Agric. Water Management 7:157-178.

Kanemasu, E.T., L.R. Stone, and W.L. Power. 1976. Agron. J. 68:569-572.

Kanwar, R.S. 1988. Trans. ASAE 31:1423-1429.

Kanwar, R.S., J.L. Baker, L. Mukgtar. 1988. Trans. ASAE 31:133- 141.

Lascano, R.J., C.H.M. van Bavel, J.L. Hatfield, and D.R. Upchurch. 1987. Soil Sci. Soc. Am. J. 51:1113-1121.

Marek, T.H., A.D. Schneider, T.A. Howell, and L.L. Ebling. 1988. Trans. ASAE 31:477-484.

Makkink, G.F. 1957. J. Inst. Water Eng. 11:277-288.

Mapp, H.P. 1987. *In* Ground water quality and agricultural practices, p. 35-45. Lewis Publishers, Chelsea, Michigan, U.S.A.

Monteith, J.L. 1965. *In* G.E. Fogg (ed.) The state and movement of water in living organisms Symp. Soc. Exp. Biol., vol. 19. p. 205-234. Academic Press. New York.

Monteith, J.L. 1973. Principles of environmental physics. Edward Arnold Ltd., London. 241 p.

Morton, F.I. 1983. J. Hydrol. 66:1-76.

Morton, F.I. 1985. *In* Advances in evapotranspiration. ASAE Publ. No. 14-85. p. 377-384. Amer. Soc. Agric. Eng., St. Joseph, Michigan.

National Research Council. 1989. Alternative agriculture,. Report of the Board on Agriculture, Committee on the Role of Alternative Farming Methods in Modern Production Agriculture. National Academy Press, Washington, D.C., U.S.A.

Nix, H. 1987. *In* A.H. Bunting (ed.) Agricultural environments. characterization, classification and mapping, p. 113-118. C.A.B. International, Wallingford, Oxon, U.K.

Oldeman, L.R. 1987. *In* A.H. Bunting (ed.) Agricultural environments. characterization, classification and mapping, p. 101-112. C.A.B. International, Wallingford, Oxon, U.K.

Ottle, C.D. Vidal-Madjor, and G. Girard. 1989. J. Hydrol. 105:369-384.

Penman, H.L. 1948. Proc. Roy. Soc. London. Ser. A, Vol. 193:120- 145.

Phene, C.J., D.W. Meek, R.L. McCormick, and K.R. Davis. 1986. ASAE Paper No. 86-1594.

Priestley, C.H.B. and R.J. Taylor. 1972. Mon. Weather Rev. 100:81-92.

Pruitt, W.O. and F.J. Lourence. 1985. *In* Advances in evapotranspiration. ASAE Publ. No. 14-85, p. 51-69. Amer. Soc. Agric. Eng., St. Joseph, Michigan.

Rangley, W.R. 1987. *In* W.R. Jordon (ed.) Water and water policy in world food supplies, p. 29-35. Texas A&M University Press, College Station, Texas, U.S.A.

Ritchie, J.T. 1972. Water Resource. Res. 8:1204-1213.

Ritchie, J.T. 1985. *In* W. Day and R.K. Atkin (eds.) Wheat growth and modelling, p. 293-305. Plenum Press, London.

Rolland, L. 1982. Mechanized sprinkler irrigation. FAO Irrig. and Drainage Paper 35. Food and Agric. Org. of the United Nations, Rome.

Rosenthal, W.D., R.L. Vanderlip, B.S. Jackson, and G.F. Arkin. 1989.

SORKAM : A grain sorghum crop growth model. Misc. Publ. No. 1669. Texas Agric. Exp. Sta., College Station, Texas, U.S.A.

Sammis, T.W., C.L. Mapel, D.G. Lugg, R.R. Lansford, and J.T. McGuckin. 1985. Trans. ASAE 28:773-780.

Schmugge, T.J., T.J. Jackson, and H.L. McKim. 1980. Water Resourc. Res. 16:961-979.

Shaw, R.H. and A.R. Pereira. 1982. Agric. Meteor. 26:51-65.

Shuttleworth, W.J. 1976. Boundary-Layer Meteor. 10:273-302.

Shuttleworth, W.J. and J.S. Wallace. 1985. Quart. J. Roy. Meteor. Soc. 111:839-855.

Snane, M.H., J.M. Gregory, and T.R. McCarty. 1984. ASAE Paper 84- 2016.

Stafford, J.V. 1985. J. Agric. Eng. Res. 11:151-172.

Staple, W.J. 1974. Soil Sci. Soc. Amer. J. 38:837-839.

Steiner, J.L., T.A. Howell, and A.D. Schneider. 1989a. ASAE Paper 89-2090.

Steiner, J.L., T.A. Howell, A.D. Schneider, and R.J. Lascano. 1989b. *In* 1989 Agron. Abstracts, p. 23. Amer. Soc. Agron., Madison, Wisconsin, U.S.A.

Steiner, J.L., R.C.G. Smith, W.S. Meyer, and J.A. Adeney. 1985. Aust. J. Agric. Res. 36:1-11.

Tanner, C.B. 1967. *In* R.M. Hagan, H.R. Haise, and T.W. Edminster (eds.) Irrigation of agricultural lands. Agron. Monogr. No. 11, p. 534-574. Amer. Soc. Agronomy. Madison, Wisconsin, U.S.A.

Tanner, C.B. and M. Fuchs. 1968. J. Geophys. Res. 73:1299-1304.

Tanner, C.B. and T.R. Sinclair. 1983. *In* H.M. Taylor, W.R. Jordon, and T.R. Sindair (eds.) Limitations to efficient water use in crop production, p. 1-27. Amer. Soc. Agron. Madison, Wisconsin, U.S.A.

Thom, A.S. 1972. Quart. J. Roy. Meteor. Soc. 98:124-134.

Thornthwaite, C.W. 1948. Geogr. Rev. No. 38:55-94.

Tripathi, R.P., H.S. Kushwaha, and A. Agrawal. 1987. Agric. For. Meteor. 41:275-288.

Turc, L. 1961. Ann. Agron. 12:13-49.

van Bavel, C.H.M. 1966. Water Resourc. Res. 2:455-467.

van Bavel, C.H.M., R.J. Lascano, and L. Stroosnijder. 1984. Soil Sci. 137:443-456.

Vermeiren, L. and G.A. Jobling. 1984. Localized irrigation. Design, installation, operation, evaluation. FAO Irrig. and Drainage Paper 36. Food and Agric. Org. of the United Nations, Rome.

White, I. 1988. *In* W.L. Steffan and O.T. Denmead (eds.) Flow and transport in the natural environment : Advances and applications, p. 59-85. Springer-Verlag, Berlin.

Wright, J.L. 1982. Amer. Soc. Civil Eng., J. Irrig. and Drain. Div. 108(IR1):57-74.

Wright, J.L. 1985. *In* Advances in evapotranspiration. ASAE Publ. No. 14-85, p. 105-113. Amer. Soc. Agric. Eng., St. Joseph, Michigan.

4

Yield To Water-Use Relationships And Use Of The Concepts In Managing A Limited Water Supply

S.D. Singh and Y.V. Singh

I. INTRODUCTION

Knowledge of yield (Y) to water-use (T, ET) relationship is indispensable in planning strategies for managing a limited water supply in an arid region, where increasing population and subsequent demand for food are facing the diminishing water resources. This relationship is important in evaluating crop water requirement, crop growth modeling, irrigation scheduling, water resource allocations, water project planning, and cost- benefit analysis of water project under planning. The dry matter (DM) or economic yield (Y) to transpiration (T) or evapotranspiration (ET) relations and yield response factors are reported to have been location specific, year specific, crop specific, cultivar specific, and management specific. In this chapter there is a fresh look into yield to water-use relations in crops adapted to dryland conditions and to irrigated agriculture in arid areas.

II. YIELD-TRANSPIRATION RELATIONSHIPS

Studies with Plants Grown in Containers

Experiments carried out in containers or in green house in early this century (Briggs and Shantz, 1914; Miller and Duley, 1923;

Shantz and Piemeisel, 1927; Kiesselback, 1929; Dillman, 1931) suggested that when yield was transpiration-limited a strong relationship existed between cumulative seasonal dry matter production and cumulative seasonal transpiration. In all these studies, dry matter production vs. transpiration curves were linear from the origin.

The reviews by Arkley (1963) and de Wit (1958) focused much attention on functional relations between dry matter and transpiration in a variety of crops grown in the containers. Similar to what was found from classic studies made by Briggs and Shantz (1914) and others the dry matter vs transpiration curves were found to be linear from the beginning in all those experimental crops that were subject to review by these two reviewers. As transpiration and evapotranspiration are closely associated, relationship between dry matter and evapotranspiration would also be linear, but this curve would never originate from the beginning, for reasons discussed in the following pages.

In voluminous data on yield to water-use relationships gathered since the classic work by Briggs and Shantz (1914) carried out in small containers, the potted plants did not represent the field conditions. Apart from this, containers were sealed to check evaporation, E, an inseparable component of ET in field agriculture. As a result, not much use of huge amount of yield to water-use data from the container experiments could be made in application to field. It was in the late fifties that de Wit (1958) re-examined the old data on transpiration ratio (an old term used to denote water requirement), and on dry matter production per unit of transpired water, and concluded that for semi-arid climate, similar to the U.S. Great Plains-the conditions under which Briggs and Shantz (1914) had worked - an equation relating yield (Y) to water-use (T) is :

$$Y = mT/Eo \text{ or } m = YEo/T \text{ or } m = YmEo\beta/ETm \qquad [4.1]$$

In this Y is yield of dry matter in kg ha^{-1}, T is transpiration in cm, Eo is average free-water (potential) evaporation in cm d^{-1}, and m is a crop factor in kg ha^{-1} d^{-1} dependent only on crop and variety. When T = Tm, then T or Tm = ETm/β, and $\beta = ETm/Tm$. As β is dimensionless, Tm assumes the same unit as ETm.

de Wit (1958) used climatic correlation factor E, for applicability of his model to semi-arid climate similar to the U.S. Great Plains. A review by Arkley (1963) proposed the use of relative humidity term, for applicability of this model to all climatic conditions. In a review on yield and water-use relations Tanner and Sinclair (1983) concluded that climatic correlation factor should be vapour pressure deficit.

With regard to m factor, a review on plant productivity in arid and semi-arid climate by Fisher and Turner (1978) concluded that de Wit (1958) model [Eq. 4.1] could be used to describe variations in water-use efficiency between plant communities. When root dry weight was included in the total dry matter production, the m factor was found to be 110-140 kg ha^{-1} d^{-1} for wheat [*Triticum aestivum* (L.) emend Thell.] and many other C_3 plant species, and 210 kg ha^{-1} d^{-1} for corn [*Zea mays* L.] and many other C_4 plant species. With both root and shoot dry matter production used in predition, m values were found to be 112, 122, 140, and 229 kg ha^{-1} d^{-1} for annual pasture in Israel, for wheat cv. 'Heron' in Australia, for shrubland [*Ceretoides lanata*] in Curlew Valley-Utah, and corn cv. '85 RM' at Mandan- N. Dakota, respectively. The m values for shrubland, annual pasture, wheat, and corn were, respectively, 34, 81, 108, and 183 kg ha^{-1} d^{-1} if only aboveground dry matter production was used in the prediction (Fisher and Turner, 1978). The variations in m factor between plant species were large in the case of latter data.

Hanks et al. (1969) verified the applicability of de Wit (1958) model and found that estimated m values were 125, 140, and 150 for wheat, sorghum [*Sorghum bicolor* L.], and millet, respectively. The m factor for wheat was similar to that from de Wit (1958), but the values for sorghum and millet were lower. Differences in m factor were reported to be due to differences in dry matter production potential of crop varieties and not due to evapotranspiration (Sorensen et al., 1979; Retta and Hanks, 1980).

The m values calculated for the studies as discussed above were found to be dependent on years of study, variety, location, etc. Our data (Table 4.1) showed that other than data aberration in cluster bean [*Cyamopsis tetragonoloba* (L.) Taub.] the computed m values were constant across the years as well as the varieties, for example in sorghum. For winter wheat, it was found to be identical to that from de Wit (1958) and Hanks (1983). In millet and sorghum, it was higher than individual values reported by Hanks (1983).

If Y in de Wit (1958) model were dry matter in grain yield, it may be possible to arrive at some acceptable values of harvest index (HI) using the data presented in Tables 4.1 and 4.2, instead of relying on a generalized HI values as of 0.36 for Kubana wheat grown in the containers (Briggs and Shantz, 1914; Dillman, 1931), and 0.42 that was suggested by Cole and Mathew (1923) for many locations in the U.S. Great Plains. From a review Hanks (1983) concluded that for many grain crops the HI was constant at about 0.40.

In this table ETm is maximum seasonal ET, Tm is maximum transpiration estimated as ETm/β, Ym is maximum yield where ET = ETm,β is dimensionless crop response factor that corresponds to the slope of yield reduction vs. Et reduction, and HI is ratio of m in Table 4.1 to m in Table 4.2. As m = Economic Ym.Eo/Tm, or total DM.Eo/Tm and as Eo/Tm is common to both numerator and denominator in estimating HI, m economic Ym/m total DM is the same as economic Ym/total DM.

At Ym level associated with ET = ETm, harvest index for winter wheat was 0.42, identical to that for the pearl millet [*Pennisetum americanum* (L.) Leeke]. Our harvest indices for pearl millet and winter wheat were the same as for wheat reported long-long ago by Cole and Mathew (1923) for many locations in the U.S. Great Plains. An HI of 0.20 may be acceptable for cluster bean, 0.22 for grain sorghum, 0.24 for mustard [*Brassica campestris* (L.) var. *dichotoma* Watt.], and about 0.34 for rainy season grain legumes such as cowpea [*Vigna unguiculata* (L.) Walp.] and mung bean [*Vigna radiata* (L.) Wilczek]. These values were estimated from the Ym and ETm, therefore were the upper limit to HI (Y/total DM) and could be used to predict the economic yield from the total DM predicted using de Wit (1958) model under well- watered environment.

Hanks (1983) assumed a constant HI at 0.4 for many grain crops; however, he indicated that HI might decrease during ex-

Table 4.1. Estimated crop factor (m) fo several crops across two years. From Niiwas (1975) and Singh (1984).

Crop	Variety	Year	Tm cm	Ym (DM) kg ha^{-1}	Eo cm d^{-1}	m kg ha^{-1} d^{-1}
Pearl millet	BJ 104	1980	17.3	8740	0.75	379
	BJ 104	1981	16.0	8510	0.72	383
Cowpea	FS 68	1981	21.3	4730	0.72	160
Mung bean	S 8	1980	19.2	3590	0.75	140
	S 8	1981	19.6	3810	0.72	140
Cluster bean	Sona	1980	30.9	6970	0.75	169
	Sona	1981	24.2	3890	0.72	116
Sorghum	CSH 4	1980	26.9	19900	0.75	562
	CSH 5	1981	26.4	20830	0.72	568
Mustard	T 59	1980–81	17.9	8340	0.65	303
	T 59	1981–82	16.8	8050	0.66	316
Wheat	Kalyansona	1971–72	65.6	13136	0.65	130
		1972–73	65.2	12800	0.66	130

Tm = ETm/β

Table 4.2. Maximum ET, maximum yield, β coefficient, estimated m factor, and harvest index (HI) for several crops grown in arid, northwestern India. From Singh (1984).

Crop	Year	β	ETm cm	Tm cm	Ym kg ha^{-1}	Eo cm d^{-1}	m kg $ha^{-1}d^{-1}$	HI
Pearl millet	1980	1.61	27.8	17.3	3590	0.75	156	0.41
	1981	1.62	26.0	16.0	3550	0.72	159	0.42
Cowpea	1981	1.20	25.6	21.3	1670	0.72	56	0.35
Mung bean	1980	1.61	30.9	19.2	1210	0.75	47	0.34
	1981	1.54	30.2	19.6	1170	0.72	43	0.31
Cluster bean	1980	1.32	40.8	30.9	1270	0.75	31	0.18
	1981	1.58	38.2	24.2	1140	0.72	34	0.29
Sorghum	1980	1.62	43.5	26.9	4340	0.75	121	0.22
	1981	1.74	45.9	26.4	4390	0.72	120	0.21
Mustard	1980–81	1.74	31.1	17.9	1950	0.65	71	0.23
	1981–82	1.76	29.5	16.8	1910	0.66	75	0.24
Wheat	1971–72	1.24	81.4	65.2	5470	0.65	54	0.42
	1972–73	1.25	81.5	65.2	5390	0.66	55	0.42

treme droughts. Harvest indices of winter wheat grown on good soils under good management with nitrogen and seeding rates maintained at the level least limiting to yield were 0.32, 0.37, 0.40, 0.40, and 0.39 for the seasonal irrigation water supplies of 10, 29, 56, 83, and 102 cm, respectively (S.D. Singh, unpublished). Thus similar to what Hanks (1983) had concluded, HI of winter wheat tended to decrease with increasing water deficits.

A decrease in harvest index as the water stress increased was brought about by similar effects of irrigation water on head-bearing percentage, fertile spikelet number $head^{-1}$, grain number $head^{-1}$, and weight $grain^{-1}$, which are evident from yield to yield components relationships shown below (S.D. Singh, unpublished). Thus, Hanks generalization held for well-watered environment, and decrease in harvest index was ought to occur with extreme to moderate seasonal water deficits.

Y (Grain wt. $g\ m^{-2}$) = 2.50** (Head no. m^{-2})-416.6 (R^2 = 0.90)
Y = 248.39** (head length, cm)-1794.0 (R^2 = 0.98)
Y = 0.13** (fertile spikelet no. m^{-2})+200.0 (R^2 = 0.96)
Y = 24.40** (grain no. $head^{-1}$)-580.3 (R^2 = 0.98)
Y = 47.78** (wt. $grain^{-1}$, mg)-976.2 (R^2 = 0.90)

Several workers (Passioura, 1977; Hodges, 1978; Rasmussen, 1979; DeLoughery and Crookston, 1979) have also reported decreasing harvest index (Y/total DM) with increasing water stress in corn, sorghum, and wheat.

Thus, determination of economic yield from aboveground dry matter predicted using de Wit (1958) model was not possible in water- stressed environment, or the model was applicable to crops in which the total dry matter production was of interest (e.g., alfalfa and maize silage) and in crops in which economic yield was a constant percentage of biological yield. In addition, constancy of harvest index as proposed by Hanks (1983) was not borne out from the data presented for several grain crops in Table 4.2.

From a review on yield and water-use relations, Hanks (1983) concluded that yield (dry matter) and transpiration were closely related. de Wit (1958) found that soil water had similar effect on both transpiration and yield (DM), therefore dry matter and transpiration were more highly correlated than economic yield and transpiration. Fig. 4.1 shows a plot of grain yield vs. T/Eo for pearl millet. It was found that correlation of yield vs. T/Eo was always better than yield vs. transpiration of de Wit (1958).

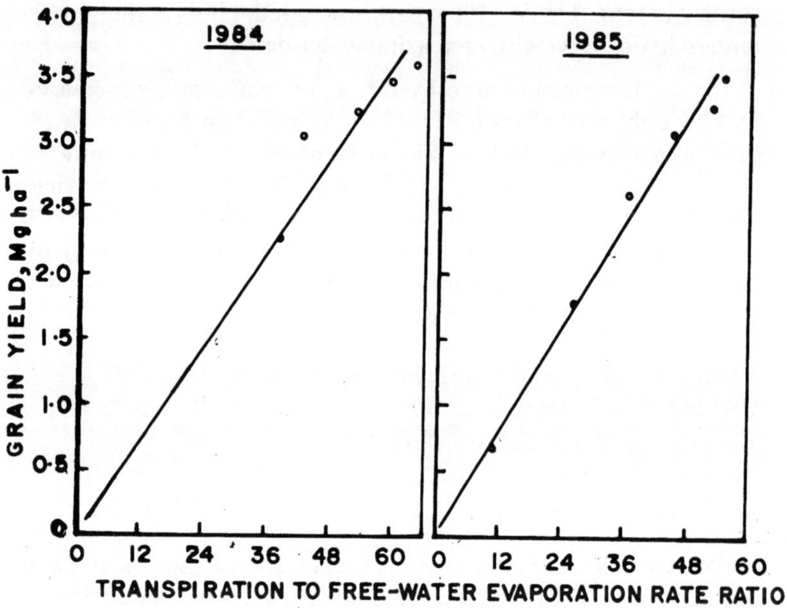

Fig. 4.1 Relation of yield of pearl millet to the ratio of transpiration to free-water (potential) evaporation corrected by a factor. From Singh (1989).

For a given crop and year (m and Eo constant) the relation of relative transpiration to relative yield could be given from equation [4.1] as equation [4.2] (Hanks, 1974; Hanks and Rasmussen, 1982).

$$Y/Ym = T/Tm \qquad [4.2]$$

In this Ym is the maximum yield when T=Tm (that does not limit growth). Shown in Fig. 4.2 is Y/Ym vs. T/Tm for pearl millet. The coefficients of determination we found were very high, and the plot of Y/Ym vs. T/Tm passed through 1:1 line. In Figs. 4.1 and 4.2 the curve passes through the origin.

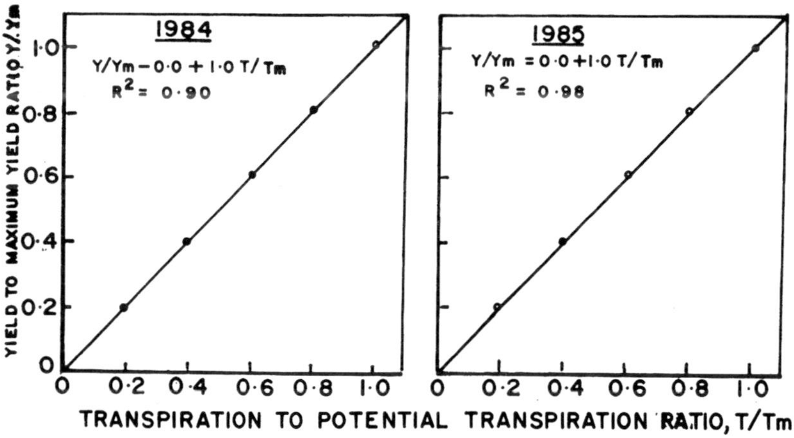

Fig. 4.2. Relation between Y/Ym and T/Tm for pearl millet. From Singh (1989).

That yield (DM) is closely related to T is especially true in water-limiting conditions (Kanemasu, 1983). Obviously, aboveground dry matter production and photosynthesis are closely related; and photosynthesis (Nc) and T are also related, via the diffusion resistances. Daily water-use efficiency (Nc/ET) of grain sorghum has been reported to decline with increasing solar radiation. A 10-d cycle of water deficit imposed in this crop gave Nc/ET ratios that were similar to ratios obtained from well- watered grain sorghum.

III. YIELD AND WATER-USE RELATIONS BASED ON FIELD-GROWN CROPS

A. Yield to ET Relationships

Yield to evapotranspiration relationships for the five important crops grown during 1980 and 1981 in arid, northwestern India are shown in Fig. 4.3. The yield-ET functional relation was similar to linear dry matter to transpiration relationship established from studies carried out in sealed containers to determine the "water

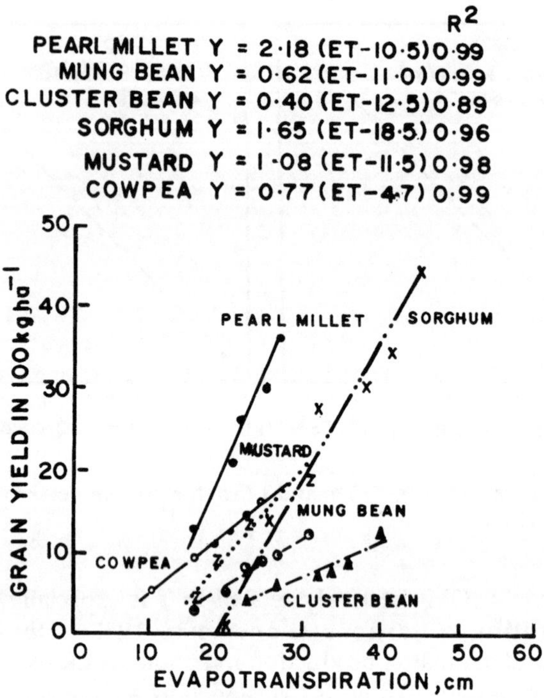

Fig. 4.3. Yield to ET relationship for the five crops grown during 1980 and 1981. From Singh (1984).

requirement" or "transpiration ratio" (T/DM). The DM-T curve passed through the origin, whereas in this case the Y-ET curve begins from the ET-axis.

The contrasting reports from container and field studies are mainly on two aspects of the yield to water-use relationships: the "form" and the "origin" of the fitted function. This contrast seems to arise from differences in: (i) crop species and varieties, (ii) definitions of "yield" and "water-use", (iii) actual maximum yield attained and the value selected in place of that to represent Ym, and (iv) selection of the data points from plots in which growth factors other than water may be limiting. For instance, in wanting supply of nitrogen, increase in water supply failed to increase the yield of winter wheat substantially, but as the N-supplies were increased, increasing water-supply enhanced the yield linearly up to 60 cm of the seasonal water (Fig. 4.4).

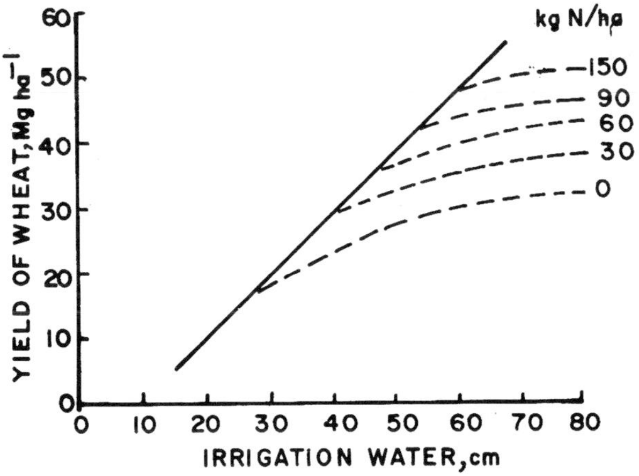

Fig. 4.4. Adequate supply of nitrogen is necessary for optimum use of scarce water. From Singh (1985a).

Thus the only important deviation is this. The dry matter-transpiration curve passes through the origin, whereas in this case the yield-evapotranspiration curve begins from the ET-axis. In crop fields, evapotranspiration includes evaporation (E), and in grain crops the interest lies in the harvestable yield, a portion of the total dry matter production. Evaporation from the soil early in the season, when the crop cover is incomplete, contributes appreciably to the total evapotranspiration, yet the yield is zero. The threshold requirements of ET, while yields were still zero, were found to be 4.7, 10.5, 11.0, 11.5, 12.5, and 18.5 cm in the cases of cowpea, pearl millet, mung bean, mustard, cluster bean, and grain sorghum (see equations on Fig. 4.3), respectively. The yield-evapotranspiration curve would therefore never begin at the origin. These results corroborated with the findings of Hanks et al. (1969) concerning grain sorghum, wheat, oats [*Avena sativa* L.], and millets. In their crops, the dry matter and evapotranspiration relationship was linear from the ET-axis. However, when evaporation was subtracted from evapotranspiration to estimate transpiration, the DM-

T curve tended to pass through the origin. Stewart and Hagan (1973) referred to a few studies conducted in the sixties. In those studies and in their own, the yield-evapotranspiration relationship was linear from the ET-axis.

Yield and evapotranspiration relations have also been found to be straight line from the ET-axis in other studies with maize (Stewart and Hagan, 1973), grain sorghum (Stewart et al., 1975; Howell and Hiler, 1975; Stone et al., 1978; Chaudhuri and Kanemasu, 1982; Owonubi and Kanemasu, 1982; Stewart et al., 1983; Singh, 1984, 1985a; Stewart, 1989), wheat (Niwas, 1975; Singh and Mann, 1979; Singh 1981), pearl millet (Singh, 1984, 1985a; Singh, 1989), cowpea, mung bean, and cluster bean (Singh, 1985a). There were a few exceptions in which the yield to irrigation relationships had also been linear from the water-axis (Leggett, 1959; Power et al., 1961), especially at the lower levels of irrigation (Niwas, 1975; Singh, 1977). Vaux and Pruitt (1983) give excellent review of crop and water-use relationships.

A convex yield-evapotranspiration function to the contrary had been reported (Musick et al., 1963; Grimes et al., 1969; Kallsen et al., 1984) in studies with tall wheat, cotton, and spring barley, respectively. Thus it seems that linearity or convexity of the yield to evapotranspiration relationship depended on the ratio of harvestable yield to aboveground dry matter. A steady increase or decrease in this ratio by increasing the ET deficits would not change the yield-evapotranspiration linearity; however, a disproportionate change in this ratio would change the relationship from linear to convex. Two such examples were reported by Musick et al. (1963) with tall wheat, and by Grimes et al. (1969) with cotton. With tall wheat, lodging in high-water treatment plots was a factor to reduce maximum grain yield, but did not alter the ETm value. Lodging at high ET values had also been found to bring departure from linearity in four varieties of spring barley [*Hordeum vulgare* L.] grown on two soils (Ali, 1979). In cotton, the lowest value of evapotranspiration found to produce maximum yield was 68 cm (i.e., ETm), while in Y- ET calculation the ET values up to 82 cm were used. Values of ET higher than ETm was possible, but the plants apparently did not use more water than 68 cm, so that yield remained either at Ym or decreased (Fig. 4.5). Thus much uncertainty still remained in quantifying the relationships for many crops, soils, and climate. However, two underlying principles were evident: (i) dry matter production was highly related to crop transpiration (de Wit, 1958). As will be seen later, grain yield and evapotranspiration are also very highly correlated, with R^2 values as high as 0.99 for pearl millet, mung bean, and cowpea; 0.98 for mustard; and 0.96 for grain sorghum. (ii) light interception (Monteith, 1977). Further, as discussed by Richie (1980), the relationships between dry matter production and evapotranspiration were found to be indirectly related to management practices (fertility, plant population, etc.). Recently as reported by Stewart (1992), ET was influenced by climatic parameters and not by management, and therefore, increase in crop productivity by manipulation in management practices would increase yield without increase in crop water use; then the water-use efficiency would increase for the given amount of water.

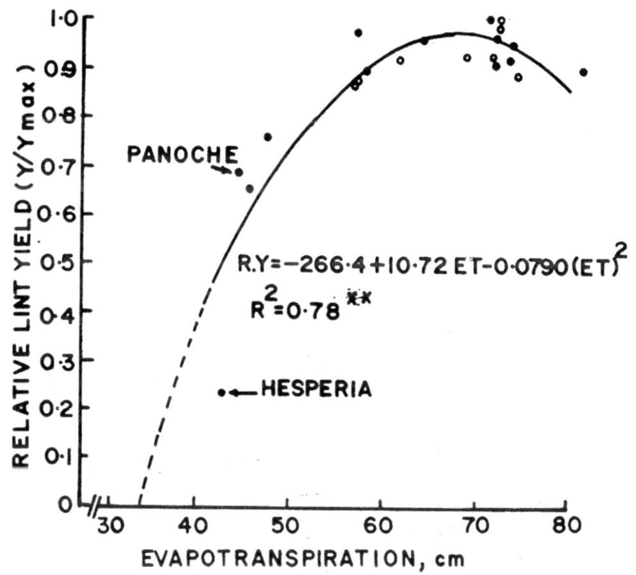

Fig. 4.5. Relation between relative cotton lint yield and evapotranspiration. Relative yield was obtained by dividing observed yield by maximum yield obtained at a specific location. From Grimes et al. (1969).

B. Yield to Irrigation Relationships

The yield-evapotranspiration linearity assumed that irrigation efficiency (ET as per cent of applied water) was 100%, water deficit sequences were optimal (water deficit timings that reduce yield the least), water distribution and infiltration over experimental fields were uniform, and there were no physical and procedural constraints. These expectations, however, are not always achieved. Nevertheless, yield-evapotranspiration relation is thought to be a standard against which planners can predict yield from available water supplies.

Evapotranspiration is the field level water parameter most directly related to yield but only indirectly related to irrigation. Irrigation (I) is the amount of water purchased and applied, therefore the water of utmost value to planners as well as to farmers. The irrigation is but one of three sources from which evapotranspiration is derived. The other two sources are profile water storage at planting time and growing season rainfall. Together these sources total crop-water-supply (CWS), i.e., CWS = soil water at planting + inseason rainfall+ irrigation. To relate yield to irrigation and to explain the concepts involved, data of an experiment conducted on winter wheat grown in arid, northwestern India were incorporated as an illustrative example. As this belt does not experience rains during cropping period, in crop water supply rainfall was not a factor. To enable planting, experimental plots had received preplant irrigation each year to assure that the crop season to begin with profile water at field capacity. In using this finding for predicting yield irrigation relationships this water was treated as if it was derived from preseason rainfall conserved in the soil.

Not all irrigation water was converted into evapotranspiration. It included non-ET water uses; I = ET + non-ET, and irrigation efficiency was the ET as per cent of applied water, i.e., I eff. = ET (from I) / I x 100. Fig. 4.6 illustrates the relationships of yield to evapotranspiration and yield to irrigation relationships within the context of functional relation between yield and crop-water-supply. The relationship between total water (available soil water at planting time in the 120 cm soil profile + irrigation) and the actual

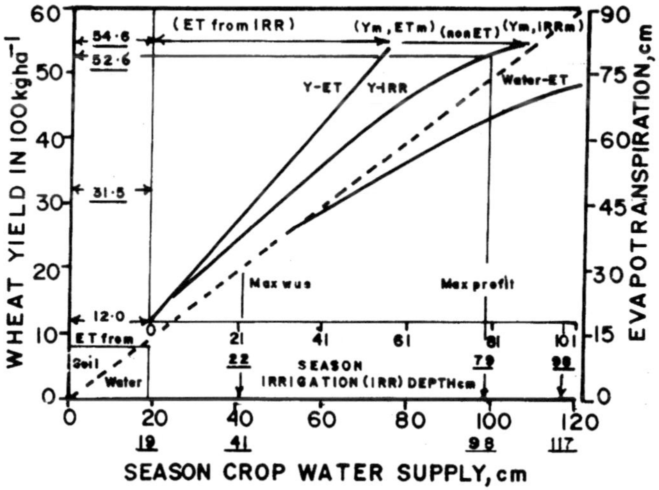

Fig. 4.6. Applied water-ET, yield-ET, and yield-irrigation water relationships plotted within yield-crop water supply functional relationship, averaged over four years from 1971 through 1975. From Singh and Mann (1979).

evapotranspiration was also shown in this figure. To quantify ET and non-ET portions of irrigation, the yield-evapotranspiration line was extended up to the point of ETm. This figure illustrated reasons for a particular form of yield and irrigation function, an exception to this, and the underlying reasons for the exception.

The dotted line in the figure depicted occurrence of an ideal situation where evapotranspiration to crop-water-supply ratio was unity, or in other words, the irrigation efficiency was 100%. This kind of situation existed until 28 cm of irrigation or 47 cm of crop-water-supply. As the season ETm approached to attain Ym, non-ET water disposition tended to increase, or in other words, irrigation efficiency went on decreasing. As a result, the relationship of evapotranspiration to total water and yield to irrigation relations became convex. The convexity of the yield and irrigation relationships was in conformity with earlier findings concerning several field crops (Howe and Rhoades, 1955; Stewart and Hagan, 1973; Singh et al., 1976).

In the low range of irrigation, the relations (combined for 1971-75) of yield to irrigation ($Y = 5.85 + 0.7023\ I$, with an R^2 of 0.95) and yield to crop-water-supply ($Y = 2.58 + 0.6722\ CWS$, with an R^2 of 0.96) were similar to linear yield to ET function. As the form of yield to irrigation relation depended on irrigation efficiency that typically reduced as ET equalled ETm to achieve Ym, the yield to irrigation relation became convex, while its linearity was a reality under dry farming conditions, or at the lower levels of irrigation.

Graphic analysis (Fig. 4.6) also revealed that where the functional relations of Y to ET and Y to CWS joined, the irrigation was zero and whatever ET derived was entirely from soil water in storage at planting. From the point of junction upward, the yield vs. CWS function became the yield vs. irrigation. The horizontal distance from the point (Ym - ETm) to the point (Ym - Im) quantified the non-ET disposition of irrigation, whereas the distance from the point (Ym - ETm) to the point of Ym at the ordinate of yield vs. I spelled out the ET derived from irrigation water alone.

IV. YIELD PREDICTION MODELS

Several investigators have developed empirical yield prediction models. Newell et al. (1976) gave an excellent review of statistical models. Usefulness of such models, however, was found to be limited to the site of study. Agronomists developed the following model [Eq. 4.3] from the measurements of ET and yield from field studies made in the U.S. Great Plains :

$$Y = a + b\ ETo \qquad [4.3]$$

where 'a' and 'b' were constants of the regression line. The value 'a' was always found to be negative. It means that some ET was needed while the yield was still zero. This was a good approximation of evaporation early in the season when groundcover was incomplete (Hanks, 1974).

From his model, de Wit (1958) concluded that crop factor m predicted field-grown crop conditions very closely. Hanks et al. (1969) later confirmed this conclusion.

If our concern was for a given crop in a given season, the de Wit (1958) model could be simplified (Hanks and Rasmussen, 1982) as given in Eq. (4.2)

In the mid-seventy, Hanks (1974) attempted to define the soil-plant-atmosphere continuum mathematically to estimate the ratio of actual transpiration (T) to potential transpiration (Tp) in maize. Yield was then related to this relationship. Four years later, Rasmussen and Hanks (1978) modified the Hanks model. They used the de Wit model to relate the relative "biological yield" (Y/Yp) to relative seasonal transpiration (T/Tp). Eq. [4.4] was the same as Eq. [4.2]

$$Y/Yp = T/Tp \qquad [4.4]$$

Table 4.3. ET, E, T, and dry matter (DM) and grain yield (GY) of pearl millet. From Singh (1989).

Year		ET	E	T	Relative yield (Y/Yp)	
					DM	GY
			--- mm ---		--- % ---	
1984	(ETm)	302	52	250	100	100
		302	52	250	100	100
		302	52	250	100	100
		302	52	250	100	100
		301	52	249	99	99
	(Rainfed)	265	45	220	88	72
1985	(ETm)	314	47	267	100	100
		314	47	267	100	100
		314	47	267	100	100
		309	47	262	98	95
		281	47	234	87	70
	(Rainfed)	172	34	138	51	2

The problem remained as how to determine transpiration separately from ET when latter was the only process that could be measured in the field. Retta and Hanks (Utah State University manual for using model PLANTGRO) development model [Eq. 4.4] that estimated transpiration and evaporation and then used to estimate the relative yield of dry matter (and grain) for a crop under the given conditions of soil, climate, crop, and irrigation management. Shown in Table 4.3 are estimated T values of pearl millet cv. 'BJ 104' using PLANTGRO model.

The seasonal transpiration was 83% and evaporation was about 17% of the growing season ET. Water stress in the rainfed plots influenced grain yield more than the aboveground dry matter production (Table 4.3). This might be one of the reasons for reported more high correlation between aboveground dry matter and transpiration than that between grain yield and evapotranspiration. Further it was found that grain yield vs. estimated transpiration was also highly significant, with coefficient of determination approaching unity, typical of a very high correlation.

Many studies had shown that water deficits during certain growth stages had more effect on grain yield than on evapotranspiration or transpiration. Jensen (1968) found that delaying irrigation in grain sorghum reduced yields by 35% but reduced water-use by 20%. In pearl millet grown at a typical site in the Rajasthan desert of high climatic variability, 12% water deficit in the rainfed plots relative to the ETm plots reduced the grain yield by 28% in 1984, and 65% water deficit in the extremely low rainfall year reduced grain yield by 98% (Table 4.3).

To account for growth stage effects of water deficit on yield, Jensen developed a model that related the relative "harvestable yield" (Y/Yp) to the product of relative transpiration (T/Tp) for each of desired growth stages. This equation is :

$$Y/Yp = \sum_{i=1}^{n} (Ti/Tp)_i^{\lambda i} \qquad [4.5]$$

In eq. [4.5] Y is "biological" or "harvestable" yield from the given level of transpiration (T), Tp is potential transpiration, Yp is potential yield obtainable from Tp, i is a physiological plant growth stage, and λi is an exponent to allow for weighting the 'ith' stage. Jensen's (1968) model approximated the Y (grain)/Yp to

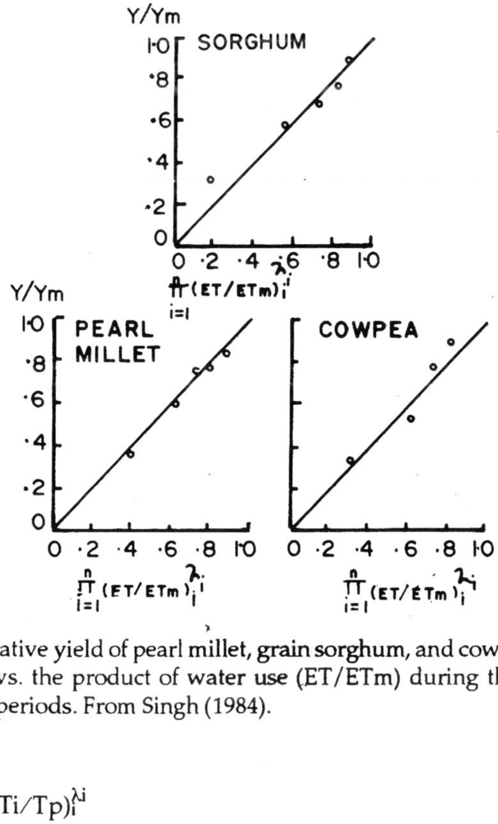

Fig. 4.7. Relative yield of pearl millet, grain sorghum, and cowpea (Y/Ym) vs. the product of water use (ET/ETm) during three growth periods. From Singh (1984).

$$\sum_{i=1}^{n} (Ti/Tp)_i^{\lambda i}$$

nevertheless it was found to have an advantage of quantifying the critical and non-critical growth stages in the life cycle of a crop.

To examine prediction precision in Jensen's model, Fig. 4.7 was constructed to develop relationship between relative yields, for example, of pearl millet, grain sorghum, and cowpea with the product of relative water use [Eq. 4.5] during three growth periods of each crop. Three periods used were : (a) early vegetative stage with $\lambda = 0.3$, (b) active tillering to boot stage with $\lambda = 0.6$, and (c) heading to grain development stage with $\lambda = 0.7$, in pearl millet. The corresponding periods were: (a) emergence to boot stage with $\lambda = 0.1$, (b) boot to milk stage with $\lambda = 0.5$, and milk to harvest with

$\lambda = 0.5$ in grain sorghum, and (i) vegetative growth with $\lambda = 0.1$, (ii) branching to flowering stage with $\lambda = 0.5$, and (iii) pod formation to harvest with $\lambda = 0.7$ in cowpea. Calibration of value [Eq. 4.5] was made by trial and error until the data for grain yield matched the estimated yield closely.

The results showed the Jensen's model adequately represented the effects of water deficit as shown by reduced water-use on yields of three determinate crops. Higher the value of λ higher was the sensitivity of a growth period to water stress.

Kanemasu (1983) proposed (i) to weight the normalized T (e.g., T/ETm) according to growth stages (i.e., with winter wheat: emergence to jointing (Stage 1), jointing to heading (Stage 2), and heading to soft dough (Stage 3), and (ii) to provide preference to longer time periods within a growth stage. Rasmussen (1979), who allowed for both of these concepts in his winter wheat study, developed the following grain yield and water-use equation :

$$Y\ (\text{kg ha}^{-1}) = 192.1\ [\Sigma\,(T/ETm)]_1^{0.172}\ [\Sigma\,(T/ETm)]_2^{0.104}\ [\Sigma\,(T/ETm)]_3^{0.646} \qquad [4.6]$$

In this the subscripts referred to the growth stages indicated above. Daily T/ETm ratios are summed during each of the stages. The value of ETm is estimated by the Priestley-Taylor equation. The maximum T/ETm summation would be the total number of days in each growth period. This equation has the advantage that neither the potential yield nor the m factor needs to be known.

There are a few other approaches to predicting yields such as the models developed by Nimah and Hanks (1973), Kanemasu et al. (1976), Childs et al. (1977), and Morgan et al. (1980). These were sophisticated models, while planners needed simple versatile model(s) for yield prediction purposes.

Stewart et al. (1977) developed models for different genotypes of maize. This model assumed that maximum yield (Ym) and maximum evapotranspiration (ETm) on good soils under good management were unique to crop cultivars as also were constant across different years.

In verification of this assumption results of some pertinent studies are presented. In field experiments by Garrity and coworkers (1982) carried out during 1977 to 1978 at the University of Nebraska, the slope of the yield reduction vs. ET deficit for 'RS 626' genotype of grain sorghum was not found to be constant, which was attributed to large variation in Ym (8.8 Mg ha^{-1} in 1977 to 7.8 Mg ha^{-1} in 1978), although the ETm for the two years was found to be the same (50 cm). On the contrary, in studies by Rasmussen and Hanks (1978) on five spring wheat cultivars the ETm values ranged from 31 to 44 cm. In studies with four genotypes of maize (Stewart et al., 1973, 1977), Ym was found to vary greatly from 11.2 to 12.0 Mg ha^{-1} at Davis, and 7.5 to 11.0 Mg ha^{-1} at Fort Collins; nevertheless, the slopes of the Y reduction vs. ET deficit in their studies were found to be nearly constant (1.03 to 1.33) across the different years.

With mounting evidence of controversy, data from winter wheat study conducted during 1971-73 at Jodhpur were utilized to develop the model [Eq. 4.7], which was incorporated in Ph.D. thesis of Niwas (Niwas, 1975) and published lately in 1979 (Singh and Mann, 1979). This model has been the basis of write-up hereafter, as an attempt to resolve the controversy on constancy across the years of ETm, Ym, and β values.

$$Y/Ym = 1 - \beta\,(1 - ET/ETm) \qquad [4.7]$$

In this Y is grain yield, Ym is maximum attainable yield on good soil under good management, ETm is minimum level of evapotranspiration that produces Ym, ET is desired evapotranspiration, and β is crop response factor that corresponds to slope of yield reduction vs. Et deficit. It may be reiterated that the model's requirements were that Ym and ETm be known and crop response factor be similar across the years. These aspects were examined in further detail.

The grain yields and seasons total evapotranspiration of several crops of study made at Jodhpur are given in Fig. 4.8 and numerical values of Ym and ETm are given earlier in Table 4.2. The figure showed that differences in Ym between years were small. This narrow yield range verified the first part of the hypothesis, i.e., for a given genotype and location a range of weather conditions

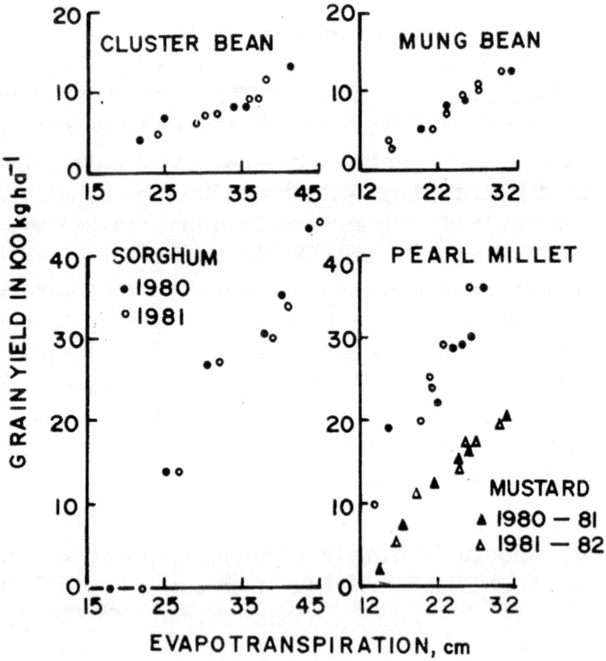

Fig 4.8. Relationship between yield and ET for two years of field studies. From Singh (1985 b)

exists in which the maximum yields on a good soil under management least limiting to yield can be anticipated. Our results do not lend support to general philosophy contained in the Hanks model, which indicated that Ym might vary from one year to another. In spite of this belief, Rasmussen and Hanks (1978) used a single Ym value in four years of their spring wheat study, which itself assumed that Ym would have been constant across the years. The Ym value they used was found to overestimate yields in 1973, whilst in 1972 the predicted yields were about 20% lower than the measured yields.

Shown earlier in Fig. 4.3 is Y-ET relationship, vital for the model, for the data averaged over two years together with regression equations in deterministic form. In study crops the relationships were linear with R^2 values close to unity. This figure showed that approximately 19 cm of water was necessary in the ET process

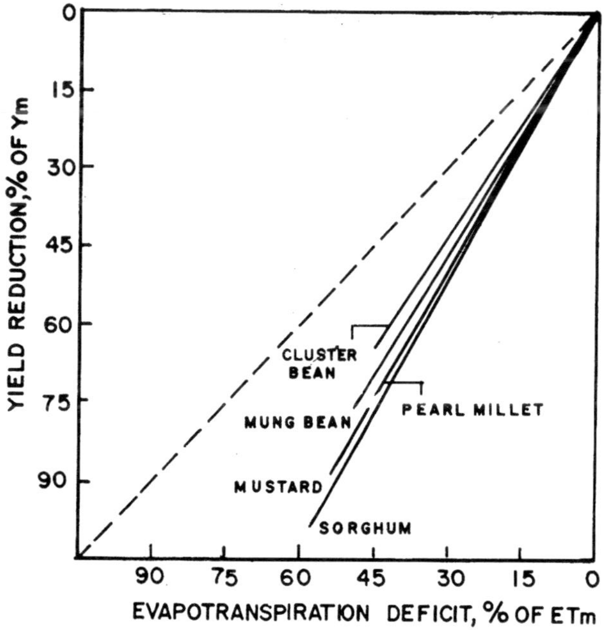

Fig. 4.9. Relationship between yield reduction and ET deficits. From Singh (1984).

before measurable yield of sorghum was obtained. In other experimental crops this requirement was a little over one-half (one-fourth in cowpea) of what sorghum required. Thus in the production process of grain sorghum in the sandy arid plains of Rajasthan, about 45% of the seasonal ETm was lost in the evaporative process without contributing anything to grain yield.

To verify whether or not the crop response factor remained constant across the years, data in Fig. 4.8 were transformed relative to Ym and ETm, taken as 100, 100, and Fig. 4.9 was constructed. Each relative value was subtracted from 100, which had the advantage of shifting the origin of the function from (100, 100) to (0,0), where season evapotranspiration deficit (ETd) was zero, and ET = ETm. This figure showed the yield possible at various ETd levels - from no reduction, Y = Ym where ET = ETm, throughout the yield range of interest. It was observed that the relationship between yield reduction and ETd was highly linear.

The yield reduction to ETd ratio of 1.0, shown by broken line on Fig. 4.9, showed that the rate of reduction in yield from Ym was equal to the associated ETd. In figure 4.9 the regression lines, drawn over the range of data points, continued to deflect farther and farther from the 1:1 line showing the rate of reduction in yield being faster at the higher than at the lower ETd intensities. This kind of trend supported the conclusion that a small supplement to rainfall, which is possible by sprinkler irrigation, could improve the production considerably. For example, a small increase in ET from 41 to 45 cm was found to increase the yield of grain sorghum from 3.4 to 4.4 Mg ha^{-1} (25 kg ha^{-1} mm^{-1}), whereas another small increase in season ET from 22 to 27 cm improved the pearl millet yield from 2.6 to 3.6 Mg ha^{-1} (20 kg ha^{-1} mm^{-1}, above 22 cm of ET).

Shown in Table 4.4 are crop response factors collected from various sources. The data led to many conclusions. (1) Other than cluster bean among the rainy season crops studied at Jodhpur, the values were similar across the years. (2) Crops such as barley and wheat with β value at 1.24 and pearl millet and mung bean with β value at 1.61 could be grouped together. (3) Maize studies revealed location specific nature of crop response factor. (4) From the values for grain sorghum cultivar 'CSH4' and 'CSH5' (Table 4.1, 4.2) it seems that they would vary from one genotype to another.

Our finding relating to Ym and crop response factor did not lend support to that of Garrity et al. (1982). In their grain sorghum study, response factor was found to differ from year to year, because of large variation in Ym (i.e., 8.8 Mg ha^{-1} in 1977 to 7.8 Mg ha^{-1} in 1978) due to high climatic variability at the study site. The reason for variability in Ym, however, was caused by a difference in the number of plants maintained per unit area - 16 plants m^{-2} in 1977 as against 13 m^{-2} in 1978. Thus, an 11% lower value of Ym in 1978 was owing to 30,000 ha^{-1} less plant population maintained in that year. In model [4.7], the first assumption was that management practices across the years be least limiting to yield.

Table 4.4 : Values of β as determined for different crops, locations, and years.

Crop	Location	Year	β	Reference
Maize	Logan, UT (North farm)	1974	1.39	Stewart et al. (1979)
Maize	Logan, UT (North Fram)	1975	1.40	Stewart et al. (1977)
Maize	Logan, UT (South Fram)	1976	1.18	Hanks et al. (1977)
Maize	Logan, UT (North Fram)	1977	1.38	Wenda (1979)
Maize	Farmington, UT	1977	1.52	Wenda (1979)
Maize	Logan, UT (South Farm)	1978	1.33	Wenda (1979)
Maize	Trenton, UT	1978	1.36	Wenda (1979)
Maize	Davis, CA	1974	1.17	Stewart et al. (1977)
Maize	Davis, CA	1975	1.18	Stewart et al. (1977)
Maize	Fort Collins, Co	1974	1.42	Stewart et al. (1977)
Maize	Fort Collins, Co	1975	1.39	Stewart et al. (1977)
Maize	Yuma, AZ	1974	1.40	Stewart et al. (1977)
Maize	Yuma, AZ	1975	1.50	Stewart et al. (1977)
Barley	Logan, UT (North Fram)	1978	1.24	Ali (1979)

Table contd

Crop	Location	Year	β	Reference
Barley	Logan, UT (South Farm)	1978	1.24	Ali (1979)
Alfalfa	Logan, UT (South Farm)	1977	1.05	Retta and Hanks (1979)
Wheat	Jodhpur, Rajasthan	1971–72	1.24	Niwas (1975)
Wheat	Jodhpur, Rajasthan	1972–73	1.25	Niwas (1975)
Pearl millet	Jodhpur, Rajasthan	1980	1.61	Singh (1984)
Pearl millet	Jodhpur, Rajasthan	1981	1.62	Singh (1984)
Cowpea	Jodhpur, Rajasthan	1981	1.20	Singh (1984)
Mung bean	Jodhpur, Rajasthan	1980	1.61	Singh (1984)
Mung bean	Jodhpur, Rajasthan	1981	1.54	Singh (1984)
Cluster bean	Jodhpur, Rajasthan	1980	1.32	Singh (1984)
Cluster bean	Jodhpur, Rajasthan	1981	1.58	Singh (1984)
Sorghum	Jodhpur, Rajasthan	1980	1.62	Singh (1984)
Sorghum	Jodhpur, Rajasthan	1981	1.74	Singh (1984)
Mustard	Jodhpur, Rajasthan	1980–81	1.74	Singh (1984)
Mustard	Jodhpur, Rajasthan	1981–82	1.76	Singh (1984)

It may be seen that crop integrates the effects of soil and climate. When a crop cultivar adapted to a region is finally selected, effects of soil-climate variables are normalized. In studies with different maize varieties no measurable differences in ET were found due to variety, but differences in aboveground dry matter and grain yield were found (Hanks, 1983). ET deficits in the range of data therefore are likely to be similar across the genotypes. If differences in yield due to variety are normalized in transformation of Y relative to Ym as 100, then response factor from research site can be used for yield prediction at a planning site possessing soil and climatic conditions similar to that of research site.

Eq. [4.7] can also be written as

$$Y/Ym = 1 - \beta + \beta\ (ET/ETm) \qquad [4.8]$$

The equation of Doorenbos and Kassam (1979) can be shown to be the same as Eq. [4.7] and Eq. [4.8].

If it is assumed that ET that corresponds to an extrapolation of Y-ET curve to zero, which can also be determined by simple expedient of letting Y equal to zero and solving for ET in Y-ET curve, is an estimate of evaporation from the soil surface then the following equations can be derived (Utah State University, Utah Manual for PLANTGRO Model) :

$$Y/Ym = 0 = 1 - \beta + \beta\ (E/ETm + T/ETm) \qquad [4.9]$$

and for the situation where Y/Ym = 0 the transpiration will be zero so that

$$E/ETm = 1 - 1/\beta \qquad [4.10]$$

If response factor is known we can then determine how much E is in relation to ETm. We can also derive an expression for Tm/ETm because for this situation Y/Ym = 1.0. Combining Eq. [4.7] and Eq. [4.10] gives

$$Tm/ETm = 1/\beta \qquad [4.11]$$

Equation [4.11] gives a simple method to estimate the relative contribution of E and T to ET.

It may be seen that the value of response factor must be greater than 1 to satisfy the field observation that there is some ET value at which yield is zero. A value of 1.0 would mean that E = 0 and ET

= T. A value of 1.24 would mean that E = 0.19 ETm, and Tm = 0.81 ETm. Thus, it provides a method for estimating T and E separately from field data, and can be used to estimate m in Eq. [4.1].

Prediction Precision of Model [4.7] Versus Jensen's Model

Table 4.5 shows observed versus predicted yields using equation [4.7] and the Jensen model for three important crops of Rajasthan desert namely pearl millet, grain sorghum, and cowpea. The coefficient of regression between observed yield and yield predicted using model [4.7] was 0.99 or say 1.0 for pearl millet, 0.995 or say 1.0 for cowpea, and 0.95 for grain sorghum. Thus, the model predicted accurately the grain yields of pearl millet and cowpea, while predicted yields of grain sorghum were 5% lower than the observed yields.

Table 4.5. Observed yields of pearl millet, grain sorghum and cowpea versus predicted yield using equation [4.7] and Jensen model.

Crop	Observed	Predicted	
		Model [4.7]	Jensen's model
		... $Mg\ ha^{-1}$	
Pearl millet	3.57	3.57	3.57
	2.97	3.05	3.09
	2.70	2.77	2.81
	2.63	2.56	2.57
	2.12	2.30	2.23
	1.29	1.29	1.77
Grain sorghum	4.37	4.37	4.37
	3.44	3.66	3.86
	3.04	3.32	3.36
	2.69	2.20	2.25
	1.38	1.33	1.40
	0.00	0.38	0.45
Cowpea	1.66	1.66	1.66
	1.46	1.50	1.48
	1.26	1.33	1.47
	1.11	1.19	1.39
	0.96	0.96	1.10
	0.51	0.54	0.39

The corresponding regression coefficients for observed versus predicted yields using Jensen's model approximated 0.81 in pearl millet, 0.95 in cowpea, and 1.08 in grain sorghum. Results showed that this model underestimated grain yields by 19% in pearl millet and by 5% in cowpea, and overestimated yields by 8% in grain sorghum. Thus model [4.7] can be more useful to economists and planners engaged in formulating strategy for managing limited waters and deficit irrigation.

Jensen's model has one advantage. It can be used fruitfully in quantification of relative sensitivity of a crop to water deficit during various growth periods. The accuracy of this model, however, depends upon how accurately the sensitivity factor λ is determined by the trial and error method. The yield estimated using Jensen's model is the product of

$$\sum_{i=1}^{n} (ET/ETm)_i^{\lambda i}$$

The assumption that yield be zero if ET is zero in any stage (Vaux and Pruitt, 1983) may be applicable to i = 1 container experiments where containers were sealed to check evaporation completely. In field agriculture, evaporation is a component of ET and it continues to occur in the pre-emergence stage as well as after physiological maturity until harvest. ET can never be zero in any of the growth periods.

V. USE OF EMERGING CONCEPTS IN MANAGING LIMITED WATER SUPPLY

1. Crop Sensitivity Rating

The slope of the yield reduction vs. ET deficit, which corresponded to yield reduction ratio (YRR), i.e. % yield reduction/% ET deficit) termed by Stewart and coworkers (1975), may be useful for crop/varietal selection.

This slope pooled over two years, was found to be 1.20 for cowpea, 1.45 for cluster bean, 1.58 for mung bean, 1.62 for pearl millet, and 1.68 for grain sorghum, showing that; among the rainy season crops; the cowpea, cluster bean, and pearl millet crops were least affected by the soil water stress. Yield reduction per unit of ET reduction was the highest in sorghum. Thus cowpea, cluster bean, and pearl millet appeared to be environmentally the best adapted. However, this alone might not be sufficient. Yield stability and productivity per unit use of a limited water supply may be other indices to evaluate crop adaption to environment and to management practices that are concomitant to better use of water.

(a) **Yield stability and engineering for higher wheat grain yield :** To illustrate, the data of experiments conducted with cowpea, cluster bean, mung bean, pearl millet, grain sorghum, mustard, and winter wheat at a typical site in the Rajasthan desert were utilized. It was found that the linear regression of grain yield ratio YR, expressed as rainfed yield/irrigated yield, on grain yield had an R^2 of 0.96 with a regression of YR = 0.95-0.170 Mg ha^{-1} for pearl millet, R^2 of 0.98 with a regression of YR = 0.81- 0.311 Mg ha^{-1} for cowpea, R^2 of 0.93 with a regression of YR = 0.92-0.561 Mg ha^{-1} for mung bean, and R^2 of 0.97 with a regression of YR = 1.03-0.571 Mg ha^{-1} for cluster bean. Such computation was not possible in grain sorghum as the rainfall amount and distribution in the 1980 and 1981 experimental seasons resulted in crop failure under rainfed system.

The regression coefficients were found to be highly significant in respect of all the four crops studied. The coefficients of determination, R^2, ranging from 0.93 to 0.98 were illustrious of a very high relationship between grain yield and grain yield ratio in all crops under study. The numerator in yield ratio defining-term was constant for the experimental crops, i.e., 1.29, 0.51, 0.34, and 0.44 Mg ha^{-1} in order of crops named above. On the contrary, the denominator had a tendency to increase with increasing ET values till the ETm-associated Ym was reached, under management practices maintained at the level least limiting to yield of each crop.

Thus lower the grain yield ratio, higher would be the crop response to irrigation if given to supplement the seasonal rainwater. Regression coefficient was the lowest for pearl millet, significantly lower ($P < 0.01$) than that for cowpea, which in turn had a significantly lower ($P < 0.05$) coefficient value than those for mung bean and cluster bean that ended up with identical coefficients.

Grain yield ratios of mustard and wheat, which were winter season crops grown in the region with irrigation, were also calculated using data points from studies carried out in the seventies (Singh and Yusuf, 1979; Singh and Mann, 1979). In those studies rainfed-like soil moisture conditions had been simulated by a preplant irrigation to supply 76 mm of water to mustard and 102 mm of water to the wheat crop.

Linear regression showed an R^2 of 0.94 with a regression of YR = 0.56-0.200 Mg ha^{-1} for mustard and R^2 of 0.94 with a regression of YR = 0.43-0.045 Mg ha^{-1} for wheat. Compared to pearl millet and three grain legumes, all grown during a single rainy season from July to September, the mustard crop exhibited yield stability similar to that of pearl millet, and like this crop had significantly higher yield stability than cowpea. The winter wheat cv. 'Kalyansona' showed significantly ($P < 0.01$) the highest yield stability among all study crops.

Table 4.6. Regression of grain weight m^{-2} in g (Y) on yield and growth components of winter wheat

Components	Y = a + bx		SE	R^2
	Constant	Coefficient		
Head number m^{-2}	–416.6	2.50**	0.42	0.90
Head length, cm	–1794.0	248.39**	20.51	0.98
Fertile spikelet number m^{-2}	200.0	0.13**	1.37	0.96
Grain number $head^{-1}$	–580.3	24.40**	1.69	0.98
Weight per grain, mg	–976.2	47.78**	9.18	0.90
Tillers m^{-2} at heading	–261.7	1.76*	0.50	0.81
Green leaves m^{-2} at heading	36.1	0.42*	0.12	0.79

Of the four rainfed crops, pearl millet was found to have the highest yield stability, highest response to water, highest potential for attainable yield on good soil under good management, and so it emerged as the best user of each added unit of water, over rainfed system, evapotranspired; via a reflection of water in growth and yield components that interplayed to bring about high yield response, and therefore, a high yield stability (Tables 4.6 and 4.7, and Fig. 4.10).

Table 4.7 Regression of fertile spikelet number per head (Y) on growth components and their nitrogen composition of winter wheat grown at Jodhpur

Components	Y = a + bx		SE	r
	Constant	Coefficient		
Green leaves 20 days after sowing (DAS)	0.9	4.44**	0.49	0.98
Green leaves 52 DAS	15.6	–0.12	1.24	–0.06
% N in stem at heading	30.2	–18.79*	4.52	–0.92
% N in green leaves at heading	30.2	–4.54	1.95	–0.80
% N in senescent leaves at heading	23.6	–5.62*	1.83	–0.87
Grain No. $head^{-1}$ on spikelet no. $head^{-1}$	7.3	0.16**	0.03	0.89

Reasons for high yield stabilities shown by pearl millet, cowpea, and wheat are straightforward. Pearl millet with C_4 pathway is characterized by high light-saturated photosynthetic rate, ranging 40-80 mg CO_2 $ha^{-1}dm^{-2}$ of leaf in an optimum day temperature range of 30-45°C, and the capacity to maintain this high rate at a low level of CO_2 concentration inside the air spaces of its leaves (Black et al., 1971). The two together allow this and other C_4 species to fix a larger amount of CO_2 for a given amount of water transpired.

Pearl millet undergoes osmoregulation, a mechanism that helps its plants in the maintenance of stomatal opening, photosynthesis, and more water uptake from the soil (Henson et al., 1982). A nonsignificant irrigation x genotype interaction for water use reported by Singh et al. (1983) further showed that most of its irrigated or nonirrigated genotypes studied by them did not differ in crop water-use. This showed that, contrary to sorghum

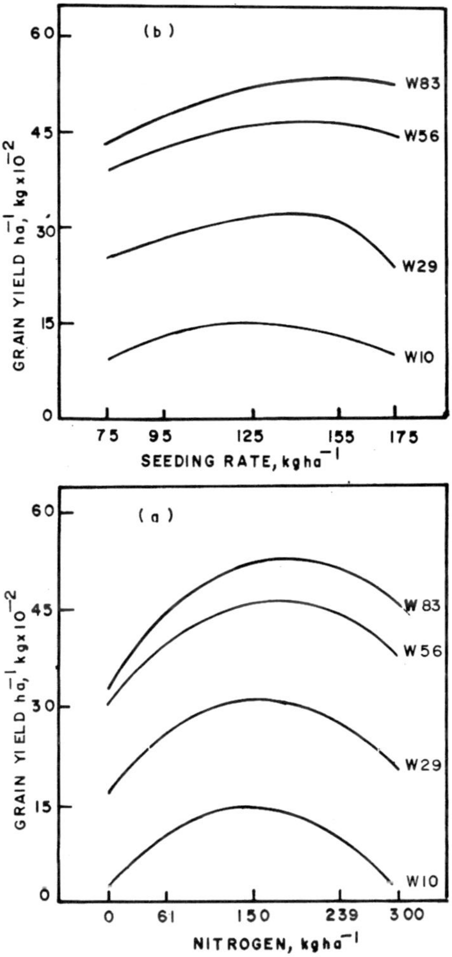

Fig. 4.10. Effect of significant interaction between (a) water and N and (b) water and seeding rate on grain yield of winter wheat grown at Jodhpur. W10, W29 etc. refer to water levels of 10 cm, 29 cm etc.

genotype, the slope of the yield reduction vs. ET deficit relationship found for 'BJ104' cultivar of pearl millet could possibly be used for yield prediction of genotypes presently grown in the region, or would be released for adaption to changed conditions of agronomic management and plant protection.

Grain yield of this crop was negatively and significantly correlated with leaf diffusive resistance (LDR) in both of their irrigated ($r = -0.85$) and nonirrigated ($r = -0.72$) treatments. In wake of these results, greater yield stability in pearl millet could be characterized by low leaf conductance (LC) or leaf permeability to water loss in transpiration.

As LDR is inversely proportional to transpiration as also to LC and as CO_2 and transpiration share the common pathway (stomata), pearl millet with high CO_2 fixation rate will have higher yield potential. On the other hand, selection of genotype of low leaf conductance for nonstressed environment (irrigated) will have greater yield stability, a conclusion supported by significant correlation of grain yield ratio with LDR in their irrigated treatment (Singh et al., 1983).

In cowpea seed yield is the expression of plant density x pod number per plant x seed number per pod x weight per sed. Experimental evidence revealed that pod number per plant and seed number per pod mainly regulated the seed yield while seed size remained reasonably constant across the row width, plant density, and irrigation (Herbert and Baggerman, 1983). Pod number per plant has been reported to be more sensitive to water stress than the seed number per pod. This was evident from an R^2 of 0.96 with regression of pod number $plant^{-1} = 1.5 + 1.01^{**}$ ETd, and R^2 of 0.89 with regression of seed number $pod^{-1} = 2.8 + 0.73^{**}$ ETd found in our study. The former regression coefficient was significantly greater than that for the latter.

Further, cowpea has been found to yield satisfactorily under a greater diversity of climate, soil, and cultural conditions than most other leguminous crops (Ligon, 1958), and this may be one reason why this crop in our study showed higher grain yield stability relative to mung bean and cluster bean. High yield stability could also be attributed to early maturity of its cultivar 'FS 68' that helped the plants in escaping environmental stress at the flowering and pod-filling stages.

With regard to wheat, high yield stability could be attributed to large yield reduction in the simulated rainfed treatment, relative to high response to water in accompaniment of fertilizer N and optimal seeding rate. To quantify the response and to engineer a blue print for high yield, analyses were made into data gathered from a study conducted in 3-factor central composite rotatable design comprising water, nitrogen, and seeding rate each at predetermined values -1.682, -1, 0, 1, and 1.692. The levels in equations for the yield and yield components as shown in Table 4.8 are the predetermined values and not the actual treatment level.

Data presented in graphic form were compiled from the response functions (Table 4.8). In cases where responses of two factors were presented, as in grain yield, the third factor was set to the central value, 0. Similarly, when the main effect of one factor, for example water, was presented, other two factors were set to zero, the middle value.

Increasing levels of water up to 83 cm tended to increase grain yield almost linearly up to 150 kg N ha^{-1} and seeding rate up to 155 kg ha^{-1} (Fig. 4.10a and b), but while water use at 120 cm lowered the grain yields use of 10 cm of water did not produce economical yield. Such trends were brought about by a steady increase in majority of yield components such as potential tillers m^{-2} recorded before commencement of heading, grain number $head^{-1}$, aboveground biomass, and harvest index up to 83 cm of water. Except green leaves m^{-2} at heading as well as heading percentage (Table 4.8) that showed increases only up to 56 cm of water (Fig. 4.11). In two instances, significant components such as fertile spikelets $head^{-1}$ and weight $grain^{-1}$ continued to increase across the seasonal irrigation amounts (Fig. 4.11b and c).

Numerical expression of yield components observed from 83 cm of water-use resulted in a yield Y (kg ha^{-1}) = 13090 kg ha^{-1} biomass x 40.4% HI = 5288 (Fig. 4.11d), as opposed to predicted yield at 5300 kg ha^{-1} (Fig. 4.10). Shown in another way, Y (kg ha^{-1}) = 322 heads m^{-2} (488 potential tillers x 66% head bearing) x 48.3 grains $head^{-1}$ x 33.7 mg wt. $grain^{-1}$ = 5241 (Fig. 4.11). This blue print

Table 4.8 Constants and coefficients of response suface for grain yield and yield components of wheat grown at Jodhpur

Team	Yield ha^{-1} kg x 10^{-2}	Grain wt. m^{-2}, g	Heading % age	Fertile spikelets $head^{-1}$	Potential tillers m^{-2}	Gains $head^{-1}$	wt. $grain^{-1}$ mg	Harvest index, %
Constant	47.18	537.39	69.1	14.8	47.4	46.51	30.51	40.0
X_1 (Water)	10.51**	128.98*	5.3*	1.0**	44.3*	4.53**	3.60**	2.0**
X_2 (Nitrogen)	2.75**	74.36*	6.6*	1.5**	48.6*	3.31**	–1.38**	–1.3*
X_3 (seed rate)	1.99**	10.48	–3.5	–0.8**	11.1	–2.04*	0.04	0.5
X_{11}	–5.01**	–59.80	–8.1**	–0.2*	–30.0	–2.76	–0.45	–1.6*
X_{22}	–4.59**	–38.04	4.2	–0.4**	–31.1	–2.14*	0.25	–0.3
X_{33}	–1.77**	15.06	0.2	0.1	0.9	0.36	0.27	–1.2*
X_{12}	1.44**	49.40	3.1	0.1	–11.7	–0.02	0.41	0.3
X_{13}	0.82*	15.91	–3.9	0.2	–2.9	–0.94	0.10	–1.5*
X_{23}	0.21	22.79	3.0	0.02	–25.8	0.22	0.23	0.7
SE : Linear terms	0.18	19.41	1.7	0.07	20.5	0.79	0.23	0.4
SE : Quadratic terms	0.17	18.00	1.7	0.07	20.0	0.77	0.23	0.4
SE : Interaction terms	0.23	25.36	2.3	0.10	26.8	1.04	0.31	0.5
R^2 (%)	91	95	88	94	–	91	87	95

* ** Significant at the 5 and 1% level of probabilities by the t-test, respectively.

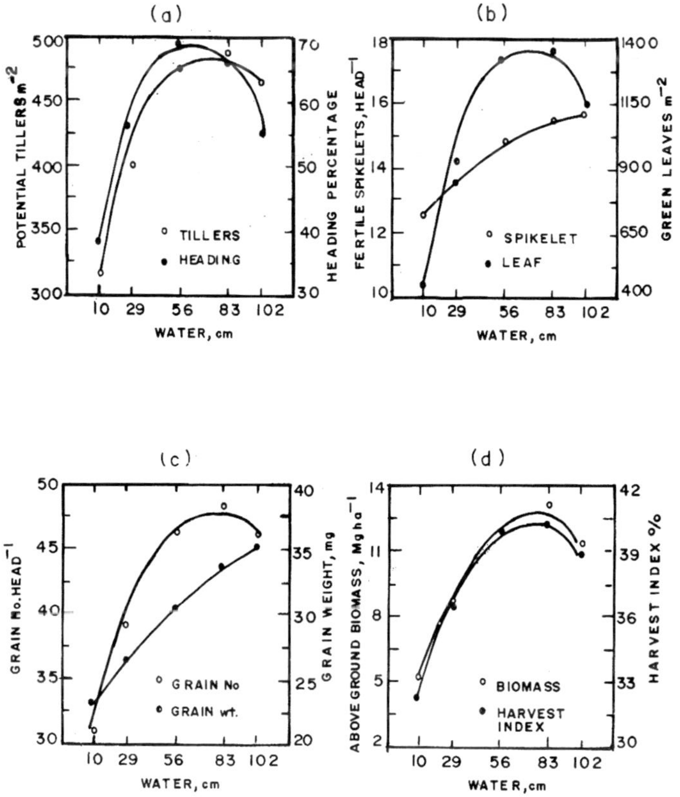

Fig. 4.11. Effect of irrigation water on (a) potential tiller no. and heading percentage, (b) spikelet no. head^{-1} and green leaf no., (c) grain no. head^{-1} and wt. grain^{-1}, and (d) above-ground biomass and harvest index. N and seed rate are set to the central value, 0.

for higher yield assumes special significance in context to India, as 16% of world population is packed in 2.4% of its land area (World Population Day - 11 July 1992, Hindustan Times 11 July 1991).

Thus a very low yield in the simulated rainfed plot, and ability of the wheat crop to respond amazingly to increasing water supplies up to 83 cm accounted for the low yield ratio implying high yield stability.

2. Inference from Yield Reduction to ETd Regression

As shown in Fig. 4.9, regression lines continued to deflect farther and farther from the 1:1 line showing the rate of reduction being faster at the higher than at the lower ETd intensities. This kind of trend supported the conclusion that a small supplement to rainfall that is possible by sprinkler irrigation could improve the production considerably. In verification of this conclusion, field trials integrating dryland farming into supplemental irrigation system were conducted during 1980 and 1981 on almost all the major crops adapted to desert agriculture, in the arid tract of Rajasthan. In this tract, two out of every five years are the years of drought and crop failure. Crops fail or give low yields for want of just one or two showers early in the season or at the critical stage of grain development.

Supplemental irrigation was integrated into dryland farming by giving life-saving irrigation(s) varying in depths through the sprinkler system of irrigation. Data presented in Table-4.9 show the significance of supplemental irrigation in improving rainwater-use efficiency, apart from substantial increase in production and security of subsistence coupled with providing a wide spectrum of choice to supplement irrigation to rain depending upon the amount of water available and yield goal in mind. Absolute water use yield figures are applicable to the Rajasthan desert, but the concept emerging provides insight for the use of integrated dryland farming-supplemental irrigation system in any of the arid regions of the world.

A small 5.9 cm of water supplemented to rain gave 104% increase in yield, and the highest added yield of 22.7 kg ha^{-1} mm^{-1} of supplemental irrigation to the pearl millet crop. Sorghum as a crop was found to yield about 24 kg ha^{-1} mm^{-1} of ET from supplemental irrigation totalling 11.1 cm in the growing season. Yield of grain legumes, by and large, showed a linear increase as the ET increased from supplemental irrigation, but there can be only a marginal increase in efficiency of water use (Table 4.9, Fig. 4.14). The results of studies carried out by Herbert and Baggerman (1983)

Table 4.9 : Yield (Y) benefits from the integrated dryland farming (DF) – supplemental irrigation (Si) system, part of data from Singh (1985a)

Pearl millet						
ET from Si, cm	0.0+	4.7	5.9	6.9	8.2	10.6
Yield, kg ha^{-1}	1290	2120	2630	2700	2970	3570
Added Y/ET, kg ha^{-1} mm^{-1}	–	17.7	22.7	20.4	20.5	21.5
Sorghum						
ET from Si, cm	0.0	5.9	11.1	17.9	20.3	24.3
Yield, kg ha^{-1}	0	1380	2690	3040	3440	4370
Added Y/ET, kg ha^{-1} mm^{-1}	–	23.4	24.2	17.0	16.9	18.0
Cowpea						
ET from Si, cm	0.0	5.4	8.5	10.2	12.4	14.4
Yield, kg ha^{-1}	510	960	1110	1260	1460	1660
Added Y/ET, kg ha^{-1} mm^{-1}	–	8.3	7.1	7.4	7.7	8.0

Table contd....

Mung bean						
ET from Si, cm	0.0	4.8	7.5	9.6	11.4	15.0
Yield, kg ha^{-1}	340	500	750	900	1050	1190
Added Y/ET, kg ha^{-1} mm^{-1}	–	3.3	5.5	5.8	6.2	5.7
Cluster bean						
ET from Si, cm	0.0	4.0	8.9	10.5	13.3	16.5
Yield, kg ha^{-1}	440	620	690	720	880	1210
Added Y/ET, kg ha^{-1} mm^{-1}	–	4.5	2.8	2.7	3.3	4.7

+ under zero Et from Si, yield is from DF plots that received 16.3, 20.4, 11.2, 15.6 and 23.0 cm of rainwater in pearl millet, sorghum, cowpea, mung bean, and cluster bean, respectively; rainwater amounts varied according to crop growing period.

with cowpea cultivar 'California' No.5' on an Oltan sand clay loam on the Texas Tech Univ. research farm in Lubbock, Texas bear these facts out. Yield from the control plot (no growing season irrigation, except 80 mm of preplant irrigation to all plots using trickle irrigation system) was 635 kg ha^{-1}, averaging 7.3 kg ha^{1} mm^{-1} of water used, which increased to 2043 kg ha^{-1} at the highest 200 mm level of irrigation applied in two equal applications, showing only a slight increase in WUE to 7.8 kg ha^{-1} mm^{-1}. Hence use of scare, expensive water resource of an arid region is contemplated in water responsive pearl millet and grain sorghum. The date and depth of irrigation to a crop would depend on the amount and distribution of rainfall. Low added out put per unit use of water input does not justify any use of a limited water supply of the region to these grain legumes.

3. Minimization of Evaporation Relative to Transpiration

As will be seen from data in Table 4.11 early in the season up to day 25 when groundcover was least, ET started with a low figure and was independent of plant density. A large part of water use was in the form of evaporation and least through transpiration. Hence we limit our discussion to leaf area concept and irrigation in terms of groundcover that influences the relative proportion of evaporation to transpiration.

(i) Optimum leaf area concept : As seen from Fig. 4.12 the leaf area index (LAI) of an annual crop, for example wheat, started form zero early in the season, increased rapidly to a maximum, remained at the maximum level for a short period, and then declined rapidly. Maintenance of 'optimum' leaf area at all times is therefore difficult. However a system it comes closest to this ideal is expected to be most efficient.

To ensure that the evaporation is less and less relative to transpiration it is desirable to have an optimum density of vegetative cover by properly adjusting the row spacing and seed rate, by selecting the most favourable time of planting, and by a fast development of crop canopy through application of nitrogen early in the season. In the winter wheat study at Jodhpur, LAI increases of the crop (seeded at 125 kg ha^{-1}) at its booting stage were found to be significantly curvilinear as season water increased in the 10 to 102 cm range, as also when water and fertilizer N were applied in a plot. When water and seeding rate were set to 56 cm and 125 kg ha^{-1}, application of 150 kg N ha^{-1} maintained an optimal leaf area index at 3.4. At this value, evaporation whatever value it had was a constant percentage of ET, and evapotranspiration exceeded the daily rate of evaporation (Fig. 4.12). Maintenance of optimal soil fertility therefore is essential in managing relative proportion of evaporation to transpiration.

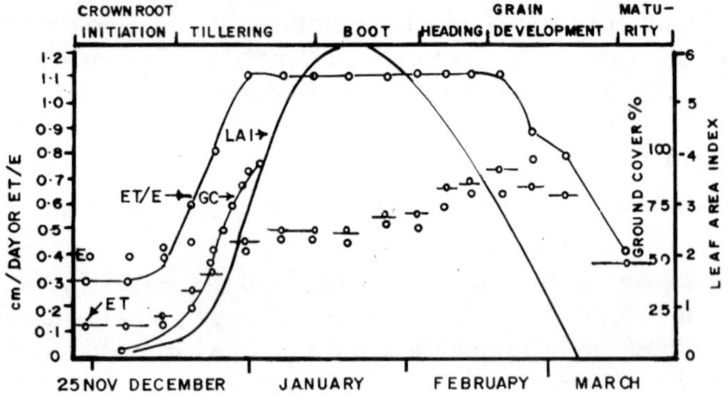

Fig. 4.12. Per cent ground cover GC, ET, E (evaporation) In Class A pan and ET/E ratio in relation to seasonal changes with time in leaf area index (LAI) of winter wheat. From Singh (1977).

As the percentage crop cover increased up to 50% through an increase in LAI from 0.3 to at about 1.2, ET tended to increase (Fig.4.12). During weeks 7 to 12, LAI was more than 3, or the groundcover was more than 50% complete. Beyond a certain point (LAI = 3.5 in this case), the groundcover did not increase because of the mutual shading of leaves to a large extent. During the period in which crop cover was more than 50% complete, the ET crop exceeded the evaporation.

The evaporation contributed a substantial part of water use again when the crop approached maturity, the LAI declined fast, as reduced canopy due to successive leaf senescence allowed a substantial transmission of radiant energy to the soil. During this time, the ET continued, but at a diminishing rate. In most grain crops, it is for this reason that the soil moisture should not be maintained at a high level during maturation.

(ii) Irrigation in terms of groundcover : It may also be mentioned (though not shown in Fig. 4.12) that frequent irrigation until the groundcover is incomplete (LAI less than 1.2) will cause low ratios of transpiration to evaporation for 2- 3 days subsequent to irrigation by excessive evaporation from the wet soil surface. Once the groundcover is complete (LAI = 1.2 to 3.5), the ratio is more than unity through the preponderance of transpiration, and evaporation whatever value it may have, appears to become a constant percentage of ET. Efficient water management requires some restrictions on frequent irrigation early in the season until the maximum energy available is used through evaporation, on the one hand, and a wise use of water is made if applied during the period of complete groundcover, on the other.

4. Removal of Non-ET Water Uses

As discussed earlier, irrigation water = ET + non-ET and irrigation efficiency is the ET expressed as percentage of irrigation water owned/purchased and applied. As we approached ETm to attain Ym, some deep percolation was inevitable, besides some available soil water remaining in storage at time of harvest. A rainfall exceeding the size of soil storage also results in non-ET water disposition, and sandier the soil, the greater are the losses through deep percolation.

From the foregoing account it seems that the optimum water use policy in arid and semi-arid regions should be to apply the concept of modest irrigation, and to terminate irrigation at an appropriate stage to ensure that losses due to deep percolation are minimal or even eliminated, and no available soil water remains in storage at harvest. The termination of irrigation is specific to soil, climate, crop, irrigation system, etc.

(a) Higher plant density : Higher plant density is a shotgun easy-to-control practice known for its potential in enhancing transpiration relative to evaporation. One should not be mistaken that a twofold increase in plant density will require twice as much water. The ET of winter wheat was found to remain about the same or increase only slightly, whereas the yield tended to increase up to the density where the compensation in individual and group performance of plants was compromising (Table 4.10)

Table 4.10. Yield Y, ET, Y/ET, LAI, grain no. head^{-1}, and grain wt. of the winter wheat crop at five plant populations; with water and N fixed at 56 cm and 150 kg ha^{-1}. From Singh (1977)

Seed, kg ha^{-1}	75	95	125	155	175
Initial stand, No.m^{-2}	178	201	249	328	356
LAI	2.5	3.2	3.4	4.1	6.2
Grain,No.head^{-1}	51	49	47	45	44
Wt.grain^{-1},mg	31	31	31	31	31
Yield,kg ha^{-1}	3554	4342	4717	4740	4500
ET,cm	59	61	62	62	70
Y/ET, kg ha^{-1} mm^{-1}	6.0	7.1	7.6	7.6	7.0

Table 4.11 Two-year mean growth rate (i.e. dry matter production, DM), ET, and DM/ET at different stages of winger wheat for three selected seeding rates. Water and N set to optimum level.

Kg seed ha^{-1}	Days after planting		
	25	60	84
	DM, kg ha^{-1}		
75	155	2262	8463
125	220	3215	9550
175	299	3333	7751
	ET, cm		
75	7.2	23.8	40.5
125	7.4	25.1	41.7
175	7.6	26.6	43.2
	DM/ET, kg ha^{-1} mm^{-1}		
75	2.2	9.5	20.9
125	3.0	12.8	22.9
175	3.9	12.5	17.9

Data as given in Table 4.11 revealed no visible differences in ET due to density early in the season, and only slight differences later in the growing season. It seems that the plant density influenced ET through its effect on leaf area index. In the winter wheat study to which these data pertain, the leaf area indices 61 days after planting when the crop had entered into its boot stage were found to be 2.5, 3.4, and 6.2 at the selected seeding rates 75, 125, and 175 kg ha^{-1}. Optimal planting density would therefore be instrumental in minimizing soil evaporation early in the season, and in prudent use of applied water during the grand growth period.

Optimal plant density may be helpful in improved water management also through a control on water flow below the root zone. Fig 4.13 shows water distribution under population densities of tomatoes, viz. 48,96, and 192 plants per 12 m length of drip irrigation lateral, each receiving an equal amount of water based on the daily crop ET demand. Water distribution in the profile pertains to the stage tomatoes were finally picked. In each case, the plants had to share the same amount of finite water supply. As may

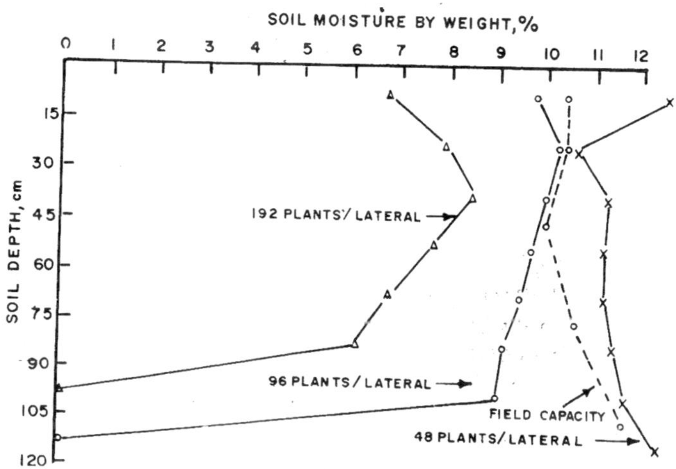

Fig. 4.13. Effect of plant densities on water distribution in the root zone of tomato. From Singh (1978).

be seen from three curves, the low population crop exerted less use thrust on added water. The net result was that soil moisture at all depths was never reduced to values where drainage could cease. On the contrary, where a large number of plants had to share the same amount of finite water supply, the soil moisture at all depths below 45 cm was reduced to values where no drainage could occur. Dry soils below 105 cm on 96 plants per lateral plot, and below 90 cm on 192 plants per lateral plot further confirmed that no water was lost form the potential root zone.

5. Maximum Yield/Optimal Yield Concept

Application of the knowledge of yield and irrigation relationships to distribution of a given water will depend on whether land or water in a region is limited or adequate. In an area where land is limited water is available on payment, the objective may be to maximize the production or profit per unit area of land. To illustrate, the levels of irrigation that maximized the production and profit per unit area of land in a winter wheat, for example, were 98 cm and 79 cm, respectively (Fig.4.6) These values were determined by equating a partial derivative of Y with respect to I (from the Y-I functional equation given earlier) to zero for "maximum" and to input-output price ratio for "optimality" computation. In algebraic terms this could be written as $dY/dI = 0$, Pi/Py set to (0.2141) in an input- output price situation. Pi is price per cm of water (including application cost) and Py is price per 100 kg of wheat grain. The respective yields found to be associated with "yield" and "profit" maximizing levels of irrigation were 5460 kg ha^{-1} (observed 4 years average maximum yield was 5430 kg ha^{-1}) and 5260 kg ha^{-1} (Fig.4.6).

In arid tropic where water is limited and expensive, the objective may be to maximize production per unit application of water. In this case, a lower average yield on a large area can result in higher total benefit than the most profitable yield from a small area. Consequently, one may limit seasonal irrigation to a level that maximizes water-use efficiency. This point of maximum water-use efficiency in Fig.4.6 could be determined by the point of contact of a tangent from origin of the CWS-axis to the function Y-CWS. The tangent, which originated from zero of the CWS-axis, touched the curve relating Y to CWS ($Y = 1.32\ CWS - 0.0067\ CWS^2 - 11.03$; R^2 =

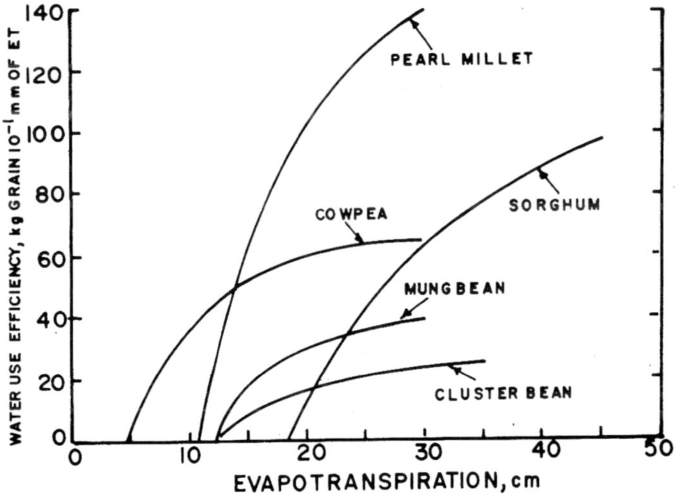

Fig. 4.14. Relationship between water-use efficiency as predicted by the Y-ET regression and seasonal ET. From Singh (1985b).

0.99; combined for 1971-75 function) at the coordinate: CWS = $\sqrt{a/c}$ and Y = (2a + b $\sqrt{a/c}$, in which a, b and c are the same as in equation relating Y to CWS. This point of contact occurred at 41 cm of CWS and 3150 kg ha^{-1} of wheat yield; 7.7 kg ha^{-1} mm^{-1} of crop water supply. At this level of water a yield of 4666 kg ha^{-1} was obtained from another field trail with wheat cultivar 'HD 2009'. It showed the significance of a proper crop cultivar in improving the level of production and water-use efficiency.

Under water deficit conditions most field crops are known to have higher water-use efficiency. To verify, water-use efficiencies of pearl millet, grain sorghum, cowpea, mung bean, and cluster bean were calculated from the yield and water-use relations shown on Fig. 4.3. As the best fit line between yield Y and ET in a crop was linear: Y = a + bET, it was possible to calculate water-use efficiency (WUE) from the relationship : Y/ET = b + a/ET in which 'a' and 'b' are constants of the regression line given on Fig. 4.3. Predicted WUE are plotted on Fig. 4.14, which showed a general tendency for it to decline, over the range of data, with a decline in ET.

This was obvious since the intercept 'a' was negative for all crops and a decrease in WUE as the ET decreased could be expected.

Continuing research on water-use optimization at Jodhpur showed that integration of the optimum supplies of relatively abundant resources (nitrogen and seed) with suboptimal supply of the scarce resource (water) can increase the production of winter wheat. With optimum combination of inputs at rates of 150 kg N and 125 kg ha^{-1} seed designated respectively as Nopt and Dopt, the WUE maximizing level of irrigation was found to be 41 cm and yield was 3150 kg ha^{-1}, or 7.7 kg ha^{-1} mm^{-1}. The Nopt and Dopt were combined with 30.4 cm of water designated arbitrarily as Wsubopt and subjected to field validation. In the field, 30.4 cm of water was scheduled as follows : 7.6 cm of water as preplant irrigation, and 3.8 cm of water applied in six equal applications at the crown-root initiation, tillering, booting, heading, and early dough stages. Observed yield from validation trial was found to be 3000 kg ha^{-1}, or 9.9 kg ha^{-1} mm^{-1}. Input mix Wsubopt Nopt Dopt was thus found to maximize water-use efficiency of wheat, confirming that the best possible water application rationality was at the level much below what was needed for maximum profit per unit area of land.

There is another way to maximize yield per unit application of water. It is like what has been shown in Table 4.12. The seasonal crop water requirement was minimized by optimally programming the date and depth of all seasonal irrigations in such a way that yield reduction was proportionately lesser than reduction of water below the gross amount needed to meet fully the crop water requirement of each hectare of the area to be cropped.

For instance, in wheat cultivar 'Kalyansona' the prevailing climate at Jodhpur set up maximum irrigation demand of 84 cm resulting in a yield of 5430 kg ha^{-1}. Optimally planned 52% and 76% less water than required 84 cm resulted in only 13% and 32% yield reductions, respectively.

Table 4.12 : Optimal irrigation programme for wheat cultivar 'Kalyansona'. From Singh and Mann (1979)

Seasonal water cm	Reduced water use %	Gross Irr. days after (cm)							I. eff. %	Yield reduction %	Yield kg ha^{-1}	WUE kg ha^{-1} mm^{-1}
		21	40	54	68	78	90	100				
84 (Imax)	nil	8	11	11	13	14	14	13	75	nil	5430	6.5
40 (Imed)	52	–	3	10	5	10	12	–	87	13	4724	11.8
20 (Ilow)	76	–	3	–	5	7	5	–	95	32	3668	18.3

6. Extensive Irrigation Concept

Extensive irrigation is synonym to deficit irrigation. It was found that, when better seed, fertilizer, water management, and crop husbandry practices were used in the production system, it was possible to achieve about two-folds increase in the total production of several crops from a small quantity of water applied over larger area rather than a large quantity of water application over a small area (Table 4.13). The only minus point it has is the enlarged-field-surface area it provides for evaporation direct from the soil. Field application of extensive irrigation practice requires control of evaporation by judicious use of cultural practices.

Table 4.13. Production for a given water quantity used over different areas. From Singh (1977).

Crop	Water used cm	On area ha	Total production, kg
Wheat	84	1	5500
		3	9100
Pearl millet	25	1	4200
		4	10300
Sorghum	28	1	4400
		2.5	6400
Sunflower	50	1	1600
		2	2700
Mustard	25	1	1100
		1.5	2000

7. Minimum Acceptable Yield Concept

Partial derivative of wheat grain yield Y with respect to water-use X in the yield to irrigation relation defined physical optimum when equated to zero ($dY/dX = 0$) and to input/output price ratio to define the optimum profit. The mathematical term $\sqrt{a/c}$ in which 'a' and 'c' are the constants in quadratic equation between yield and water-use defined the water level that maximized water-use efficiency (WUE). As we know, the level of water what was needed to maximize profit was lower than the yield maximizing level whilst WUE maximizing level was still lower. In coming down from the physical optimum what would be the minimum tolerable limit to water-use below that one should not plan to use water was defined in this section. Prima facie, we have now wide spectrum of choice to programming irrigation bearing in mind the objective of achieving the physical optimum, the economic optimum, the maximum WUE, or the minimum acceptable limit to water-use.

Table 4.14. Simple iteration to define minimal tolerable limit of water (cm) for "minimal acceptable" yield (kg ha^{-1}) of winter wheat.

Water W	Yield Y	CSS of deviations from $\overline{W}$	CSS of deviations from $\overline{Y}$	R^2, i.e., $\left[\frac{TCSS-(CSScol.3+4)}{TCSS}\right]$
86	5366	1089	1382976	0.9639
84	5362	961	1373584	0.9642
82	5352	841	1350244	0.9648
74	5247	441	1117249	0.9709
72	5206	361	1032256	0.9731
70	5157	289	935089	0.9757
62	4902	81	506944	0.9868
60	4822	49	399424	0.9896
58	4736	25	298116	0.9922
50	4329	9	19321	0.9995
48	4212	25	484	0.9999
46	4088	49	10404	0.9997
44	3958	81	53824	0.9986
36	3375	289	664225	0.9827
34	3213	361	954529	0.9751
32	3045	441	1311029	0.9659
24	2311	841	3530641	0.9081
22	2111	961	4322241	0.8875
20	1906	1089	5216656	0.8643

$\overline{W}$ = 53, $\overline{Y}$ = 4190, and TCSS = 38475632

The minimum tolerable limit to water-use in, for example, winter wheat crop was determined from the yield to water-use relation given in Table 4.8, with nitrogen and seeding rate set to nonlimiting level at 0. The physical optimum found to maximize yield was 86 cm of water. This 86 cm was allowed to diminish at the interval of 2 cm. By plugging each diminishing value into yield to water-use regression equation the reduction in yield was iterated until 34 number of iterations (Table 4.14) when the water reached the value where other low-water requiring crops, e.g., barley and oats, may inter into competition for water with wheat.

To maximize $Y - \bar{Y}$ while at the same time keeping W (water) as small as possible, the deviations squares of W from its mean and Y from its mean were estimated. This gave large positive deviations on $Y - \bar{Y}$ and large negative deviations on $W - \bar{W}$. The R^2 (estimated from the total corrected sum of squares (TCSS) of deviations of Y from its mean minus corrected sum of squares (CSS) of deviations from mean of W and Y divided by TCSS) would be highest when W is at its mean or very close to its mean because, when plugging $\bar{W}$ into the regression equation, the predicted Y would be close to $\bar{Y}$. The minimum tolerable limit to water where R^2 was the highest was 48 cm of water, with a yield of 4212 kg ha^{-1} (Table 4.14). The value was very close to $\bar{W}$. It showed again that the minimum tolerable limit to water would be close to the mean, or $\bar{W}$ in almost every case.

VI. SUMMARY

Different aspects of yield and water-use relations are discussed. Yield to evapotranspiration relationship was found to be linear. The linearity from the ET-axis, unlike linearity of dry matter production-transpiration curve from the origin, was a reality under dry farming conditions or at the lower levels of irrigation. Under these conditions, yield-irrigation relation was similar to linear yield-ET relation. As the season potential evapotranspiration approached to attain the maximum attainable yield, non-ET water dispositions tended to increase, or in other words, irrigation efficiency (ET expressed as per cent of water applied) went on decreasing. As a result, the relationships of ET to crop water supply and yield to irrigation-water became convex.

Applicability of several yield prediction models developed all over the world from time to time has been examined using the data from studies made on the crops adapted to dryland conditions and to irrigated areas in the arid, northwestern India.

Some concepts emerged from the yield to ET, yield to irrigation- water, and yield to total crop water supply relationships. Important among them are: crop response factor as a tool to crop sensitivity rating, optimal leaf area concept, maximum yield concept, optimal yield concept, water-use efficiency maximizing concept, minimum acceptable yield concept, extensive irrigation concept, etc. Applicability of each of these concepts to managing a limited water supply in the arid region has been interpreted.

REFERENCES

Ali, M.A. 1979. M.S. Thesis, Utah State University, Logan.

Arkley, R.J. 1963. Hilgardia. 34:559-584.

Black, C.C., G.E. Edwards, R. Kanai, and H.H. Mollenhaver. 1971. *In* 2nd International Congress on Photosynthesis, p 1745- 1757.

Briggs, L.J. and H.L. Shantz. 1914. J. Agric. Res. 3: 1-63.

Chaudhuri and E.T. Kanemasu. 1982. Canadian J. Plant Sci. 62: 599-607.

Childs, S.W., J.R. Gilley, and W.E. Splinter. 1977. Transactions ASAE 20: 858-865.

Cole, J.S. and O.R. Mathews. 1923. USDA Bull. 1004. U.S. Government Printing Office, Washington, D.C.

DeLoughery, R.L. and R.K. Crookston. 1979. Agron. J. 71: 577-580.

de Wit, C.T. 1958. Versl. Landbouwkd. Donderzock. no.64.6.

Dillman, A.C. 1931. J. Agric. Res. 42:187-238.

Doorenbos, J. and A.H. Kassam. 1979. FAO Irrigation and Drainage Paper No. 33. FAO, Rome, 193 p.

Fisher, R.A. and N.C. Turner. 1978. Annu. Rev. Plant Physiol. 29:277-317.

Garrity, D.P., D.G. Watts, C.Y. Sullivan, and J.R. Gilley, 1982. Agron. J. 74:815-820.

Grimes, D.W., H. Yamada, and W.L. Dickens. 1969. Agron. J. 61:769-773.

Hanks, R.J. 1974. Agron. J. 66:660-665.

Hanks, R.J. 1983. *In* H.M. Taylor, W.R. Jordon, T.R. Sinclair (eds.) Limitations to efficient water use in crop production, p. 393-411, ASA, CSSA, and SSSA. Madison, WI, U.S.A.

Hanks, R.J., V. Sorenson, and A. Retta. 1977. Utah Sci. 38:38-43.

Hanks, R.J., H.R. Gardner, and R.L. Florian. 1969. Agron. J. 61.30-34.

Hanks, R.J. and V.P. Rasmussen, 1982. Adv. Agron. 35:193-215.

Henson, I.E., V. Mahalakshmi, F.R. Bidinger, and G. Alagarswamy. 1982. Plant, Cell Environ. 5:147-154.

Herbert, S.J. and F.D. Baggerman. 1983. Agron. J. 75:982-986.

Hodges, T. 1978. Ph.D. Thesis, Kansas State University, Manhattan.

Howe, O.W. and H.F. Rhoades. 1955. Proc. Soil Sci. Soc. Amer. 19:94-98.

Howell, T.A. and E.A. Hiler. 1975. Transactions ASAE 18:873-878.

Jensen, M.E. 1968. *In* T.T. Kozlowski (ed.) Water deficits and plant growth. vol.2, p. 22, Acad. Press, New York.

Kallsen, C.E., T.W. Sammis, and E.T. Gregory. 1984. Agron. J. 76:59-64.

Kanemasu, E.T. 1983. *In* H.M. Taylor, W.R. Jordan, T.R. Sinclair (eds.) Limitations to efficient water use in crop production, p.413-417, ASA, CSSA, and SSSA, Madison, WI, U.S.A.

Kanemasu, E.T., L.R. Stone, and W.L. Powers. 1976. Agron. J. 68:569-572.

Kiesselback, T.A. 1929. Proc. International Congress of Plant Science, Sec. of Agronomy, p. 87-105.

Leggett, G.E. 1959. Wash. Agric. Exp. Stn. Bull. 609.

Ligon, L.L. 1958. Okla. Agric. Exp. Stn. Bull. B-518.

Miller, M.G. and F.L. Duley. 1923. Miss. Agric. Exp. Stn. Res. Bull. 76.

Monteith, J.L. 1977. Phil. Trans. of the Royal Soc. London (B) 281:277-294.

Morgan, T.H., A.W. Biere, and E.T. Kanemasu. 1980. Water Resour. Res. 16:59-64.

Musick, J.T., D.W. Grimes, and G.M. Herron. 1963. Production Res. Rep. No. 75, ARS, USDA, in cooperation with Kansas Agric. Exp. Stn.

Newell, R.E., M. Tanaka, and B. Misra. 1976. Climate and food workshop : A report. Bull. Amer. Meteorl. Soc. 57(2):192-197.

Nimah, M.N. and R.J. Hanks. 1973. Soil Sci. Soc. Amer. Proc. 37:522-532.

Niwas, Ram. 1975. Ph.D. Thesis, University of Jodhpur, Jodhpur.

Owonubi, J.J. and E.T. Kanemasu. 1982. Canadian J. Plant Sci. 62:35-46.

Passioura, J.B. 1977. J. Aust. Inst. Agric. Sci. 43:117-120.

Power, J.F., D.L. Grunes, and G.A. Reichman. 1961. Proc. Soil Sci. Soc. Amer. 25:207-210.

Rasmussen, V.P. 1979. Ph.D. Thesis, Kansas State University, Manhattan.

Rasmussen, V.P. and R.J. Hanks. 1978. Agron. J. 70:940-944.

Retta, A. and R.J. Hanks. 1980. Irrig. Sci. 1:135-147.

Richi, J.T. 1980. *In* Moving up the yield curve : Advances and obstacles. Ass special publication no. 39, p. 1-23.

Shantz, H.L. and L.N. Piemeisel. 1927. Colo. J. Agric. Res. 34- 1093-1190.

Singh, Piara, E.T. Kanemasu, and Phool Singh. 1983. Agron. J. 75:886-890.

Singh, S.D. 1977. *In* Desertification and its control, p. 237-255, ICAR, New Delhi.

Singh, S.D. 1978. Agron. J. 70:951-954.

Singh, S.D. 1981. Agron. J. 73:387-391.

Singh, S.D. 1984. *In* Proceedings Indo-US Workshop on Arid Zone Research, 9-14 January 1984, Jodhpur. Published by DST, Govt. of India, New Delhi, p. 485-491.

Singh, S.D. 1985a. *In* S.D. Singh (ed.) Irrigated agriculture in arid areas, p. 442-454, WAPCOS, New Delhi.

Singh, S.D. 1985b. *In* S.D. Singh (ed.) Irrigated agriculture in arid areas. p. 477-489. WAPCOS, New Delhi.

Singh, S.D. and H.S. Mann 1979. CAZRI Jodhpur (India) Res. Bull. 1, 88 p.

Singh, S.D. and M. Yusuf. 1979. Canadian J. Plant Sci. 59:437- 444.

Singh, S.D., M. Yusuf, R.C. Bhandari, and H.S. Daulay. 1976. Indian J. agric. Sci. 46:567-575.

Singh, Y.V. 1989. Ph.D. Thesis, University of Jodhpur, Jodhpur.

Sorensen, V.M., R.J. Hanks, and R.L. Cartee. 1979. Agron. J. 72:266-270.

Stewart, B.A. 1989. *In* Proceedings of an IBSRAM inaugural workshop, 18-22 February 1985. ICRISAT Centre, India. Patancheru, A.P. 502 324, India : ICRISAT.

Stewart, B.A. 1992. International symposium on management for sustained productivity, Ludhiana, India, 10-12 February 1992.

Stewart, B.A., J.T. Musick, and D.A. Dusek. 1983. Agron. J. 75:629-634.

Stewart, J.I. and R.M. Hagan. 1973. Irr. and Dr. Div., ASCE 99 (IR4):421-439.

Stewart, J.I., R.M. Hagan, W.O. Pruitt, and W.A. Hall. 1973. U.S.Dept. of the Interior, Bur. of Recl., Engineering and Research Centre, Denver, Report 14-06-D-7329. 164 p.

Stewart, J.I., R.M. Hagan, W.O. Pruitt, R.E. Danielson, W.T. Franklin, R.J. Hanks, J.P. Riley, and E.B. Jackson. 1977. Water Resources Center, Univ. of California, Davis. 190 p.

Stewart, J.I., R.D. Misra, W.O. Pruitt, and R.M. Hagan. 1975. Trans. ASAE. 18:270-279.

Stone, L.R., R.E. Grwin, Jr., and M.A. Dillon. 1978. Kansas Water News 21:12-21.

Tanner, C.B. and T.R. Sinclair. 1983. *In* H.M. Tylor, W.R. Jordon, and T.R. Sinclair (eds.) Limitations to efficient water use in crop production, p. 1-27, ASA, CSSA, and SSSA, Madison, W.I., U.S.A.

Vaux, H.J. Jr. and W.O. Pruitt. 1983. *In* D. Hillel (ed.) Advances in Irrigation. 2:61-97.

Wenda, W.I. 1979. M.S. thesis, Utah State University.

5

Managing Salt Affected Waters For Crop Production

Raj K. Gupta and P.S. Minhas

I. INTRODUCTION

Water is a limiting resource in arid and semiarid regions. Therefore, all out efforts have been made the world over for inter-basin transportations of surface waters through canal networks and exploitation of groundwaters wherever available. However, competition for imported good quality waters for industrial and drinking purposes is so intense that their use for agriculture becomes last priority. As the use of groundwater in irrigated agriculture is constrained by limited aquifer availability as well as quality, irrigation with salt affected waters to meet crop water requirements becomes inevitable.

Faulty irrigation management of salt affected waters often leads to excessive accumulation of salt and other toxic elements to levels that may affect the crop yields, quality of produce, and the choice of crops to be grown. At times problem becomes so severe that cultivators are forced to leave their lands fallow. Besides the irrigation practices, major ion chemistry of irrigation water as well as agro-ecological factors influence the processes of salt build up in the crop root zone. Therefore, all these factors are important for the development and management of salt affected waters for sustained agricultural production.

The purpose of this chapter is to evaluate the interactions of the parameteric factors of water quality on physico-chemical properties of soils and the influence of such changes on crop productivity. Problems emanating from the long-term use of low quality irrigation waters can be broadly grouped into three categories : (1) salinity hazard (quantity of dissolved salts), (2) alkalinity/sodicity hazards (high proportions of Na and CO_3+HCO_3 ions in relation to Ca), and (3) toxicity hazard (specific solutes causing damage to plant tissues or nutritional imbalances, e.g., B, F, NO_3, Se, etc.). Another related question addressed here is : what strategies can be adopted for management of multiquality irrigation waters that has relevance for many areas receiving limited quantities of canal water supplies.

II. EFFECTS OF LONG-TERM USE OF LOW QUALITY WATER

1. Salinity Hazards

Irrigation with saline waters and upward movement of soluble salts during evaporative process from saline-water table are the main cause of salt accumulation in the root zone of crops. For the extraction of soil water to meet the crop evapotranspiration (ET) plant roots have to overcome both matric and osmotic forces. An increase in the amount of soluble salts in soil water causes the total soil water potential to fall and forces the plants to spend additional energy for water extraction. When plants are unable to adjust osmotically, the growth reduces due to subdued uptake of water by crop resulting in symptoms such as wilting and small thick leaves with a dark bluish green colour. In severely affected fields, bare spots, patchy sparse crop stands with stunted growth are not uncommon. Salinity build up resulting from long-term use of saline waters depends upon several factors such as soil type, water table, drainage conditions, water management, rainfall pattern, and the cropping sequence.

Distribution of salts and water in the root zone is highly dynamic varying both in time and space depending upon the dynamic balance of the forces governing the rate and direction of soil water flow. When changes in salt storage of soil profile are negligible, a sort of steady state equilibrium exists that can be represented by the mass balance equation :

$$c_i d_i = c_d \ d_d \qquad [5.1]$$

where c and d represent salt concentrations and equivalent depths per unit surface area of irrigation (i) and drainage (d) water, respectively. This expression also defines the leaching fraction (LF), taken as the fraction of the applied irrigation water that leaves the root zone as drainage. Leaching fraction can alternately be expressed as

$$LF = d_d / (d_i + d_r) \qquad [5.2]$$

where d_r is depth of rainwater. Substituting the value of d_d from the equation (5.2) in expression (5.1) yields the following equation

$$c_d = c_i \ d_i / [L (d_i + d_r)] \qquad [5.3]$$

that relates the mean concentration of salts in the drainage water leaving the root zone at steady state as a function of leaching fraction, depth of the irrigation water and its salt content. Above relation that assumes no precipitation or the dissolution of salts, has been used to calculate the leaching requirement of crops. It is the minimum value of leaching fraction that a crop will require for the use of given salinity water without any reduction in yield. Using the above relation the average root zone salinity can be calculated if the water extraction patterns of the crops are known. Salinity of soil water (ECsw) at the different points of the root zone can be calculated using the equation :

$$ECsw\,(n) = ECw / [1 - \sum_{n}^{n=1} f\,(n)\ (1 - LF\ )] \qquad [5.4]$$

where ECsw(n) represents the ECsw at the nth layer, f(n) is the fraction of moisture extraction down to nth layer, and LF is the leaching fraction. The linear average of the ECsw was calculated as :

$$\overline{ECsw} = \sum_{n}^{n=1} ECsw\,(n)/n \qquad [5.5]$$

Salinity of saturation paste (ECe) extract is taken as ECsw/2 and the concentration factor, X = ECe/ECiw.

Achieving the leaching requirements has been considered essential in saline water irrigated agriculture (U.S. Salinity Laboratory Staff, 1954). However, such a concept has limited application in the continental monsoonal type of climate. Because steady state conditions are never achieved under the practical farming conditions.

The soil texture and the amount of rainfall received during the monsoon determine whether the salts accumulated in upper soil layers will leach out of the root zone, partially or completely. Measurements of the average soil salinity of the surface 0-30 cm of a sandy loam soil in Agra district of the Indian state of Uttar Pradesh under mustard-sorghum and wheat-pearl millet crop rotations in which winter crops of wheat / mustard received 3-5 irrigations with saline water (ECiw = 5.9 dS m^{-1}) revealed (Fig. 5.1) that the soil salinity build up occurs primarily during the winter season. The steady state conditions, a pre-requisite for use of

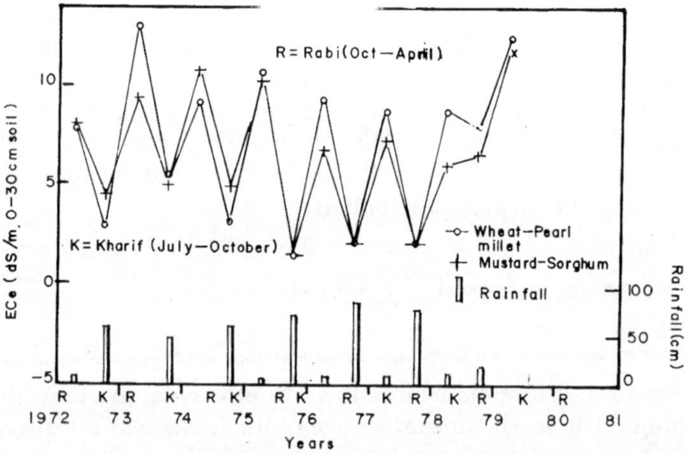

Fig. 5.1. Changes in soil salinity at the harvest of crops in rotation (wheat-pearl millet and mustard-sorghum) irrigated with saline water (ECiw 5.9 dS m^{-1}).

equation (5.5) for computation of average salinity was not attained even after 8 years of saline irrigation. A plot of salinity profiles obtained at the harvest of wheat (Fig. 5.2) usually shows 'inverted salinity profiles' where the salinity decreases with soil depth. In contrast, steady state conditions expectedly should have resulted in profiles where salinity increases with soil depth. In figure 5.2, steady state salinity profiles have been predicted using water uptake distribution of 60:25:10:5 from different quarters of the root

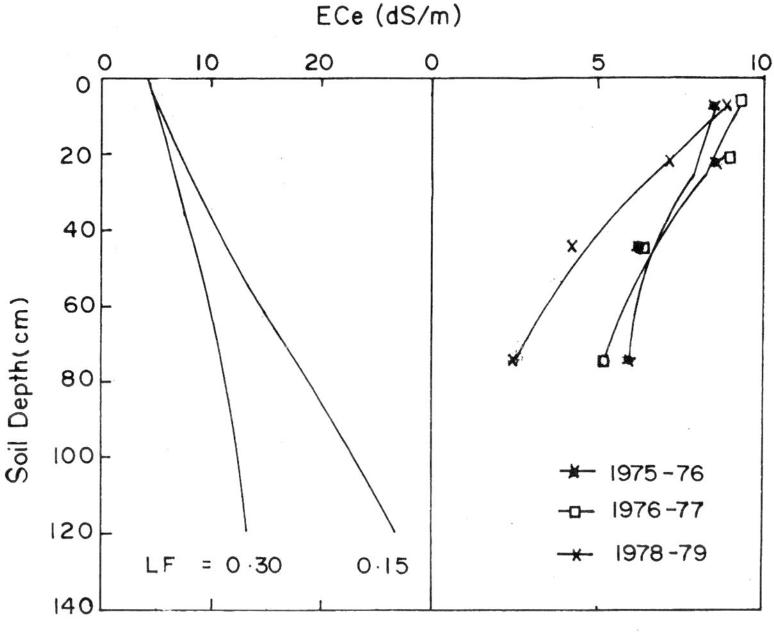

Fig. 5.2 Salinity profiles predicted for steady state conditions (ECiw 8 dS m^{-1}, LF=0.15 and 0.30) and those observed at the harvest of wheat.

zone, which is in conformity of the average pattern of water uptake observed for the saline water irrigated crops (Minhas et al., 1989b, 1990a and b).

Soil salinity at the harvest of winter crops depends on the weighted average salinity of the irrigation water and the rainfall during the growing season (Minhas and Gupta, 1990). Rains received during the monsoons decide the initial salinity of the soil before sowing of winter crops. Further, salt accumulation from the long-term use of water of comparable salinity in sandy loam soils was found to be faster, and its leaching with monsoon rains slower for the waters having higher SAR (Dhankar et al., 1986; Singh et al., 1990).

Use of a given salinity water (ECiw) builds up higher salinity (ECe) in fine-textured soils as inferred from the concentration factor, ECe/ECiw. On compilation of published results the average values of concentration factor are reported (Manchanda et al., 1989) to be 0.76, 1.12, and 1.80 for soils with clay contents of <10, 10-20, and > 20%, respectively. By multiplication of this factor with the ratio of saturation to field capacity moisture contents (θs/θfc) it was possible to transform the value of 'X' (calculated value of Ca^{2+} concentration in the soil solution in meq L^{-1}) into salinity of the soil water (Minhas and Gupta, 1990). A plot of soil water salinity versus clay content (Fig. 5.3) indicates the salinity of soil water occurring upon irrigation with water of a given salinity is independent of soil texture. This is contrary to existing knowledge

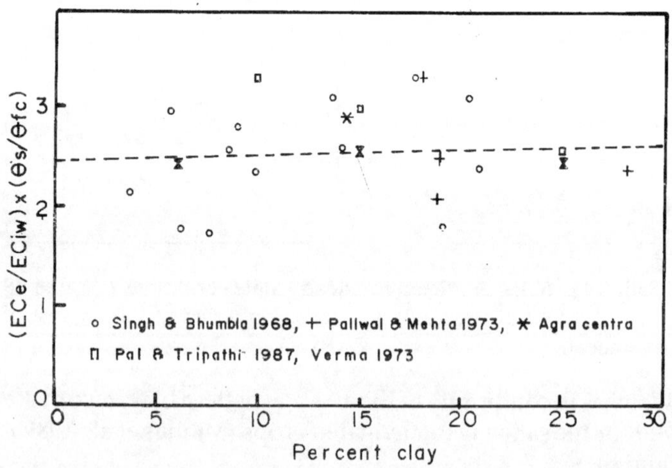

Fig. 5.3 Soil water salinity in texturally different soils irrigated with saline water.

about the acceptability of high salinity waters on coarse than on fine soils, as higher leaching fractions in the former result in lower salinity development during irrigation period, on the one hand, and higher leaching of accumulated salts with similar amount of monsoon rains, on the other. Compilation of rain-induced salt removal data when presented in the form of hyperbolic functions (Minhas and Gupta, 1990) further bears this fact out. They found the rainwater requirements per unit depth of fine, medium, and coarse-textured soils to remove 80% of the salts accumulated the preceding irrigation period were 18.5, 9.6, and 7.6 mm, respectively.

2. Alkalinity/Sodicity Hazards

Irrigation with waters containing high sodium (> 75% of total cations) and plenty of CO_3^{2-} and HCO_3^- ions relative to Ca^{2+} and Mg^{2+} leads to development of high exchangeable sodium (ESP) and high pH in soils. These together adversely affect the soil physical properties including water infiltration and soil aeration. The soil a few cm below the surface remains almost dry even though water may be stagnating at the surface. On drying it becomes very hard that affects the root development. The plants show the symptoms of scorching and leaf burning. The waters with low contents of Ca^{2+} (< 2 meq L^{-1}) and high amounts of CO_3^- and HCO_3^- ions injure the plants by burning their stems and leaves.

Changes in the exchangeable cation composition of the soil following irrigation with sodicity-inducing-waters are generally described on the basis of cation exchange equilibria. A useful index for evaluating the sodic hazard of waters is the sodium adsorption ratio (SAR), which is related to exchangeable sodium ratio by the Gapon equation:

$$ESR = \frac{\text{Clay–Na}}{\text{CEC–Clay Na}} = K_G \frac{Na}{\sqrt{[(Ca + Mg)/2]}} = K_G \,.\, SAR \quad [5.6]$$

where exchangeable cations are in meq 100^{-1} g and solution concentrations are in meq L^{-1}. K_G = Gapon's constant for Na-(Ca+Mg) exchange equilibria. Instead of ESR, the exchangeable sodium percentage describing the per cent of the cation exchange capacity (CEC) occupied by Na^+ ions (Clay-Na/CEC) has been widely used.

$$ESP = 100 \times ESR/(1 + ESR) \qquad [5.7]$$

In the presence of alkalinity causing anionic species (CO_3^{2-} and HCO_3^-) soil solution concentration in between the irrigation cycles leads to removal of a part of Ca^{2+} ions caused by precipitation of $CaCO_3$ and an increase in the relative proportion of sodium or SAR of soil solution. Eaton (1950) presumed complete precipitation of Ca^{2+} and Mg^{2+} ions when concentrations of HCO_3^- plus CO_3^{2-} exceeded the divalent cations. He proposed the concept of residual sodium carbonate (RSC) for evaluating the sodicity hazard of irrigation waters. In fact, RSC is an equivalent expression for the residual alkalinity (meq L^{-1}) in waters :

$$RSC = (CO_3^{2-} + HCO_3^-) - (Ca^{2+} + Mg^{2+}) \qquad [5.8]$$

It was later recognised that precipitation of Ca^{2+} though not be complete yet it increases the relative concentration of Na^+ relative to Ca^{2+} in soil solution. Therefore, the concepts of SAR and RSC have been refined from time to time by making adjustments for precipitation and dissolution reactions of $CaCO_3$ in soils. Bower (1961) suggested the use of a modified version of the Langlier's saturation index (SI) as

$$SI = 8.4 - pHc \qquad [5.9]$$

where pHc is the theoretically calculated pH of the irrigation water in contact with lime and in equilibrium with Pco_2 of the soil solution. The equilibrium pH (pHeq) is taken as 8.4. A positive value of saturation index indicates the tendency of $CaCO_3$ to precipitate from water whereas its dissolution will occur in waters having negative values of SI. The value of pHc is derived from the following equation.

$$pHc = (pK_2 - pK_s) + p(Ca^{2+}) + p(Alk) \qquad [5.10]$$

Here $p(Ca^{2+})$ and p(Alk) are negative logarithms of the molar concentrations of Ca^{2+} and of the equivalent concentrations of methyl orange alkalinity ($CO_3^{2-} + HCO_3^-$), respectively. The pK_2

and pK_c refer to the negative logarithms of the second dissociation constant of $H_2CO^-_3$ and the solubility constant of $CaCO_3$, respectively. The concept of pHc was basically developed to account for calcite precipitation / dissolution reactions. However, both Ca^{2+} and Mg^{2+} are invariably included for calculating the pCa term. There is no theoretical basis for the inclusion of Mg^{2+} because of preponderant solubility of $MgCO_3$. The presence of Mg^{2+} ion in quantities higher than Ca^{2+} in soil solution and underneath groundwater aquifers also indirectly suggests to the lower selectivity of Mg^{2+} and higher solubility of $MgCO_3$.

The SAR of irrigation water can be adjusted for the expected calcite precipitation and used to predict the sodic hazard of irrigation water (Bower and Maasland, 1963) as :

$$ESP_{surface} = adj.SARiw = C.SARiw\,[1 + (8.4 - pHc)] \qquad [5.11]$$

where C is the empirical coefficient. Later, Bower et al. (1968) observed that increasing the leaching fractions (LF) decreased the precipitation of $CaCO_3$. To account for the effects of leaching fraction on $CaCO_3$ dissolution, Bower proposed an empirical relation for predicting the ESP at the bottom of the root zone or adj.SAR of drainage waters as :

$$ESP_{bottom} = adj.SARdw = (1/LF)^{1/2}\,SARiw\,[1+(8.4-pHc)] \qquad [5.12]$$

Equation [5.12] predicted much higher sodication of soils than actually observed. To improve the predictions, Rhoades (1968) further modified this equation to account for the effects of mineral weathering in soils. With the Incorporation of weathering coefficient (Y), equation [5.12] becomes :

$$ESP_{bottom} = adj.SARdw = \frac{Y^{(1+2LF)}}{(LF)^{.5}}\,SARiw\,[1+(8.4-pHc)] \qquad [5.13]$$

where empirically determined mineral weathering coefficient, Y, usually equals 0.65. In a laboratory column study, if the values of weathering coefficient are 0.80 and 0.75 for waters having 7.5 < pHc < 7.5 the predicted SARdw is close to observed value (Sharma, 1978). The pHeq = 8.4 is not constant (Suarez, 1981) because of its dependence on the solution composition, leaching fraction, and PCO_2 in the root zone. It was thus shown that the observed and predicted sodicities (SARdw) can be better related through the use of Eq [5.14].

$$SARdw = (Na_{iw}/LF) / [(Mg_{iw}/LF) + X\,PCo_2)^{.33}]^{.5} \quad [5.14]$$

where X is the calculated value of Ca^{2+} concentration in the soil solution (meq L^{-1}) that depends upon the ratio of HCO^-_3/Ca^{2+} ions and the ionic strength. The value of X represents the modified CaX values, later used by Ayers and Westcot (1985) to adjust the SAR of waters and termed as adj.RNa :

$$adj.RNa = \frac{Na}{\sqrt{(CaX + Mg)/2}} \quad [5.15]$$

The SAR.adj concept proposed by Suarez (1981) takes into account the HCO^-_3 ions only whereas many alkali waters do contain some amounts of CO^{2-}_3 ions. Alkali waters if have CO^{2-}_3 ions are more hazardous to soils than the HCO^-_3 type waters (Kanwar and Kanwar, 1971; Gupta, 1983; Siyag et al., 1988).

In combination with appropriate coefficients, equations [5.11] and [5.15] can be used to predict ESP build up of soils irrigated with alkali waters. The coefficient, 2, in the equation [5.11] was found to fit to the data on the Punjab waters (Bower and Maasland, 1963). A scrutiny of published results (Sharma and Mondal, 1981; Bajwa et al, 1983, 1986; Bajwa and Josan, 1989 a,b,c), has indicated a range of these coefficients for rice- wheat and pearl millet/maize-wheat rotations as summarised in Table 5.1.

Table 5.1. Values of empirical constants for predicting ESP (0–30 cm) from adj.SAR of irrigation waters.

Crop rotation	Value of empirical constant			
	Adj. SAR			Adj. RNa
	(1)	(2)	(3)	
Rice-wheat	1.5	1.8	2.6	2.6
Millet/maize-wheat	0.7	0.8	1.3	1.0

(1), (2) and (3) represent adj.SAR calculated using pCa+Mg, pCa, and the latter corrected for rainfall, respectively.

In early stages of irrigation with alkali waters, large amounts of divalent ions are released into the soil solution from exchange sites and other reserves (easily weathered minerals) besides their content in irrigation water. When divalent ions released from soil sources become zero or negligible (constant), the steady state is considered to be achieved. All above equations for predicting adj.SARdw or ESP are based on the assumption of attainment of steady state conditions. For the transient field conditions, prediction of sodication/alkalization rates from the SARdw below root zone is difficult because of deterioration of upper soil layers does not permit deeper penetration of water. Also alternating cycles of irrigation during winter and rainwater during monsoon season induce cycles of precipitation and dissolution of salts to hinder the attainment of steady state conditions. In a series of field microplot experiments (Bajwa et al. 1983, 1986; Bajwa and Josan, 1989a), ESP profiles as predicted for steady state conditions could not be ob-

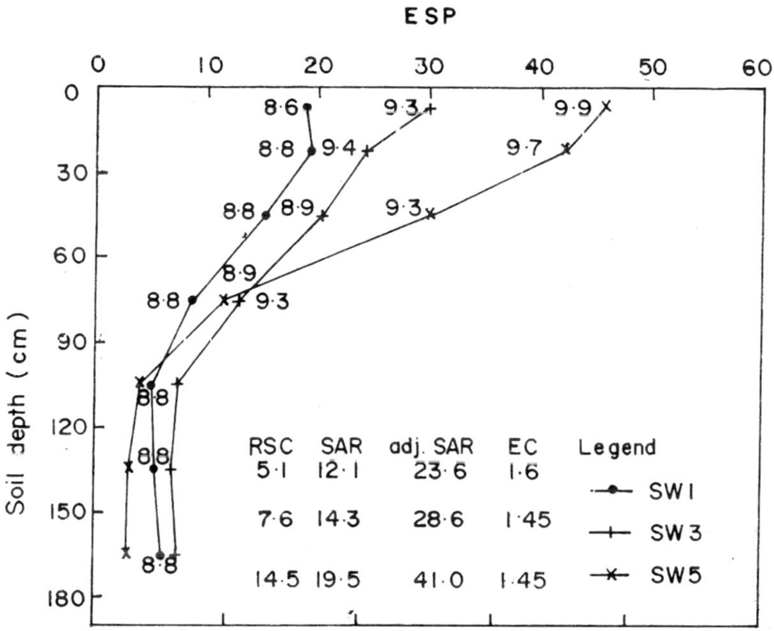

Fig. 5.4. Build up of sodicity (ESP) and alkalinity (pH) in sandy loam soil after 9 years of alkali water irrigation in millet- wheat rotation.

tained even after the use of alkali waters for nine years (Fig. 5.4). The sodicity and pH of soil especially of upper layers, continued to increase over the years whereas lower layers were less affected. Because of reduced infiltration (RIR = 0.3 at an ESP of ≈ 20), the opportunity for alkali waters to penetrate deeper reduces. Sodication process mainly of surface layers continues further preventing attainment of steady state conditions. Results presented (Table 5.1) suggest that adj.SAR concept can serve as useful index to predict the sodication hazard of alkali waters for the most important surface layer (0- 30 cm).

The concept of SAR/adj.SAR is widely used for estimating/predicting the sodic hazard of irrigation waters. However, SAR of alkali waters alone provides no information on the potentiality of an irrigation water to result in high soil pH. Use of alkali waters gives rise to sodic soils having high pH but the sodic soils by definition do not necessarily have a high pH, but may even show an acidic reaction (Beek and Breeman, 1973). Alkalinity being held constant, residual alkalinity may be modified by changing the ratio of mono/divalent cations. Eaton (1950) associated the residual alkalinity concept with a possible maximum sodicity while it is fundamentally related to pH. The presence of residual alkalinity implies that SAR will increase upon evaporation and subsequent precipitation of divalent cations.

Analogous to salinity hazards, the adverse effects of sodicity build up by alkali water use are higher in the fine than coarse-textured soils. However, the sodication rates of soils having low cation exchange capacity will be faster (Aggarwal and Ramamoorthy, 1974; Bajwa et al., 1986). Most soils have the higher selectivity for Ca^{2+} than for Mg^{2+} implying that at comparable SAR values, higher ESP would be observed for Mg-Na than Ca-Na system (Pal and Poonia, 1979). Calcareousness and the soil texture also influence the ionic preferences of the two divalent cations (Paliwal and Gandhi, 1976; Yadav and Girdhar, 1981). The magnitude of the association constant of the anions with Ca^{2+} also affect the sodicity build up in soils (Siyag et al, 1988). ESP build up in soil columns leached with comparable volumes of different salt solutions of sodium was observed to follow the order : CO^{2-}_3 > SO^{2-}_4 = HCO^-_3 > Cl^-.

3. Infiltration Problems

Long-term use of high SAR and/or high alkalinity waters has been known to induce permeability problems in soils, which in turn reduces replenishment of the soil water storage thus diminishing water supply to the plant roots besides leading to poor soil aeration and gas exchanges.

Most of the work for determining the effects of water quality on water transmission characteristics has been limited to laboratory conditions (Shainberg and Letey, 1983) and a very few attempts have been made to relate the infilteration rates to ion chemistry of waters under field conditions, although infiltration rates have been shown to be more sensitive than hydraulic conductivity (Oster and Schroer, 1979; Aggassi et al., 1981). Under field conditions the reduction in infiltration is caused by greater opportunities for the clay movement as well as rearrangements of the resultant subaggregates during separated successive wetting and drying cycles. Oster and Schroer (1979) observed a multiple linear regression of infiltration rate (IR) on SAR (in the range of 5-30) and concentration of electrolyte (C, 2-30 meq L^{-1}) for various waters as the best fit with an R^2 of -0.86 and regression of IR = 6.8-1.1 SAR + 0.79 C. Sharma and Minhas (1986) in their study on the influence of the two types of water on the relative infilteration rate (RIR) of the illitic calcareous sandy loam soil found the following relationships of RIR to EC and to SAR/adj.SAR.

(1) $Cl^- - SO_4^{2-}$ waters with no residual alkalinity

RIR = 0.887 + 0.037 EC - 0.015 SAR (R^2 = 0.74**)

(2) HCO_3^- waters with residual alkalinity

RIR = 0.690 + 0.077 EC - 0.015 adj.SAR (R^2 = 0.80**)

For the RIR, the IR (3 mm h^{-1}) determined with a water of ECiw = 0.3 dS m^{-1}, SAR < 1.0 for the normal soil was taken as reference. The SAR and adj.SAR were negatively related to RIR and or IR, similar in magnitude for water with and without residual alkalinity. Inverse relationship of SAR/adj.SAR to IR/RIR has also been reported by Manchanda et al. (1985) and Bajwa et al. (1986;

1989a), with extent of decrease in IR dependent on soil type, length of use of alkali water, cropping system, etc.

They found that long-term use of an alkali water (ECiw 4.0 dS m^{-1}, SAR > 30, and RSC 10 meq L^{-1}) on a sandy loam soil reduced the IR to 5 mm d^{-1}. Alkali water irrigation to maize-wheat rotation on a sandy loam (11% clay) and a loam (24% clay) soil for a 5- year period decreased IR progressively with an increase in adj.SAR of alkali waters. Reductions in RIR were found to be more in the case of loam soil as well as when ECiw increased at the same level of RSC. Irrigation with alkali water (ECiw 3.2 dS m^{-1}, RSC 4 meq L^{-1}, and adj.SAR 30.5) to rice-wheat and maize/millet-wheat rotations practised on a loamy sand soil (5 to 8.8% clay) reduced the IR to 46 and 29% (referenced to canal water) after 5 years of irrigation under the two cropping systems, respectively. Corresponding reductions after 10 years of alkali water irrigation under the same cropping systems came down to 25 and 18%. The use of various EC (6 and 12 dS m^{-1}) and SAR (> 20 and RSC 0) waters for 8 consecutive years adversely affected the infiltration of water into a sandy loam soil (Figure 5.5).

Regression coefficient for the EC to RIR relationship was positive, showing that the presence of salts in water mitigated the adverse effect of any high SAR on RIR thereby. The value of coefficient 108% higher with than without residual alkalinity suggested that for similar ECiw and SAR, waters having high alkalinity prove more harmful in reducing the IR of the soils. IRs of soil were lower when salinity of the irrigation water was increased (Singh et al., 1990).

4. Toxicity Hazards

Groundwaters having high contents of toxic ions such as B, F, NO_3, Se, etc. also become problematic for irrigating crops. High accumulation of B in plants causes necrosis and sometimes may reduce leaf size to the extent of significantly reducing photosynthesis. Even the small amounts of B in irrigation waters tend to cumulate in soils (Singh and Kanwar, 1963) as its displacement in soils is much more difficult. B may require 3 times more water than required for removal of similar amounts of soluble salts (Hoffman, 1980). Excess of fluorides, usually present in HCO_3 type waters,

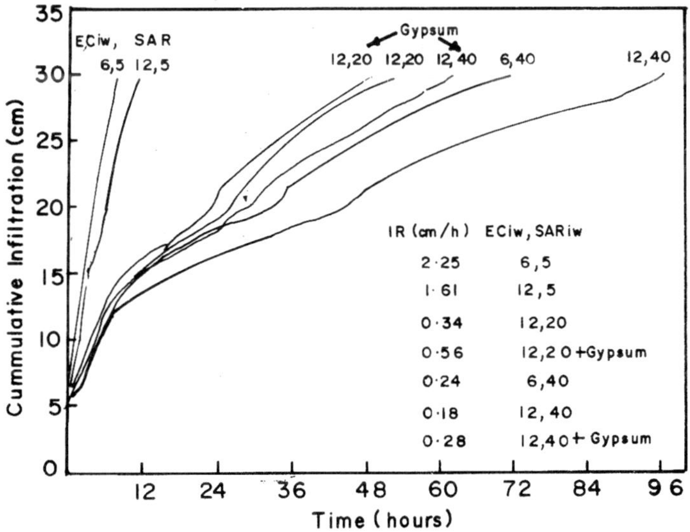

Fig. 5.5. Infiltration as affected by irrigation with various EC and SAR waters.

causes serious health problems such as fluoresces, decay of teeth, and weakening of bones. When irrigation waters contain excess of soluble NO_3, vigorous plant growth, delayed maturity, increased succulence resulting in lodging, poor grain filling, and shriveled grain do occur. Accumulation of organic acids such as malate and oxalate in high amounts in forages and leafy vegetables may prove toxic to animals and human belngs.

III. MANAGEMENT OF POOR QUALITY WATERS

Concerns about the proper use of poor quality waters for crop production in areas where their use is inevitable are not new. Researchers have made concerted efforts for integrating the effects of agroecological factors and water quality characteristics on soils and crops for evolving appropriate crop / agronomic and irrigation management practices believed to modify the water quality hazards for the sustained production of crops.

A. Crop Management

1. Selection of crops : Crops differ considerably in their tolerance to salinity/sodicity. The intergenic differences can be exploited for selecting the crop capable of producing acceptable yield under a given root zone salinity. The crop tolerances to irrigation water salinity can be obtained from tables compiled by Maas (1986) and Minhas and Gupta (1990). In general, the oilseed crops that require less water have higher salinity tolerance than many cereals and pulses. Further, the crops grown during high evaporative conditions rapidly deplete soil water and this results in higher salinity around the roots than in the bulk soil (Sinha and Singh, 1976). As a result their tolerance to irrigation water salinity is reduced.

Similarly, the relative sodicity tolerance of some agricultural crops has been compiled by Gupta and Abrol (1990). The sodicity tolerance limits of crops have been observed to differ for the two conditions of soils undergoing sodication process caused by irrigation with alkali waters and being desodicated with amendments like gypsum. Relative yields of wheat as a function of ESP (0-15 cm soil depth) for an alkali soil undergoing reclamation (Abrol and Bhumbla, 1979) and sodication (Bajwa et al., 1983; Bajwa and Josan 1989a and b) have been compared in Fig. 5.6. The lower tolerance of wheat crop on soils under uphill sodication process is evident.

2. Selection of varieties : Like the intergenic differences in crop tolerance to salinity/sodicity, wide variations do exist in the inherent salt tolerance of the crop cultivars, which must be considered for their selection. It has generally has observed that cultivars having high yield potential, as for example, rice variety 'Jaya' and wheat 'HD-2304', show considerable yield reductions caused by salinity, yet they produce higher yields than local varieties such as 'Damodar' in rice and 'Kharchia' in wheat having higher salt tolerance. This explains the reason why the high yielding varieties are the preferred choice of growers. Results of recent breeding trials suggest that it is possible to breed cultivars that have both high relative salt tolerance and high yield stability in saline environment. Of these two traits of a cultivar, latter has been preferred for selection of cultivars for commercial cultivation (Gupta et al., 1989). For example, when wheat cultivars 'RAJ 2560'

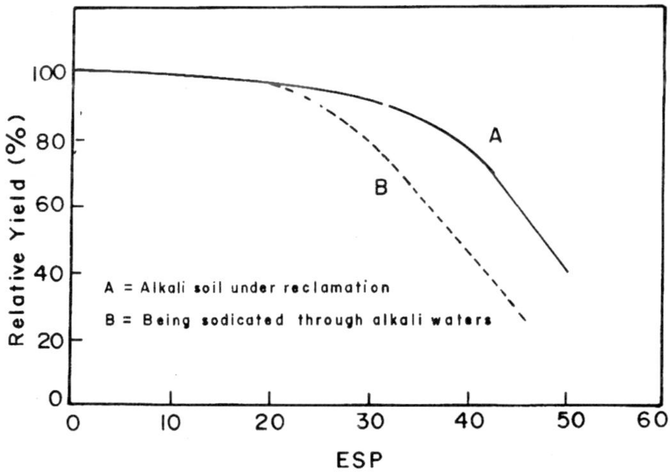

Fig. 5.6. Tolerance of wheat to soil sodicity (ESP, 0-15 cm)

and 'WL 2382' were tested over four seasons (1985-1989), the former was found to be more stable for salinity-environment interactions as is obvious from the lower scattering of data points (Fig. 5.7)

Table 5.2 : Critical growth stages of crops. From Gupta et al. (1989).

Crops	Salinity sensitive growth stages
Cotton [*Gossypium hirsutum* L.]	Seedling establishment
Wheat [*Triticum aesivum* (L.) emend Thell].	Seedling and flowering
Barley [*Hordeum vulgare* L.]	Seedling establishment
Maize [*Zea mays* L.]	Seedling and silking
Safflower [*Carthamus tinctorius* L.]	Seedling, rosetting / flowering
Indian mustard [*Brassica juncea* (L) Czern. & Coss.]	Seedling and flowering
Mung bean [*Vigna radiata* (L.) Wilczek]	Seedling and pod development
Black gram [*Vigna mungo* L. Heppes]	Seedling and pod development

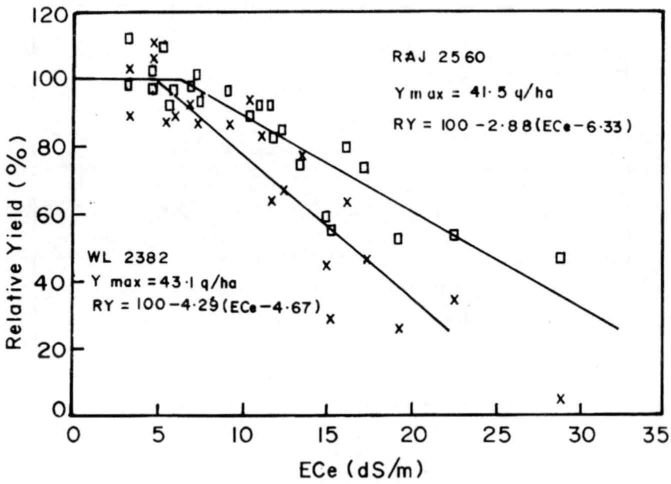

Fig. 5.7. Salinity response of two wheat varieties.

3. Identification of sensitive growth stages : All crops do not tolerate salinity equally well at different growth stages. Important growth stages of the crops recognised to be sensitive to salinity are given in Table 5.2. In most of the crops, the germination and early seedling establishment stages are most sensitive to salts. Therefore to enhance the plant stands, strategies as discussed later should prove beneficial. The other critical period for crops is during phasic changes from vegetative to reproductive stage, i.e., heading and flowering to fruit setting. In general, tolerance of the crops to saline water usage increases with the crop ageing.

B. Irrigation Management

1. Irrigation frequency and leaching requirements : Irrigations with saline waters should meet the water requirement of crops and the leaching requirement to maintain a favourable salt balance in the root zone. In the intervening period between irrigations, evapotranspiration by crops reduces the soil water content that in turn decreases the matric (due to attraction of soil matrix for water and water molecules for each other) as well as solute (due to increase of soil solution concentration) potentials. Consequently the water uptake by the crops and final yields are adversely affected. The irrigation with saline water should, therefore, be frequent as to reduce the cumulative water deficits (both matric and osmotic) between irrigations. However, frequent irrigations tend to increase the salt load, limit the uptake of water from upper soil layers where a larger fraction of the root system occurs, and enhance the evaporative losses from the soil surface. With high frequency irrigations more salts accumulate in the upper portion of the root zone as compared to low frequency heavy irrigations.

Results of several experiments compiled by Minhas and Gupta (1990) have shown that an increase in depth of irrigation though slightly enhances the wheat yields, there has been no interactive effect with salinity of irrigation water. It appears that higher quantities of saline irrigation suffers from the disadvantage of displacing the nonsaline soil water, carried over from the monsoon rains, beyond the reach of plant roots. Similarly, several other reports (Shalhevet et al., 1983, 1986; Hoffman et al., 1983) do not support the proposition that irrigation interval should be shortened while irrigating with saline waters. However, positive responses to light and frequent irrigations with saline waters (ECiw = 11 to 12 dS m^{-1}) of wheat grown on coarse textured soils had been observed where soils were initially saline (Sharma et al., 1979 ; Minhas et al., 1989a).

Leaching requirement (LR) concept requires the application of additional water (over that required to replenish the evapotranspirational requirements) to maintain the salt balance. LR for the crops irrigated with given salinity water (ECiw) is usually calculated from the following equation (Rhoades ad Merrill, 1976) ·

$$LR = ECiw / 5(ECe - ECiw)$$

where ECe is the salinity of the saturation extract of the soil at 10% yield reduction. This formulation, however, remains controversial since there is ambiguity in defining the appropriate values of ECe. whereas Bernstein and Francois (1973) suggested the use of ECe values at 50% yield reductions, ECe at zero yield has also been recommended by van Schilfgaarde et al. (1974). Experiments conducted in the All India Co-ordinated Research Project on Saline Waters have shown no additional advantage of applying extra water to wheat, even at higher levels of irrigation water salinity (Minhas and Gupta 1990). Similarly, salinity control could not be achieved by applying 50% extra water (ECiw = 3.2 dS m^{-1}, SAR = 21 (m mole $L^{-1})^{.5}$ RSC = 4 meq L^{-1}) to meet LR in rice - wheat and maize / pearl millet-wheat rotations (Bajwa et al., 1986). Rather, salinity build up was 30- 50% higher that decreased the yield of crops.

Thus, an alternative strategy may be to irrigate normally up to the time when salts become excessive. This may be followed by application of water for leaching if required to lessen the stress at later sensitive stage of crops, e.g., at the time of third irrigation coinciding with flowering in wheat (Minhas et al., 1989a). In areas characterised by irrigation with highly saline waters and monocropped in winter only, the monsoon rains, if occur can be utilised to maintain salt balance. A leaching strategy that seems to work well is to apply saline water for boosting the antecedent soil moisture content and reduce salinity before the onset of monsoon. The refill of the top soil prior to the onset of monsoon will enhance salt leaching by ensuing rains, and therefore, help in cropping in the rainy season also, as the front may move below the potential root zone.

2. Method of irrigation : The distribution of water and salts added to soil with each irrigation varies with the method of irrigation. Any of the methods followed should create and maintain favourable salt and water regimes in the root zone such that water is made readily available to plants for growth without much sagging effect on yield.

The low-cost surface irrigation methods including the border strips, check basins, and furrows are most commonly practiced in irrigated agriculture. These irrigation methods even after following the best design criteria generally result in excessive irrigation and non-uniformity in water application. Consequently the on-farm irrigation efficiency is low (50-70 %). However, properly designed and operated surface irrigation methods can maintain the salt balance and minimize salinity hazards of saline waters. These twin objectives can be met through appropriate choice of specifications designed for surface methods (Rajput and Aggarwal, 1983).

Modern irrigation methods such as sprinkler and drip are more efficient as the quantity of water to be applied can be adequately controlled but the initial investment and maintenance costs of such systems are high. Sprinklers can ensure uniform distribution of water even on undulating and sandy terrains and help in better leaching of the salts. Saline water use through sprinklers, however, may cause leaf burning and toxicity when used in some sensitive crops. The general guideline of relative susceptibility of crops to foliar injury from saline / sodic sprinkling waters based on the direct accumulation of salts through the leaves have been reported by Maas (1986).

The use of irrigation waters through drip system has revolutionised the production of some high value crops in arid regions. Drip system of irrigating crops has been observed to enhance the threshold limits of their salt tolerance by modifying the patterns of salt distribution and maintenance of constantly higher matric potentials (Shalhevet, 1984). Due to enhanced leaching and accumulation of the salts at the wetting front and the soil between the drip laterals, the salt accumulation below the drippers remain very low whereas the water contents are maintained at higher levels under the drippers. As the crop roots are known to follow the path of least resistance, most roots are found below the surface drippers. Hence the drip system seems to be the best system of saline water application because it avoids leaf injury to plants as is with splinkers and maintains optimum condition for the uptake of water by the plant roots. Some of the alternatives to drips like the earthen pitchers (Mondal et al., 1987) and the specially designed earthen cups (Yadav, 1986) have been tried for irrigation of vegetable crops but their long-term usage for field scale still awaits detailed investigations.

3. Conjunctive use of multiquality waters : In many situations, the canal water supplies are either unassured or in short supply such that farmers are often forced to pump groundwaters of varying quality for crop production. This calls for using the limited quantities of nonsaline (canal) waters most judiciously in combination with poor quality waters. For the combined use of saline and nonsaline waters, two options are available to the farmers :

(a) Mixing of water supplies : Mixing of poor quality water with the canal water can dilute such water to acceptable quality and can broaden the choice of crops. Mixing improves the stream size that would enhance the uniformity in irrigation by the surface methods if practiced on sandy soils. Though simple, it is considered as a forcible solution to the water quality problems if facilities for mixing are available and the mixing ratio is known. To achieve this, a prior information on the threshold salinity values of the crops to be grown in sequence and salt build up in the soil with use of a given quality water during the cropping seasons is essential. Mixing does not reduce the total salt load but may allow for more area to be planted. However, Rhoades (1984) has disfavoured mixing because in this process usable and the unusable waters are mixed into one blend that has to be again separated by plants for their use. The plants are forced to expand bioenergy (that would have otherwise been used in their biomass production) to extract (distill out) the usable fraction from the saline soil water solution until the fraction of the mix made up of excessively saline portion is left. This fraction is just as unusable at this point as it was before mixing. Thus on the whole mixing does not stretch the overall water supply for crops as far as saline waters are concerned

Though blending of saline and canal waters may not always be beneficial to crop production, the process, however, will lead to improvement in the quality of alkali waters. Both canal and groundwaters are usually in equilibrium with calcite but at different PCO_2. The equilibrium PCO_2 in groundwaters is much higher. Interestingly the relationship between concentration of Ca^{2+} in solution phase and PCo_2 is nonlinear and is governed by the relation :

$$m^3Ca = \frac{Kh.PCO_2\, K_1\, K_{Cal}}{4\, K_2\, \Gamma Ca^{2+}\, \Gamma^2 HCO_3} \qquad \{5.17)$$

In equation (5.17), Kh, K_1 and K_2 denote Henry Law's constants for CO_2, the first and second dissociation constant of H_2CO_3 acid, respectively. K_{Cal} refers to calcite solubility and Γ is activity coefficient of respective ions. It seems that mixing of surface water with groundwaters of high residual alkalinity and low calcium concentration would result in undersaturation with respect to calcite (Wingley and Plummer, 1976). Consequently the blended water on irrigation will have greater tendency to pick up calcium through dissolution of native calcite from soils. There is, however, no direct evidence available at present to support the above proposition.

(b) *Cyclic use of multiquality waters :* This strategy involves the substitution of nonsaline (canal) water for saline water at the most sensitive growth stages/crops grown in sequence and saline water is used at other stages such that the effects of the resultant soil salinity build up can be minimized. In most of the crops, the germination and seedling establishment has been identified as the most sensitive stage to salinity. A failure at this stage will lead to poor crop stands and considerable reductions in yields. Minhas et al. (1989b, 1990 a & b) have observed that mung bean, sorghum, and Indian-mustard tolerate the use of higher salinity waters, once the nonsaline water is substituted for presowing irrigation to leach out salts from the seeding zone. This substitution enhances the germination, crop growth, and better the use of soil water (Table 5.3). Thus, it is better to use nonsaline water, if available, in early growth stages and switch over to poor quality waters later in the season when the crops can tolerate higher salinity.

Table 5.3. Yield and water extraction by crops following the use of different salinity waters. From Minhas et al. (1989a, 1990 a&b).

EC iw (dS m^{-1})	Seed yield kg ha^{-1}	Water extracted cm	Dry forage yield kg ha^{-1}	Water extracted cm
Mung bean				
0.3 (throughout)	2520	44.8	9700	31.0**
4.7 (throughout)	270	22.6	6500	24.7
4.7 (Plnsw)	1560	38.1	8500	31.7
Mustard				
0.3 (throughout)	2320	36.9		
12.3 (throughout)	1050	18.1		
12.3 (Plnsw)	1800	27.9		

Plnsw = Presowing irrigation with nonsaline water, ** up to the last irrigation only.

If facilities for blending exist and different qualities of water are simultaneously available on demand, then the question arises as to which option should be preferred. Minhas and Gupta (1990) analyzed the data on crop yields obtained from alternate use modes of saline and canal water and compared these with predicted yields under mixing for the same level of ECiw (weighted average salinity). As presented in Figure 5.8, the results provide a useful evidence in favour of cyclic use modes. Rhoades et.al. (1988) have also advocated a similar strategy for the combined use of river and drainage waters. Mieri et. al. (1986) compared the performance of potatoes (*Solanum tuberosum* L.) and peanuts (*Arachis hypogaea* L.) with network blending and soil dilution, under drip system of irrigation. It was found that the crop responded to weighted mean water salinity regardless of the blending methods. In a recent farm survey in Hissar district of the Indian state of Haryana, Bouwmans et al. (1988) observed that alternate use of canal and saline waters for irrigation by farmers resulted in higher production of rainy season crop (cotton (*Gossypium hirsutum* L.) and millets) than that from mixing of saline and canal waters. On the contrary, mixing proved quite beneficial in winter crops (wheat and mustard [*Bratsica campestris* (L.) var. *dichotoma* Watt.])

In other experiments on effect of cyclic use modes for the supplementation of alkali with canal waters on soil properties and yield of rice (*Oryza sativa* L.) and wheat [*Triticum aestivum* (L.) emend Thell.] grown on a sandy loam soil (Bajwa and Josan, 1989c) irrigation with an alkali water increased the pH and ESP in surface soil and reduced the rate of infiltration to 14% (Table 5.4). The yields of rice and wheat tended to decrease with time and averaged 62% and 57% of the yield obtained from the use of canal water, respectively. However the use of alkali and canal water reduced the sodication of soils and equalised the yields of both crops to that from canal water. After accounting for rainfall and canal water in estimating the adj.SAR,ESP of surface soil was about 1.4 times the adj.SAR for the cyclic use modes. A factor of 1.8 was earlier (Table 5.1) observed for alkali waters. Less reduction in yield and reduced sodication, therefore, reflect to the advantage of using alkali and canal waters cyclically.

C. Seeding Practices

Failure to achieve satisfactory germination and required plant population is the major factor limiting crop production in saline water irrigated areas. Crops like wheat, barley (*Hordeum vulgare* L.), and safflower (*Carthamus tinctorius* L.) are sensitive at the germination and seedling stages. Once the plants are established they can withstand fairly high levels of salinity. To achieve this, use of high seed rates at times is recommended. Since it does not work out as desired, the practices that would ensure suitable environment in the seeding zone during germination and seedling emergence periods merit consideration.

When saline water is applied for presowing irrigation, seeds are exposed to high soil water salinity (ECsw > ECiw) resulting from movement of salts to the soil surface in evaporative drying during the period seeding to seedling emergence. This makes the seed germination and emergence even more critical, especially in summer crops when evaporation from the soil surface is quite high. Under such conditions a safe device for better establishment may be to go in for a post-sowing irrigation that pushes the salts deeper and maintains better moisture regimes around the germinating seeds (Minhas et al., 1988), with care taken to avoid the formation of crust. The seed yield of Indian mustard in postsowing irrigation with saline water following dry seeding was found to sustain up to ECiw of 11 dS m^{-1} (Table 5.5). Until recently this has been a standard practice amongst growers of Kharchia and other local varieties of wheat in scattered pockets of sandy arid plains of the Rajasthan desert containing various degrees of groundwater salinity.

Modifications in the shape of the seedbeds can also reduce the salt accumulation near the seed. Bernstein and Francois (1957) proposed the double row flat topped bed configuration for saline irrigation wherein most of the salts move to the centre of the beds and the shoulders are left relatively salt free to enhance the seedling establishment. Seeding on sloping beds also seems better (Bernstein and Ayers, 1955). Here in this case the seeds are placed on the sloping side and seed row is placed just above the water line and irrigation is continued until the wetting front has moved well

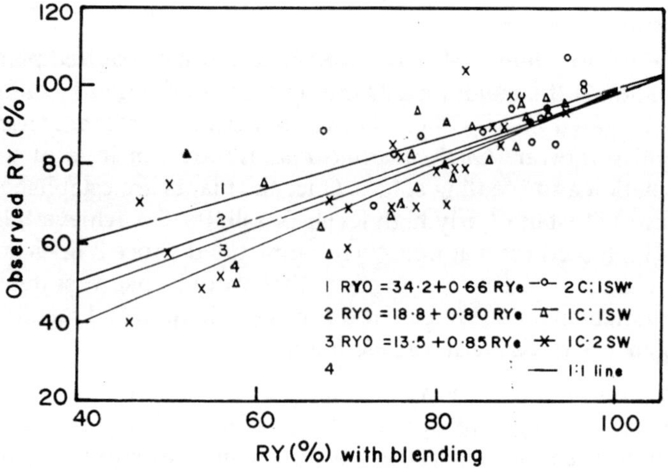

Fig. 5.8. Relative yield with blending and in alternate use modes at same average salinity of irrigation water.

past the seed row. The other alterations are to apply irrigation in alternate row (Bernstein and Fireman, 1957) and to seed on the north-east side of the ridges (Bains and Singh, 1966), because the salts move along with water to the drying surfaces and accumulate over there. For the larger seeded crops the seeds can be planted in the furrows itself (Rai and Mehrotra, 1971). One advantage in such a practice is that the seed placement takes place in a wet and a less saline zone. During ridge making more saline surface soil goes to the ridges and a postsowing irrigation leaches the salts of furrow soil more efficiently than that of the ridge soil (Minhas et al., 1988). Efforts have also been made to induce salt tolerance in plants by treating the seeds with salt solutions and growth harmones (Puntamaker et al., 1972; Darra et al., 1973 ; Chippa and Lal, 1976 and 1986). However, the success of such treatments is often doubtful (Manchanda and Bhandari, 1976).

D. Fertility Management

Most soils are deficient in N that must be supplemented through inorganic or organic fertilizer sources. Urea is by far the most widely used N source for crops. Following application, it is

Table 5.4. Effect of cyclic use of alkali and canal waters on soil properties and yield of rice and wheat. From (Bajwa and Josan (1989b)

Quality/mode	adj. SAR*	pH2	ESP	RIR %	Average Rice	Yield Wheat
						--- mg ha^{-1} ---
Canal water (CW)	0.3	8.2	4	100	6.78	5.43
Alkali water (AW)	22.0	9.7	46	14	4.17	3.08
2 CW – 1 AW	8.9	8.8	13	72	6.67	5.22
1 CW – 1 AW	12.8	9.2	18	59	6.30	5.72
1 CW – 2 Aw	18.5	9.3	22	34	5.72	4.85

* After accounting for 828 and 434 cm of irrigation and rainwater, respectively.

	ECw (dS m^{-1})	Ca	Ca+Mg	RSC (meq L^{-1})	SAR	adj. SAR
CW	0.25	1.6	2.1	0	0.3	0.4
AW	1.35	0.4	0.9	10.1	13.5	26.7

Table 5.5. Effect of seeding methods for use of saline waters on crop stand and yield of Indian mustard. From Minhas and Gupta (1990)

Seeding methods	ECw (dSm^{-1}) of irrigation water							
	3	7	11	16	3	7	11	16
	Plant stand, No. m^{-1}				Seed yield, kg ha^{-1}			
T1	10.4	9.8	2.4	1.4	1670	1740	590	190
T2	10.6	10.7	9.3	6.3	1560	1620	1540	1290
T3	10.0	9.7	7.4	2.5	1510	1570	1380	750

T1 = Seeding after conventional presowing irrigation

T2 = Dry seeding followed by post-sowing irrigation

T3 = 1/2 pre–and 1/2 posst-sowing irrigation

first hydrolyzed to ammonia and carbon dioxide by the enzyme urease and the process has the most commonly expressed disadvantage of loss of N via NH_3 volatilization. Increased soil pH resulting from the use of alkali waters has been shown to positively influence NH_3 volatilization from urea and other ammonia sources (Singh and Bajwa, 1985). Similarly, Sen and Bandyopadhyay (1987) and McClung and Frankenberger (1987) have demonstrated that nitrification is more sensitive to salinity than the ammonification process. Following the application of fertilizer N, a sudden rise in microbial activity takes place. A large pool of NH_4^+ is formed and if this is not nitrified or bound with organic forms; volatilization of ammonia takes place, more extensively in salt affected soils or upon the use of poor quality waters resulting in low N-use efficiency by crops. Split application of fertilizer N, deep placement, use of slow release N-fertilizers, application of urease inhibitors, and use of organic-N sources are reported to reduce the N-losses and increase N-use efficiency.

Contrary to benefits illustrated above, the role of fertilizer N in reducing/ overcoming the adverse effect of salts have not been found to be encouraging (Dhir et al., 1977). However on the basis of multilocational trials conducted in India, better strategy for improving N-use efficiency seems to be to substitute a part of inorganic fertilizer requirements through organic materials. In doing so, organic materials temporarily immobilize the ammonical N. Subsequently the organically bound N becomes available to crops gradually. Increased crop response to combined use of N-fertilizers and organic materials as a result of reduction in the volatilization losses of ammonia and resulting enhancement in N-use efficiency under saline environment bears this fact out.

Higher responses to P application under saline water irrigation indicate that unlike N, application of phosphorus mitigates the adverse effects of salinity. However the degree of response to P is tied-up with the type of salinity. For example, in irrigation of wheat and barley crops with chloride dominated waters the yield response to P application was higher as compared to sulfate dominated saline water (Manchanda et al., 1982, Chauhan et al., 1990).

E. Special Treatment

(1) Use of amendments : Continuous use of water containing high RSC increases the soil pH and ESP that in turn decrease the soil permeability to water.In the absence of the adequate supplies of calcium, high sodium water can bring nutritional imbalance within the plants. Therefore, to lessen the adverse effects of the long-term use of alkali/sodic water on physical and chemical properties of soils, the use from time to time of calcium containing amendment such as gypsum is recommended. Another way is to apply acids or the acid forming substance such as sulfuric acid or pyrites, which on reaction with soil $CaCo_3$ release Ca^+ in solution. Still another alternative is to reduce the RSC of the water itself by passing it through the gypsum beds.

Application of gypsum is preferably recommended when RSC of irrigation water exceeds 2.5 meq L^{-1} (Bhumbla and Abrol,1972). However the amount required precisely will depend upon such factors as the level of the existing deterioration of the soil, cropping intensity, and the water requirements of the crops chosen for cultivation. It may be pointed out that 86 kg of pure gypsum is required to neutralize each meq L^{-1} of RSC in 1000 m^3 or 10 cm of irrigation water and the agricultural grade gypsum is usually 70-80 % pure.

From the foregoing discussion it may not be taken for granted that response to gypsum is obtained in all soils, cropping systems or rainfall situations. Field trials have shown that gypsum helps maintain the yield of the crops receiving alkali waters (RSC > 5 meq L^{-1} especially when paddy is grown in rotation and rainfall of the area being double cropped is < 50 cm. An improvement in soil properties and significant increase in yields of rice and wheat crops irrigated with water of RSC 6.8 meq L^{-1} ECiw 0.85 dSm^{-1} (Bajwa and Josan, 1989c). Conversely, no response to gypsum in wheat-fallow rotation has been reported on well-drained light textured (sandy loam) soils irrigated with waters having RSC up to 10 meq L^{-1} (Gupta, 1980; Manchanda et al., 1985).

With regard to mode and time of application, Bajwa et al. (1983) observed that the application of gypsum at each irrigation was more effective for increasing the maize (*Zea mays* L.) yields in maize-wheat sequence irrigated with RSC water of 8 meq L^{-1} as compared to application of single dose annually. Recently, Bajwa and Josan (1989c) reported that in rice and wheat crops irrigated with RSC water of 6.8 meq L^{-1} ECiw = 0.85 dSm^{-1} the response to gypsum, either applied annually as one does or at each irrigation, was similar. With higher RSC (10.3 meq L^{-1}) water, although the improvement in wheat yields was similar for the two modes of gypsum application rice responded to gypsum applied at each irrigation. Pyrite has also been used for amending the deleterious effects of high RSC waters. Its application once before the sowing of wheat has proved better than split application at each irrigation or mixing it with irrigation water (Chauhan et al., 1986).

In some crops, for example maize and rice, since the use of amendment at each irrigation proves better, passages of water through gypsum bed for improving its quality will prove economically beneficial. Pal and Poonia (1979) observed that an alkali water (ECiw 1.83 dSm,$^{-1}$ RSC 15 meq L^{-1}) could be reclaimed by passing through the gypsim bed of assorted sized fragments (2-50 mm). Dissolution of gypsum is affected by factors such as size distribution of gypsum, flow velocity, salt content, and chemical composition of water (Bashir et al., 1979; Kemper et al., 1975; Singh et al., 1986).

(2) Role of organic materials : Organic manures have often proved beneficial for the saline and the gypsum amended alkali soils. However, on the soils undergoing sodication process through irrigation with alkali waters, the use of organic manures may not be advisable (Gupta et al., 1984), because it enhances dispersion of soils due to greater interparticle interaction forces at high pH. Sharma and Manchanda (1989) studied the effect of an alkali water (ECiw 4 dSm^{-1}, SAR 26 and RSC 15 meq L^{-1}) on the growth of pearl millet [*Pennisetum americanum* (L.) Leekej] and forage sorghum (*Sorghum bicolor* L.) with and without gypsum and farmyard manure (FYM) on a non-calcareous sandy clay loam soil previously deteriorated with water used in fallow-wheat rotation for six years. It was found (Table 5.6) that the use of FYM alone further reduced the crop yields as also the permeability of the soils. Therefore, organic materials should better be added after gypsum application to soils irrigated with alkali waters.

(3) Desalinisation and rainwater management : The process of desalinisation can be adhered to both the water and soils. For conversion of brackish waters into reliable water with dissolved solids in the range of 1.5 - 6.0 dSm^{-1} the reverse osmosis and electrodialysis are the two widely accepted technologies. However, this feasibility is not economically viable. Desalinasation of soil is usually accomplished by impounding of water at the surface of the salty soil. The amount of water required for bringing the salinity of soils to acceptable levels has been observed to depend upon their initial salinity, quality of water used, the soil depth to be reclaimed, the water application techniques, dispersibility of soils during leaching, etc. (Hoffman, 1980 ; Minhas and Gupta, 1990).

In monsoonal type of climate where most of desalinisation is accomplished through rains concentrated in a short span,e.g., July to September in most parts of India, conservation and proper use of rainwater is of utmost importance. In the arid, northwestern India where rainfall is scanty (< 30 cm annually) and groundwaters are saline, farmers involved in desert farming prefer to grow salt tolerant winter crops like kharchia wheat, barley, and mustard with highly saline waters ($6 < ECw < 20$ dSm^{-1}) on light textured soils and skip crops during rainy season. Analysis shows that rainwater can effectively remove salts from equivalent depth of soil in sandy loam soils (Minhas and Gupta,1990).

The general objective in rainy season should be to maximize the infiltration of rainwater into the soil and minimise its losses due to runoff and evaporation during the fallow periods. For this the fields should be properly leveled and bunded where possible and be ploughed in between the rains for increasing the intake of rainwater and controlling the unproductive losses of water through weeds and evaporation. Creation of a soil mulch has been observed to enhance the leaching of salts by 10-13 % (Minhas et al., 1986 ; Minhas and Khosla, 1987). Use of organic mulches have also been observed to enhance leaching of salts by rainfall (Carter and Fannning, 1964; Benz et al., 1967) but shortages of such mulching materials are serious impediment in adoption of this practice on field scale.

Table 5.6. Effect of alkali waters and levels of gypsum and farmyard manure (FYM) on the yield of wheat, Pearl millet, and sorghum. From Sharma and Manchanda (1989).

Teatment	Wheat grain kg ha^{-1}	Pearl millet, kg ha^{-1}		Sorghum fodder, kg ha-1
		grain	Stalk	
Canal water	480	390	2830	2540
RSC water, G–0^{+}	510	230	1350	1610
RSC water, G–50	510	330	2040	2360
RSC water, G–100	480	370	2030	2490
RSC water, G–180	–	360	1940	2340
RSC water, G–50+ FYM, 12 t/ha^{-1}	520	380	2090	2360
RSC water, FYM, 25 t/ha^{-1}	–	190	1190	1380
LSD (P = 0.05)	ns	70	280	310

+ Addition of gypsum in the soil cumulatively at rates to neutralize 0, 50, 100, and 180% residual sodium carbonate content of water (G–0, G–50, G–100, and G–180)

F. Ionic Constituents / Type of Salinity

(1) Cl^- : SO_4^{2-} ratio : Besides the amounts, proportion of different dissolved ionic constituents also influences the salinity hazards of irrigation water. Under fallow-wheat cropping when irrigation waters, similar in salt content but varying in their Cl^- : SO_4^{2-} ratios were used continuously for three years on a sandy loam soil, the threshold salinity limit of irrigation water for wheat was higher for the sulfate dominant than the chloride dominated waters (Figure 5.9). Obviously plant roots were exposed to much higher salinity when soils were irrigated with Cl than SO_4^{2-} dominated waters. More ion pairs are formed with the sulfate waters that increase the solute potential and consequently the availability of water to plant roots. Manchanda et al. (1982) reported chickpea (*Cicer arietinum L.*) to be more sensitive to Cl^{-1} than SO_4^{2-} salinity. Comparable yields were obtained at ECe of 3.2 and 6.8 when salinity was predominantly due to Cl^{-1} and SO_4^{2-} ions, respectively. However in addition to the resultant osmotic effects of the salinity, the plant growth may also be affected by the toxic effects of chloride and/or beneficial effects of sulfur nutrition. Recently, Manchand (1990) has also reported that pulse crops such as pea (*Pisum sativum* L.) mung bean [*Vigma radiata* (L.) (Wilczek)] gram, lentil (Lense ulinaris Medik) suffered more due to toxic accumulation of chlorides when irrigated with Cl^- than SO_4^{2-} dominant (70%) waters of similar salinity. In the long-run, the ECe of soil irrigated with sulfate dominated water will approach inamenable level of about 3 dSm^{-1} even following rainfall leaching during the rainy season. Because the highly soluble salts leach easily with rainwater. Chloride salts being more soluble, leach first, and the precipitated SO_4^{2-} (e.g., relatively insoluble gypsum) continues to dissolve afterwards.

(2) Mg^{2+} : Ca^{2+} ratio : Most saline waters have Ca/Mg ratio in the range of 1 to 6 or even more as in the costal belts. As the proportion of Mg^{2+} relative to Ca^{2+} increases, it enhances the sodication of soils and causes dispersion of clay particles (Paliwal and Gandhi, 1976; Girdhar and Yadav, 1980; Yadav and Girdhar, 1981). Deo and

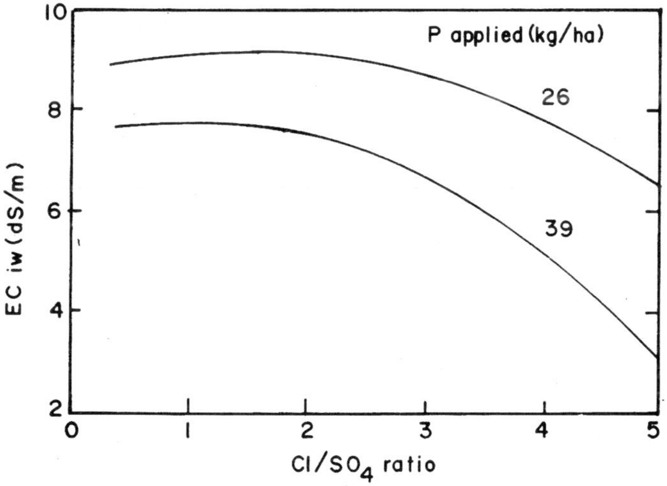

Fig. 5.9. Effect of Cl : SO4 ratio in irrigation waters on threshold salinity of wheat.

Kanwar (1969) reported that Mg behave alike Ca^{2+} at the Mg^{2+} / ΣM^{n+} 5 and alike Na or K at lower ratios. The adverse effects of increased Mg/Ca ratio in irrigation waters have been evaluated in limited number of experiments. But it seems to become critical at a ratio > 3 for optimum growth of plants. An increase in Mg/Ca ratio of irrigation waters from 3 to 9 produced deleterious effects and reduced yields. In addition to reduced permeability of soils, higher Mg/Ca ratio influences the Ca nutrition of plants. The exact mechanism of the latter, however, is not clearly understood. Higher Mg/Ca ratio may affect the uptake and translocation of Ca due to its antagonistic effect or competition for absorption sites. Also Mg-rich water induce higher uptake of P that may further reduce translocation of Ca in plants. The toxicities of other ions in the root environment are generally ameliorated due to higher utilization of Ca. Limited research evidences suggest that Ca^{2+} / ΣM^{n+} of 0.1-0.15 or greater is required for healthy growth of roots. Thus insufficiency of Ca, at increased Mg/Ca ratio in soil solution, has been found to cause reductions in yields of barley (*Hordeum vulgare* L.) and sugarbeet (*Beta vulgaris* L.) (Carter and Webster, 1972; Mostafa and Ulrich, 1976). Application of gypsum should be made particularly when such waters are used for irrigation in black soils.

IV. CONCLUSIONS

Considerable improvements in the state-of-the art for minimizing the hazards upon the use of salt affected waters in agriculture have been made in the recent past. Proper assessment of water quality characteristics usually provides the basis for decisions on various management options to be adopted. Management strategies include proper selection of crop and varieties, irrigation management (method, frequency, leaching management, and conjunctive use modes), and cultural practices (seeding practices, fertility management, need for amendment, etc.). Mechanism for integration of interactive and interdependent processes of various water quality indices with soils is a complex problem and beyond subjective human judgments. Development of computerized models to represent the field situation should serve as promising tools for arriving at the management practices based on the experiences gained under specific set of conditions and extrapolation of experimental findings to different locations and seasons.

REFERENCES

Abrol,I.P. and D.R. Bhumbla. 1979. Soil Sci. 127: 79-85.

Aggarwal,R.P. and B.Ramamoorthy. 1974. Ann.Arid Zone. 13 : 309.

Aggassi, M.,I.Shainberg, and J. Morin. 1981. Soil Sci.Soc.Amer. J. 45: 848-851.

Ayers, R.S. and D.W. Westcot. 1985.FAO irrigation and drainage paper No. 29, Rev. 1.174 p.

Bains, S.S. and K.N. Singh. 1966. Nature 212:1391-1392.

Bajwa,M.S. and A.S. Josan. 1989a. Agric.Water Management. 16:217-228.

Bajwa,M.S. and A.S. Josan.1989b. Agric. Water Management. 16:53- 61.

Bajwa,M.S. and A.S. Josan 1989c. Exptl. Agri. 25: 199-205.

Bajwa,M.S., G.S. Hira, and N.T. Singh. 1983. Irrig. Sci. 4: 191- 199.

Bajwa,M.S.,G.S. Hira, and N.T. Singh 1986. Irrig. Sci. 7: 27-34.

Bashir,A.,W.D.Kemper,G.Haider, and M.A. Niazi. 1979. Soil Sci. Soc. Amer. J. 43 : 698-702.

Beek, G.G.E.M. and N.Breeman. 1973.J.Soil Sci. 24: 129-136.

Benz,L. C., F.Sandoval, and W.O. Willis. 1967. Soil Sci. 104: 63- 68.

Bernstein,L.and R.S.Ayers. 1955. Cali.Agri.p.8.

Bernstein,L.and L.E. Francois. 1973.Soil Sci.Soc.Amer.Proc. 37:931-943.

Bernstein,L. and M.Fireman. 1957. Soil Sci. 83:249-263.

Bhumbla, D.R. and I. P. Abrol. 1972. Indian Fmg. 21 (14) : 15-16.

Bouwmans,J.H. J.W. vanHoorn, G.P. Cruiseman, and B.S. Tanwar. 1988. Agric. Water Management. 14: 537-545.

Bower,C.A. 1961. Proc. Tehran symp. salinity problem in arid zones, UNESCO, p. 1-8.

Bower C.A., G.Ogata, and J.M. Tucker. 1968.Soil Sci. 106:29-34.

Bower C.A. and M. Maasland.1963. Proc.Symsp.waterlogging and salinity in West Pakistan. p. 47-61.

Carter, M.R. and G.R. Webster. 1972. J. Soil Sci. 30: 161-174.

Carter, D.L. and C.D.Fanning 1964. Soil Sci. Soc. Amer. Proc. 28: 564-567.

Chauhan,C.P.S., R.B. Singh, P.S. Minhas, A.K.Agnihotri, and R.K. Gupta 1990. Agric. Water Management. 20:223-231.

Chauhan, R.P.S., C.P.S. Chauhan, and V.P.Singh. 1986. Indian J. agric.Sci. 56:717-21.

Chippa,B.R. and P.Lal. 1986. Proc. Int. Sem. on water management in arid and semi-arid zones. HAU, Hissar. Vol II. p. 176-184.

Chippa, B.R. and P.Lal. 1976. J. Indian Soc. Soil Sci. 20: 330-336.

Darra, B.L., S.P. Seth., H.Singh, and R.S. Mendiratta. 1973. Agron. J. 65:292-295.

Deo,R. and J.S. Kanwar. 1969. J.Indian Soc. Soil Sci. 11: 217- 226.

Dhankar, O.P., V.K. Phogat, D.R.Sharma, and O.P.Sangvan. 1986. Proc. Intl. sem. water management in arid and semiarid zones. HAU, Hissar, India. Nov. 27-29,1986. Vol II. p. 1-10.

Dhir, R.P.,S.N. Bhola, and A.S.Kolarkar. 1977. Indian.J.agric.Sci. 47:244-248.

Eaton, F.M. 1950. Soil Sci. 69: 123-33.

Girdhar,I.K. and J.S.P. Yadav. 1980. Proc Intnl. symp.salt.affected soils. CSSRI, Karnal. p. 210-218.

Gupta,I.C. 1980. Proc. Intl. sym.salt affected soils.CSSRI,Karnal. p. 382-388.

Gupta. I.C.1983. Curr. AGri. 7:97-113.

Gupta,R.K.,D.K.Bhmbla, and I.P.Abrol. 1984. Soil.Sci. 137: 245- 251.

Gupta,R.K. and I.P.Abrol. 1990. Adv.Soil Sci. 11: 233-275.

Gupta,R.K., P.S. Minhas, and I.P.Abrol.1989. Reg.sem. on desertification control in Asia and Pecific, DESCONAP, July 9- 16,1989,Khon Kaen, Bangkok, Thailand.

Hoffman, G.J. 1980. Proc. Inter-American salinity and water management tech. conf. Juarez. Maxico. p. 49–64.

Hoffman, G.J.,J.A. Jobes, and W.J. Alves. 1983. Agri. Water Management. 7: 439-456.

Kanwar,B.S. and J.S. Kanwar. 1971. Indian J. agric Sci. 41: 54- 66.

Kemper, W.D., J. Olsen, and C.J. DeMooy. 1975. Soil Sci. Soc. Amer. Proc. 39: 458-463.

Kiesseling, T.C.,P.S.C. Rao, and R.E. Jassup. 1978. Soil Sci. Soc. Amer J. 42: 234-236.

Maas, E.V. 1986. Appl. Agri. Res. 1: 12-26.

Manchanda, H.R. 1990. Indo-Pak workshop on soil salinity and water management. Feb. 10-14, 1990. p. 2-3.

Manchanda, H.R.,S.K. Sharma, and D.K.Bhandari. 1982. Plant and Soil. 66: 233-241.

Manchanda, H.R., R.N.Garg, S.K. Sharma, and J.P. Singh. 1985. J. Indian Soc. Soil Sci. 33 : 876-883.

Manchanda,H.R.,I.C. Gupta, and B.L.Jain. 1989. In Reviews of research on sandy soils of India. Int. symp. on managing sandy soils. CAZRI Jodhpur. Feb. 6-10, 1989. p. 362-383.

McClung, G, and W.T.Frenkenburger. 1985. Soil Sci. 139:405-411.

Meiri, A., J. Shalhevet, L.H.Stolzy, G.Sinai, and R.Steinhardt. 1986 BARD Tech. Rep. No. 1-402-81. Volcani Center, Bet Dagan. Israel. 172. p.

Minhas, P.S. and B.K. Khosla. 1987. J.Agric. Sci. (Camb) 109:415- 419.

Minhas P.S., B.K.Khosla, and S.S. Prihar. 1986. Soil and Tillage Res. 7: 301-313.

Minhas P.S., D.R. Sharma, and B.K. Khosla. 1988. J. Indian Soc. Soil Sci. 36 : 584-587.

Minhas, P.S.,D.R. Sharma, and B.K. Khosla 1989a. Int. symp. management of sandy soils, CAZRI, Jodhpur. Feb. 6-10, 1989.p. 437-443.

Minhas, P.S.,D.R. Sharma, and B.K. Khosla. 1989b. J. Indian Soc. Soil Sci. 37 : 140-146.

Minhas,P.S.,D.R. Sharma, and B.K. Khosla. 1990a. Irrig.Sci. 11:57-62.

Minhas,P.S.,D.R. Sharma, and B.K. Khosla. 1990b. Indian J. agri.Sci. 343-346.

Minhas,P.S. and R.K.Gupta. 1990. Irrigation water quality-assessment and management. ICAR Bull.

Mondal, R.C., S.K. Gupta, and S.K. Dubey. 1987. CSSRI,Karnal. India. Bull. No. 11, 15p.

Mostafa. M.A.E. and A. Ulrich. 1976. Soil Sci. 121 : 16-20.

Oster, J.D. and F.W. Schroer. 1979. Soil Sci. Soc. Amer. J. 43:444- 447.

Pal, R. and S.R.Poonia. 1979. J. Indian Soc. Soil Sci. 27 : 5-10.

Paliwal,K.V. and A.P. Gandhi. 1976. Soil Sci. 122:85-90.

Punatamkar,S.S.,P.C. Mehta, and S.P.Seth. 1972. J.Indian Soc.Soil Sci. 20:281-285.

Rai,B. and C.L. Mehrotra. 1971. Proc All India Symp. on Soil Salinity. U.P. Inst. of Agri.Sci.Kanpur. p. 212-214.

Rajput,R.K. and M.C. Aggarwal. 1982. Proc. nat. symp. agronomy in 2000 AD - Looking ahead. PKV, Nagpur. p. 71-97.

Rhoades. J.D. 1968. Soil Sci. Soc. Amer. Proc.32:648-652.

Rhoades, J.D. 1984. Proc ASCE Irrigation and Drainage Spec. Conf. Water Today and Tomorrow. July 24-26, 1984. Flagstalf, Arizona, p. 231-236.

Rhoades, J.D.,F.T.Bingham,J.Letey.A.R.Dedrick, M.Bean,G.J.Hoffman, W.J. Ives, R.V.Swain,P.G.Pacheco, and R.D. LeMert. 1988. Hilgardia. 56 (5) : 1-16.

Rhoades, J.D. and S.D.Merrill. 1976. FAO Soils Bull. 31: 69-109.

Sen,H.S. and S.K.Bandopadhyay. 1987.Soil Sci. 143:34-39.

Shainberg, I. and J.Letey. 1983. Hilgardia. 52 : 1-57.

Shalhevet,J. 1984 *In* Soil salinity under irrigation-principles and processes. I. Shaignerg. and J. Shalhevet (eds.) Springer-Verlag. Berlin. p. 298-318.

Shalhevet,J.,A.Vinten, and A.Meiri. 1986.Agron.J. 78 : 539-545.

Shalhevet, J., B.Heurer, and A. Meiri. 1983. Irrig. Sci. 4: 83- 93.

Sharma,D.C.,S.S. Puntamkar,S.V.Jain, and S.P.Seth. 1979. Indian J.agric.Sci. 47: 485-488.

Sharma,D.R. 1978. J. Indian Soc.Soil Sci. 27: 204-207.

Sharma,D.R. and P.S. Minhas. 1986. Annual Report.CSSRI, Karnal, India. p. 73.

Sharma,D.R. and R.C. Mohdal. 1981. J. Indian Soc. Soil Sci. 29:270-273.

Sharma,S.K. and H.R. Manchanda. 1989. Indian J. agric Sci. 59: 118-119.

Sinha, B.K. and N.T. Singh. 1976. Agron J. 60: 346-348.

Singh, B. and M.S. Bajwa. 1985. Fert. Res. 8: 231-240.

Singh,M.,S.R.Poonia, and R.Pal. 1986. Agric.Water Management 11:293-301.

Singh, R.B., P.S.Minhas, C.P.S. Chauhan, and R.K.Gupta 1990. Agri. Water Management. (submitted).

Singh, S.S. and J.S.Kanwar. 1963. J. Indian Soc. Soil Sci. 11:283-287.

Siyag,R.S.,M.S.Lamba, R.Pal., and S.R.Poonia. 1988. J. Agri. Sci. (Camb.). 111: 159-163.

Suarez, D.L. 1981. Soil Sci. Soc. Amer J. 45: 469-475.

US Salinity Laboratory Staff. 1954. Diagnosis and improvement of saline and alkali soils. USDA Handbook 60.160p.

van Schllfgaarde,J., I. Bernsteln, J.D.Rhoades, and S.Rawllns. 1974. J. Irrlg. Drainage Div. ASCE. 100 (IR 3) : 321-338.

Wingley, T.M.L. and L.N.Plumner. 1976.Geochem. Comochem. Acta. 40: 889-995.

Yadav,J.S.P. and I.K. Girdhar. 1981. Soil Sci. 131: 194-198.

Yadav,R.K. 1986. Indian Hort. 30: 2-5.

6

Deficit Irrigation And Strategy For Optimizing Yield

David E. Miller

I. INTRODUCTION

One of the most common plant stresses is consequent to deficient water. Water is a constituent of protoplasm and is an essential reagent in physiological processes. It is a solvent in which salts and gases enter and move through the plant apart from water being an essential factor in maintaining turgidity for growth, form, and position of roots and shoots and for opening of stomata (Kramer, 1959). After absorption from soil it passes through the plant and is lost to atmosphere in transpiration. If this necessary evil exceeds absorption, internal water stress develops that adversely affects the physiological activities.

The main purpose of irrigation therefore is to provides a crop with a near optimum supply of soil water so as to minimize such injury. Irrigation water applied to a field may become unavailable to a crop because of deep percolation, runoff, evaporation, and weed transpiration. As the amount of water available for irrigation diminishes relative to growing demand, it is essential to minimise these unproductive water losses. In addition, use of irrigation water may be extended through deficit or limited irrigation that seeks to apply irrigation water at rates lesser than crop potential evapotranspiration, and minimizes yield reduction by imposing water deficits at the growth stage (s) least sensitive to water stress

(Singh, 1981). The subject of this chapter is how to maximize water use efficiency, i.e., production per unit application of water or water used in the ET process, through judicial use of limited or deficit irrigation without undue loss in productivity.

II. THEMES

Deficit irrigation increases the risk of yield reductions from water stress (Musick, 1989). However, the modest yield losses may be acceptable if the water saved can be used beneficially to offset the value of the reduced yields. Irrigation may be practised to achieve one of three possible goals (Vaux and Pruitt, 1983). These are (1) maximum yield per unit of land, (2) maximum yield per unit of water, and (3) maximum profit. A grower is usually concerned about maximizing profit. When water is plentiful and inexpensive this is nearly the same as irrigation for maximum yield. However as irrigation rates approach those needed for maximum yields, water use efficiency declines If water is scarce or very expensive the interest shifts toward obtaining maximum yield per unit of water. This shift usually involves deficit irrigation and the imposition of water stress on the crop. The effect of water stress on yield and quality is a function of both the degree and the duration of stress, the marketable plant product, and the stage of growth when stress occurs (Grimes et al., 1970). One must consider the economics of deficit irrigation by comparing the savings that result from reducing water applied to the cost to achieving the reductions and the value of lost crop yield that may accompany reduced irrigation.

The maximizing of yield per unit of irrigation water often requires improvements in the irrigation system to reduce conveyance losses, increase application uniformities, and to reduce drainage. Again, the cost of the savings must be weighed against the benefits.

III. RULES FOR DEFICIT IRRIGATION

Deficit irrigation has been studied at Bushland, Texas, for over 30 years. In a recent review of that work, Musick (1989) presented seven rules for deficit irrigation management. Although these rules were developed largely for grain sorghum (*Sorghum bicolor* (L.) Moench), wheat (*Triticum aestivum* L. emend Thell.), and corn (*Zea mays* L.) production in the high plains of Texas, they apply well to this chapter. These rules are presented with Dr. Musick's permission: (1) Practice deficit irrigation only on relatively deep soils that have moderately high water holding capacities, (2) use drought resistant crops, (3) Consider the contribution of precipitation to crop water requirements, (4) Consider the crop growth stage at which irrigation deficits are imposed, (5) Consider whether or not preplant irrigation is needed, (6) Manage surface irrigation systems to reduce water infiltration rates and runoff, thus reducing required application amounts, and (7) Modify cultural practices.

Rule 1. Practice Deficit Irrigation only on Relatively Deep soils that have moderately High Water Holding Capacities

The amount of water stored in the soil profile and available to a crop to supplement low irrigation rates during high water use periods is an important factor in limited irrigation. Moderate-to-high amounts of stored water allow water deficits to develop gradually and thus improve the plants ability to stand water stress. Small amounts of plant-available water allow rapid development of stress within the plant and increase the risk of yield reductions. Low storage may be due to shallow root- restricting layers, coarse-textures, or a subsoil depleted of available water. Yields of sorghum have tended to decrease by irrigation at less than potential ET when the soil profile was initially dry (Fereres et al., 1978). A soil with a coarse surface texture but underlain by fine material may store enough profile water to be productive under deficit irrigation.

When rainfall provides some of the plant water requirement, water stored in the profile acts as a buffer to water stress because the plant can draw on this water during rainless periods (Prihar and Gajri, 1982). With irrigated corn on a clay loam soil in Nebraska, where summer rainfall is important, irrigation could be reduced to one-half to one-third of peak evapotranspiration (ET) without reduction in grain yields (Fischbach and Somerhalder, 1974). We have found marked differences in crop responses to daily deficit sprinkler irrigation due to soil texture. On a loam soil with good water storage, daily irrigation rates could be reduced to about one-half estimated ET rates without serious yield reduction of dry beans (*Phaseolus vulgaris* L.), wheat, potatoes (*Solanum tuberosum* L.) or sugarbeets (*Beta vulgaris* L.) (Miller and Burke, 1983; Miller and Hang, 1980, 1982). Miller and Martin, 1983). Although total yields of Russet Burbank potatoes (the dominant cultivar in the Pacific Northwest) were not greatly affected by deficit irrigation, the quality of tuber was reduced. On a sandy soil with low water holding capacity, irrigation rates could not be reduced to much below 100% ET without incurring losses in yield and quality.

Rule 2. Use Drought Resistant Crops

Turner (1986) refers to drought resistance as the ability of a crop to grow satisfactorily in areas subject to deficient water. Mechanisms or adaptations have evolved in higher plants that favour survival and growth with inadequate or irregular water supplies. These mechanisms have been classified as drought escape, avoidance, and tolerance (Levitt, 1972, 1978; Radin, 1983).

(i) **Drought escape** : Plants are able to complete their life cycle during periods when water is available. Short life cycles are characteristic. For example in Israel, with a Mediterranean climate, spring-type wheat grows during the mild rainy winter and matures at the beginning of the dry summer. It is usually able to complete its life cycle with an adequate water supply (Shimshi, 1973). Turner (1986) reviewed work that compared earliness of various cultivars of wheat, barley (*Hordeum vulgare* L.), and triticale (*Triticum-Secale*) with deficit irrigation. Each day of earliness increased yield by about 30 kg ha^{-1}. However, drought escape by earliness may restrict the potential for high yields when water is adequate.

(ii) **Drought avoidance** : Plants are able to maintain a favourable water status either by maintaining adequate water uptake or by decreasing water loss in the presence of high evaporative demand. Inasmuch as drought is an environmental condition, the plants do not actually avoid it but avoid the low tissue water potentials associated with drought.

(a) *Maintaining a water supply :* An extensive root system is not necessary for optimum plant growth if adequate water and nutrients can be obtained from a restricted root zone (Singh et al., 1989). However, crops with restricted roots are vulnerable to water stress should rainfall or the irrigation system fail. In the absence of root restricting profile layers, deep rooting crops such as sunflowers (*Helianthus annuus* L.), sugarbeets, alfalfa (*Medicago sativa* L.), wheat, sorghum, and cotton (*Gossypium hirsutum* L.) may avoid or reduce drought damage by continued expansion of the root system into moist soil. The benefit of deep rooting in reducing water stress with limited water supplies has been documented in several studies (Kaspar et al., 1984; Sullivan and Ross, 1979). For most species of interest, this provides only supplemental water that may be minimal for survival. If the water content is so low that flow to the roots is negligible, the available water also will be quite low. Water needed to sustain active transpiration usually cannot be supplied by root expansion only. As indicated by Power (1983) if the root zone is not recharged, the increased rooting depth increases the water supply only until the root zone has been depleted of available water. One advantage is that the plants are maintained sufficiently healthy that they can respond positively to water when it again becomes available through rain or irrigation.

Stanhill (1986) has indicated that another important avoidance characteristic is the ability to quickly and fully shade the soil surface to reduce evaporation losses.

(b) *Reducing plant water loss :* Some plants can decrease water loss through reduced leaf expansion, leaf curling, leaf orientation to minimize radiation interception, leaf shedding, radiation reflection, and high water flow resistance through the plant. Closing of stomata is a common mechanism to reduce water loss. This closing reduces evaporative cooling and also restricts CO_2 diffusion into the leaves, reducing, photosynthesis. Pineapple (*Ananas comosus* L.) has overcome this last problem by keeping stomata open at night and the CO_2 absorbed is synthesized the following day (Stanhill, 1986).

(iii) **Drought tolerance :** Water uptake is a passive process with water flowing down a water potential gradient from the soil, into and through plant roots and shoots and into the atmosphere (Hsiao and Bradford, 1983). The capacity to develop low osmotic potentials through active accumulation of solutes and thus maintain turgor is essential to survival and growth when soil water potential is low. Cary (1971) observed that drought tolerant plants can grow and exchange air with the atmosphere, even with very low plant water potentials. Species and cultivars within a species vary in the degree of adaptation to low soil water potentials. For example, in crested wheat grass [*Agropyron desertorum* (Ffisch.) Schult.] plant water potentials decrease to lower values than in potatoes (Ruf et al., 1963). Tomatoes (*Lycopersicon esculentum* Mill) can adjust osmotically during water stress (Taylor et al., 1982), as can wheat, cotton, millet (*Panicum* sp.), and sorghum (Turner, 1986). Osmotic adjustment in sugarbeets is as effective as stomatal closure in cowpeas [*Vigna unguiculata* (L.) Walp.] as a drought survival mechanism (McCree and Richardson, 1987). Radin (1983) indicated that most plants have a limit of about 0.5 to 0.7 MPa osmotic adjustment, but some species can adjust much more. At low soil water contents, soil water potential decreases rapidly with small changes in water. Accordingly the amount of additional water the plant can obtain because of osmotic adjustment is small.

> Thus, osmotic potential adjustment may be beneficial in the short-term but will have little effect in a prolonged drought (Kirkham, 1989). The water that becomes available because of osmotic adjustment may be adequate for survival but not for producing a good crop.

With the exception of escaping water stress by early maturation, increased rooting volume and turgor maintenance are the only processes that allow plants to withstand drought without a reduction in photosynthesis and crop growth (Turner, 1979).

Rule 3. Increase the Contribution of Precipitation to Crop Water Needs

Precipitation increases in importance for meeting crop water needs in areas where deficit irrigation is practised and rainfall is partially relied on to meet expected yield goals (Stewart and Musick, 1982). The contribution of precipitation in meeting crop water needs and thus reducing irrigation water requirements can be enhanced by : (1) using precipitation, if onset is timely, for stand establishment without preplant irrigation, (2) partial wetting of the profile during irrigation to allow some storage capacity for precipitation, (3) reducing or eliminating precipitation runoff (conservation tillage, furrow dams, land levelling), (4) skipping irrigation during above normal precipitation periods, and (5) timely termination of irrigation to fully utilize available profile water by the end of the growing season.

Early irrigation cutoff increases the storage capacity for precipitation occurring between harvest and planting the ensuing crop. As for instance, in a three-year study at Bushland (Texas) on a Pullman clay loam the precipitation storage efficiency ranged from 40 to 50% when the soil profile was dry after harvest consequent to termination of irrigation at the boot stage to 10% or less when the soil profile was wet after harvest (Fig.6.1). The need for leaving room to store rainfall following an irrigation has also been emphasized in the midwesterns USA. Thus, the semiarid regions receiving annual precipitation ranging from 225 to 510 mm offer the greatest potential for conjunctive use of irrigation and rainfall

(Stewart and Musick, 1982). The reduced loss of water to deep drainage reduces loss of nutrients by leaching especially in light soils.

It is important to consider the crop to be grown in relation to the expected precipitation patterns. For example, at Bushland (Texas) winter wheat matures from mid-May to mid-June, when precipitation is almost double that during maturation of grain

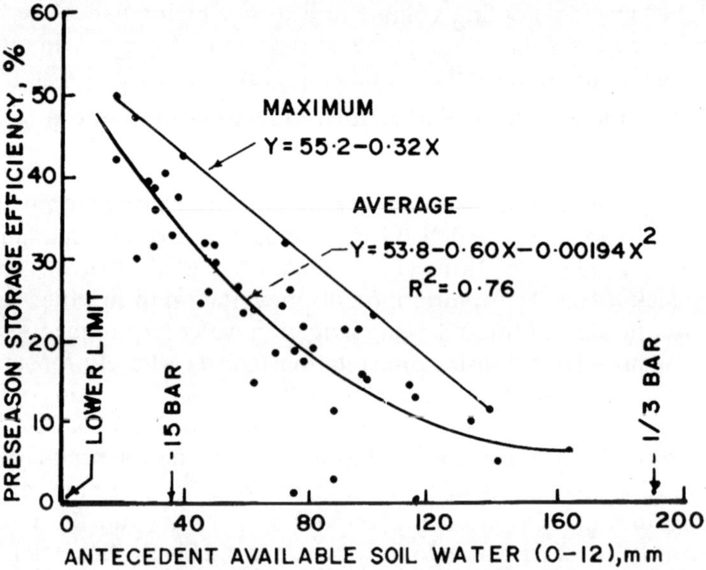

Fig. 6.1. Effect of antecedent available soil water on the precipitation storage effciency on a Pullman clay loam soil. From Stewart and Musick (1982).

sorghum in mid-August to mid-September. Irrigation during grain filling is much less likely to give a response to wheat than to sorghum. In Israel, spring wheat is grown during winter and mid-winter rains are usually sufficient to mature the crop. Response is obtained from an irrigation at seeding to provide adequate water for germination, emergence, and early development. Because of the different expected rainfall distributions a single irrigation should not be applied to sorghum in Texas but is recommended for wheat in Israel.

Rule 4. Consider the Growth Stage at Which Irrigation Deficits are Imposed.

Water is essential for the growth of plants from germination through physiological maturity but the sensitivity to water deficit changes during the growing season. At all growth stages water must be sufficient to allow plants to survive in a healthy enough condition that they can respond to adequate water when it does become available. Injury due to water stress at any growth stage should not be severe to cause damage to final yields.

Seed germination and early development of roots and shoots are dependent upon available soil water, for normal growth as well as to insure low soil resistance to seedling emergence and root extension. Plants subjected to early water stress may not develop a deep, dense root system (Ritchie et al., 1972). An early irrigation 30 days after planting of dryland winter grain crops on loamy sand and sandy loam soils was found to stimulate early deep rooting, promote plant growth, and give significant increase in yields (Prihar and Gajri, 1982). An important part of the yield response was the early deep rooting that eventually resulted in increased extraction of water stored in the lower profile. Early irrigation has also been reported to stimulate deep rooting of millet, barley, sudan grass (*Sorghum vulgare* Pers. *sudanense* Hitche.), and triticale. Others have found early irrigation to be beneficial for early germination and development of wheat (Shalhevet et al., 1976), cotton (Shimshi et al., 1973), and corn (Mantell and Bielorai, 1973), as well as for providing stored water for later use.

The need for a full soil profile at the start of the irrigation season for deficit irrigation has been emphasized by Miller (1977) and Fonken et al. (1974). The computer simulation results of Heermann et al. (1974) indicate that irrigation rates can be reduced as soil water storage increases because the stored water supplements periodic irrigation deficits.

Irrigations early in the season when the groundcover is incomplete provide conditions favourable for loss of precious water by nongrowth related evaporation more than transpiration. Hence the water-use efficiencies, the major concern of irrigators in arid regions, of early irrigations may be very low (Musick, 1989; Rule

5). The value of the response thus be weighed against such factors as the availability of water for early irrigation, expected rainfall, and the productive use of the water saved if an early irrigation is reduced or omitted.

Increased water-use efficiency can be achieved though selection of the crops to be grown according to the expected water supplies (rain and irrigation) and by consideration of the stage of plant growth at which water stress is imposed. Crop production should be timed so that the most sensitive stages of plant development will be completed when deficient water is least likely (Loomis, 1983). A major factor is the marketable plant product - vegetative growth, a seed, or a fruit. Because of the great variation among species in the harvested plant part and in the sensitivity to water stress at various growth stages, it is appropriate to discuss crops in general groups. For added detail on specific crops, the readers are referred to several good reviews (Salter and Goode, 1967; Hagan et al., 1967; Yaron et al., 1973; Stewart and Neilsen, 1989).

(i) **Crops grown for seed or grain :** Many experiments have been conducted on influence of irrigation regime on the yield and quality of wheat, corn, barley, and other seed producing crops. A detailed review of work prior to about 1962 was made by Salter and Goode (1967). Most of these studies employed intermittent watering with various degrees of soil water depletion between applications or the omission of some scheduled irrigations.

Growth stages for this group can be classified roughly as early vegetative, reproductive, and seed fill (Stegman, 1983b). The decrease in yield and quality of seed due to water stress is markedly influenced by the growth stage at which the stress occurs. It is generally accepted that water stress causes the most crop injury and yield reduction when it occurs during reproduction, especially during pollination. Excess water stress at this time can irreversibly damage crops to such an extent that yields are reduced, regardless of later water regimes. Flowering and pollination are usually associated with high rates of water- use so internal water stress can develop rapidly if soil water is deficient (Musick and Grimes, 1961). The early vegetative and seed maturation stages of development are only slightly sensitive to water stress. There is a wide variation among species. For example when water is plentiful, corn will usually show a higher yield response to irrigation than will sorghum, but with limited water, sorghum is much more tolerant of deficits than is corn.

The amount of yield reduction due to shortage of water depends as much on the timing and degree of water stress as on the total shortage. The results of a pot study with southern peas (*Vigna sinensis* L.) made by Hiler et al. (1972), in which three degrees of plant water stress were imposed at three stages of growth, bear this fact out.

Ziska and Hall (1983) observed that withholding irrigation from cowpeas during vegetative growth reduced water-use by 11 to 33% but had negligible effect on seed yields. Robins and Domingo (1956) showed that in dry beans the flowering stage was most critical to water deficits. Flowering stage was also found (Singh et al., 1987) to be critical to water stress in chickpea (*Cicer arietinum* L.) that reduced grain yields by 33%. Yield of soybean (*Glycine max* L.) tended to reduce when water stress occurred during early flowering to early bean development period (Eck et al., 1987).

Hiler et al. (1974) have given crop susceptibility factors to water stress, illustrating the variation among species and in growth stages within a species to sensitivity to water stress (Table 6.1).

Their crop susceptibility factor (CS) is the fractional reduction in yield resulting from a water deficit during a given growth stage:

$$CS_i = (X - X_i) / X$$

where X is the marketable yield under well-watered conditions, and X_i is the yield when water stress occurred only during the ith growth stage. With the exception of peanuts, the crops in Table 6.1 depend upon flowering for the yield product. These data support the general belief that the most critical time for preventing water stress is during reproduction.

Table 6.1. Crop susceptibilities (CS) for various irrigated crops. From Hiler et al. (1974).

Crop	Growth stage	CS
Corn	Vegetative	0.25
	Silking and tasseling to soft drough	0.50
	After soft drough	0.21
Grain sorghum	Vegetative-six to eight leaves	0.25
	Boot to heading	0.36
	Heading to soft drough	0.45
	After soft dough	0.25
Peanuts	Vegetative to early pegging	0.36
	Peak pegging and nut development	0.24
	Late nut development and maturation	0.12
Rice	Vegetative	0.17
	Reproductive and ripening	0.30
Southern peas	Vegetative-preflowering	0.13
	Flowering and early pod formation	0.46
	Pod development to maturation	0.43
Soybeans	Vegetative	0.12
	Early-to-peak flowering	0.24
	Late flowering and early pod development	0.35
	Late pod development to maturation	0.13

Yield responses of sorghum to irrigation is usually low during early vegetative growth, increases during boot stage through flowering, and declines to low levels during grain filling (Musick and Dusek, 1971). In Kansas a single irrigation of sorghum during boot or flowering increased yields sufficiently that additional irrigations were not justified (Stone et al., 1978). The timing of water deficits according to growth stage can result in a wide range of sorghum yields for a given seasonal water use, as illustrated by data in Table 6.2.

Table 6.2. Water-use efficiency by sorghum from one 10 cm irrigation applied at different growth stages, when the total number of seasonal irrigations ranged from one to four. From Musick (1989).

Number seasonal irrigations	Stage of development when irrigated			
	6 to 8 leaf	Mid-to late boot	Heading to flowering	Milk to soft dough
	----- $kg\ m^{-3}$ ------			
1	0	2.23	2.12	0.10
2	0.40	2.03	2.41	0.54
3	0.69	1.70	2.37	0.60
4	0.92	1.78	2.37	0.66

Corn responds to irrigation during the reproductive stage more than at any other time (Denmead and Shaw, 1960; Stone et al., 1978). Robins and Domingo (1953) found that depletion of soil water to the wilting range for only 1 or 2 days during tasseling or silking reduced corn grain yields by 22% and extending the stress to 6 to 8 days reduced yields by 50% compared to full irrigation. Later irrigations did not prevent this loss. Wilson et al. (1979) recommended that full irrigation always be applied during pollination. If water is severely limited, that available should be applied during pollination instead of dividing it among other times.

Cotton is grown both for seed and its fibre. Flowering to boll setting is a critical time while moderate water stress before flowering does little damage (Fowler et al., 1984; Shalhevet et al., 1976). Early irrigations may be detrimental by encouraging vegetative growth at the expense of lint production. If a preseason irrigation is given, the next one can usually be delayed until the beginning of flowering. Water stress during flowering affects cotton fibre production adversely (Fowler et al., 1984), but apparently has little effect on fibre quality. An example of the relative sensitivities of the vegetative and flowering stages to water stress as reflected in lint production is given by the data in Table 6.3 from a study in New Mexico.

Table 6.3. Effect of irrigation treatments during two crop development stages on cotton yield and water-use efficiency. From Fowler et al. (1984).

Irrigation regime (soil water available at irrigation)	Lint yield	Water-use efficiency
% of available	kg ha^{-1}	kg m^{-3}
	Preflower	
50	783	0.08
25	856	0.12
0	829	0.12
LSD 0.05	NS	0.02
	Flowering	
50	1034	0.13
25	786	0.10
0	678	0.09
LSD 0.05	120	0.02

Yields from irrigations before flowering at the 50% and 100% available soil water depletion were similar, accompanied by delayed maturity and decrease in water-use efficiency. The same degree of water deficit during flowering reduced lint yields by 35%. The water-use efficiency was thus highest with initial deficit irrigation and adequate water during flowering.

Bielorai et al. (1983) have discussed the variability in response to water stress occurring among cultivars and suggested that in scarce water supply selection of cultivars should base on water-use efficiency instead of maximum yield. In their views a high water-use efficiency can be achieved by scheduling irrigations according to leaf water potential. Grimes (1977) has demonstrated the large difference between waters required to achieve maximum yield and that maximum water-use efficiency (Fig. 6.2). In other studies (Singh, 1977; Singh et al., 1976) maximum economic return was obtained at about 82 to 90% of yield maximizing level of irrigation where some slight water deficit was incurred. Maximum efficiency of water-use occurred at an irrigation level lower than the maximum economic return point.

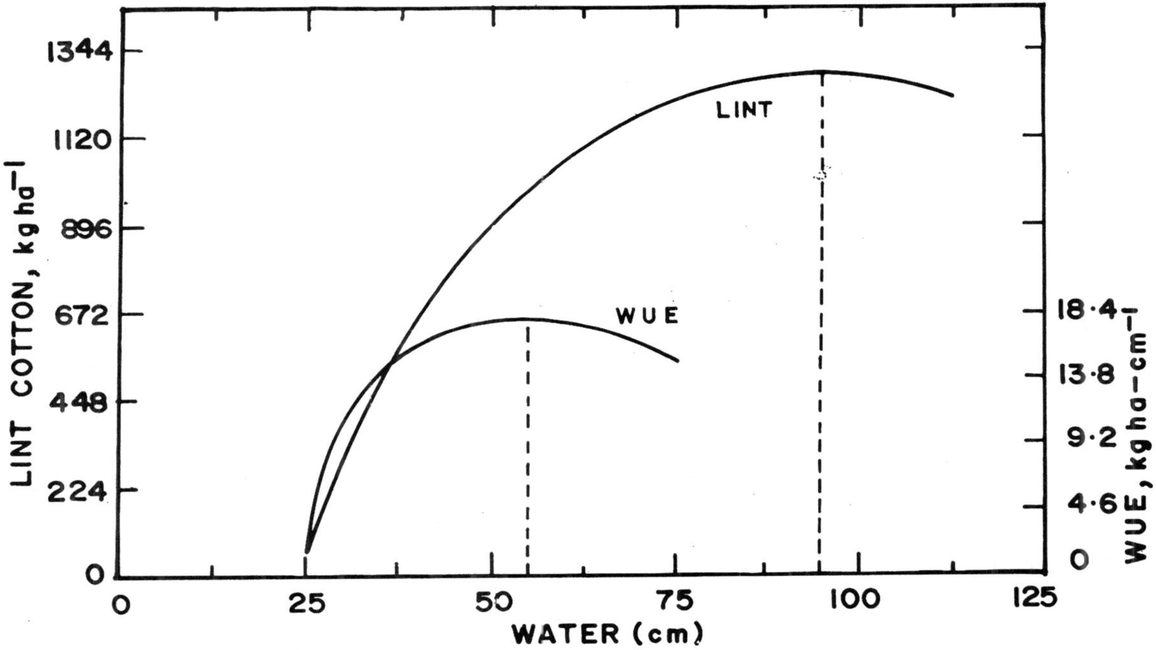

Fig. 6.2. Effect of water application on lint-cotton yield and water-use efficiency (WUE). The water-use includes soil water availabel at planting plus irrigation water. From Grimes (1977).

In wheat the most critical time to avoid water stress is from booting to early grain filling (Schneider et al., 1969). Moderate earlier stress does no harm if soil water is adequate during the critical period. Almost a similar finding has been reported by Singh (1981). In accordance with the "critical stage" concept, the three growth periods of winter wheat could be rated, in order of decreasing sensitivity, as: booting-heading, flowering to grain development, and vegetative stage. With this knowledge, an optimal plan of seasonal irrigation, similar to the one developed earlier by Singh (1977), can be developed that would incur the anticipated water deficit in the growth stage(s) or time periods the least sensitive to water stress. A critical analysis of data concerning timing water deficits revealed that without prior water deficit in the vegetative stage, wheat yields were sensitive to water deficit during the critical booting-heading period but were relatively insensitive when the plants were conditioned to some 15% moisture stress in the vegetative stage. This led to the conclusion that in relation to wheat grain yields if water is limited, the water deficits should be spread nearly evenly over the previous growth stages and the critical growth stage.

Timing of irrigation is as important as the quantity applied. Other workers support these conclusions (Robins and Domingo, 1962; Salter and Goode, 1967; Day and Intalap, 1970; Ehlig and LeMert, 1976). Water deficit, in the soft dough stage can cause shrunken kernels (Robins and Domingo, 1962; Shimshi, 1973), for a short period after anthesis significantly reduces the grain weight per ear (Wardlaw, 1971), should be avoided from the boot stage until the grain is mature (Salter and Goode, 1967), while irrigations after soft dough do not increase yields (Chauhan et al,. 1970; Misra et al., 1969). It seems that the sensitivities of various growth stages to water stress have been variable, depending upon soil types, weather factors, variety, plant type (tall or dwart), and maturity period. With regard to soil types, for instance, Miller and Hang (1982) found that in a medium- textured soil daily sprinkler irrigation equal to near the upper limit of available soil water early in the season could reduce irrigation rates to 40% of estimated ET, without yield reduction. Because the water stored in the soil supplemented the low irrigation rates afterwards resulting, therefore, in the highest water-use efficiency with irrigation rates 40% of estimated ET. On a sandy soil, irrigations at rates lesser than estimated ET reduced yields.

Sunflower shows promise as a crop suitable for limited irrigation, along with winter wheat, grain sorghum, and cotton. Because it has water stress tolerance, adaption to cooler temperatures more than sorghum or cotton, short growing season, and compatibility for growing in sequence with sorghum and wheat (Musick and Rogers, 1980). In addition, being a deep rooted crop it fits well in rotation with shallow rooted crops in that it can extract water and nutrients gone below the shallow root zone and would otherwise be lost to crop recovery. The literature reviewed by Unger (1989) emphasizes the importance of having adequate soil water during flowering and seed filling. Near-maximum yields can be obtained with less than full irrigation if water is plentiful during this critical period. The ability of sunflower to extract water stored deep in the profile is an important factor in its ability to produce a good yield with limited irrigation.

(ii) Forage crops : Forage crops apparently do not have growth stages that are especially vulnerable to water stress. Inasmuch as forage is harvested, rapid growth should be encouraged, and this requires adequate soil water. Many forage crops suffering from water deficits can resume normal growth and production when more water becomes available.

Alfalfa can maintain itself over a wide range of soil water conditions. Because of its deep root system, this crop can use water from deep in the profile. For production of a quality product and for rapid regrowth, it is usual to irrigate at least after each cutting on medium-textured soils. Coarse-textured soils will require more frequent irrigation. In Israel it has been found to be profitable to not irrigate alfalfa from June until August, producing seed rather than forage. The water saved is beneficially diverted to other crops, such as cotton, with peak water demands during this period (Mantell and Bielorai, 1973).

The value of grain sorghum and corn grown for forage depends partially on the immature grain. Thus, the reproductive stage is critical in producing a high quality product. Water stress during the vegetative period will reduce plant size and dry matter yields.

(iii) Vegetable crops : Vegetable crops are sensitive to water stress because the marketable product is usually a fresh fruit, tuber, or vegetative growth, products in which the water content at harvest is an important quality item. They are more sensitive to water deficits than crops grown for dry matter (Shimshi, 1973). These crops can tolerate mild stress and then resume near normal growth when the stress is alleviated.

Potatoes are considered to be high water users and the marketable product, the tubers, is mostly water. Deficit irrigation during tuber development will cause small tubers and reduce yields. Both total yield and quality (marketable yield) are affected by water stress and this effect varies greatly with the cultivar. This is shown by the results from a study at Prosser, WA (Miller and Martin, 1987), on a sandy soil, in which three important northwestern USA potato cultivars (Russet Burbank, Nooksack, Lemhi) were evaluated for their response to water stress due either to declining irrigation rates during the last half of the growing season or to an interruption to irrigation during tuber initiation or tuber bulking. Among these, Nooksack was generally the least sensitive to water stress, and produced a satisfactory yield and per cent U.S. No.1 tubers under conditions that seriously injured Russet Burbank.

Tomato production is sensitive to water deficits in the flowering stage. Stress at this time can cause shedding of young fruit (Shimshi, 1973).

(iv) Fruit trees : Irrigation requirements for fruit trees differ from those of field crops in several important aspects : (1) several years usually are required from planting until a marketable yield is produced; (2) water is a major component of the commercial product, the fleshy part of the reproductive organ; (3) there is a long-term cumulative response of fruit trees to water regime; and (4) the crop is relatively high- valued compared to the cost of irrigation. Proper irrigation according to specific requirements of the trees under specific climatic and soil conditions will have a marked effect on the yield and quality of product (Bielorai et al., 1973).

When an orchard is first established, transpiration is very low because of the small crop canopy. Most water lost from the soil is by evapotranspiration from among the trees. Considerable savings in irrigation water can be achieved by eliminating this superfluous ET. One way to do this is by using drip irrigation (Proebsting et al., 1977). With drip irrigation only a small volume of soil near the tree is kept wet, eliminating any applications between trees. Roots are effectively restricted to the wetted volume and this has caused apple (*Malus sylvestris* Mill.) trees to set fruit one or two years earlier than when sprinkler irrigation was used. After a full canopy had developed differences in irrigation water requirements due to the method of application were small (Proebsting et al., 1984).

A successful flowering and pollination period is essential to a fruit crop. With deciduous trees, this stage of growth occurs before leaf development, while ET is low and there usually is soil water available from winter precipitation. Water stress during blossoming and fruit setting is less likely than during fruit development and maturation.

A recent development in fruit production is the manipulation of the root/shoot ratio through use of regulated deficit irrigation (Chalmers et al., 1981, 1986). By withholding or reducing irrigation during the early season, when fruit growth is slow, excessive vegetative growth is controlled. Resumption of full irrigation when the fruit grows rapidly assures a high quality fruit. The period of deficit irrigation stimulates later fruit growth with the result of more fruit on a smaller tree, compared with full season irrigation.

It is essential that soil water be easily available during the time of rapid fruit growth and maturation. Fruits that mature under a water deficit are small with low water content and high soluble solids (Proebsting et al., 1984). This negative effect of water stress on fruit size and water content may be more important in the market place than the total yield.

Some fruit trees can survive an extreme water shortage, although marketable fruit may not be produced. For example, peach

(*Prunus persica* Batsch.) and pear (*Pyrus communis* L.) trees, with and without heavy spring pruning, were not irrigated during one summer at Prosser, WA, an area with hot, dry summers. Without pruning some of the peach trees died while the pears survived, but with severe injury. Heavy pruning reduced water demand sufficiently that all trees survived (Proebsting and Middleton, 1980). The damage from such severe stress will be evident for several years after normal irrigation is resumed. In the same area, following a year in which only 15% of normal irrigation was applied, prunes (*Prunus domestica* L.) required two years and cherries (*Prunus avium* L.) three years with adequate irrigation to attain normal growth and production (Proebsting et al, 1981).

Evergreen fruit crops such as citrus require water throughout the year. They are also grown widely in arid and semiarid regions so that more consideration must be given to an adequate soil water supply during blossoming and fruit setting than with deciduous trees. Water requirements for citrus vary widely both among species and with differences among locations. The cumulative response of citrus to wet and dry irrigation regimes was demonstrated in Israel (Bielorai et al., 1973) for a young grove. Yields at all irrigation levels increased each year but the rate of increase was much higher with adequate than with inadequate irrigation.

Water requirements of citrus fruit trees were reviewed by Kriedemann and Barrs (1981) and of deciduous fruit trees by Chalmers et al. (1983).

Rule 5. Consider the Need for Preplant Irrigation

Preplant irrigation accomplishes several objectives. Three are important: (i) Storing water in the soil profile for later crop use; (ii) Germinating weed seeds so the seedlings are killed in the preparatory tillage before planting; and (iii) Providing adequate seed zone water for germination, emergence, and early crop development.

Our primary concern is to describe the arid land irrigation and its scientific management. Any rule or strategy advocated for managing a deficient water supply must take this factor into account. As planting in many arid areas having either a summer or winter rainfall pattern is done by and large with the onset of shower in amount sufficient to provide adequate seed zone water for germination and emergence, rule 5 may not apply in the context of this chapter. Rule 5 once again does not fit into a farming system generally practised in temperate region receiving rainfall during winter followed by cropping the following summer season on soil water stored in the profile the preceding season.

Need for preplant irrigation arises at best in those areas of United States where rainfall occurs in winter and the soil profile gets dry approaching planting the following season (Musick, 1987) as well as in Indian part of the Thar desert where some irrigation facilities are available and arid land irrigated agriculture is practised during mild winter season; when preplant plus seasonal irrigations meet the crop water requirements.

Thus where preplant irrigation is essential, the application efficiencies must be as high as possible. This may require an irrigation system with which small amounts of water may be uniformly applied. Small basins in light soils (Ram et al., 1976), sprinklers, and the recently developed "surge flow" technique (Stringham and Keller, 1979) are suitable. The cost of acquiring and using a suitable irrigation system must be weighed against the value of water saved through increased application efficiencies, both preplant and during the growing season.

Rule 6. Manage Water to Reduce Deep Percolation and Runoff with Furrow Irrigation

This rule applies to areas using furrow irrigation that is known to have usually low application uniformities as a result of different intake opportunity times between the top and bottom of a field and the necessity at times to have runoff if the bottom of the field is adequately irrigated. It is difficult to irrigate uniformly with a small amount of water, either across or down a field, especially during the first irrigation after the soil has been disturbed by tillage.

Some practices that have been developed (Musick and Walker, 1987) to reduce the amount of water applied by furrow irrigation are: (1) use of wide-spaced or alternate furrows; (2) tractor wheel compaction of furrows; and (3) surge flow irrigation. In addition, tail-water runoff may be reduced by increasing infiltration rates near the bottom of the field by tillage. Reducing the length of time when tailwater runoff will decrease water loss at the expense of yield at the lower end. The water saved in this way may be used more efficiently to irrigate additional land. At times it may be economical, where possible, to install a feedback pump system that returns the tailwater to the top of the field.

Stewart et al. (1983) developed a limited irrigation-dryland (LID) system on a Pullman clay loam soil at Bushland (Texas) that prevented irrigation tailwater and storm runoff. The upper one-half of a 575 m field of sorghum was fully irrigated. Tailwater from this section was utilized on the next one-fourth of the field, and the lower one-fourth was dryland with furrow dams to retain precipitation. During a three-year test tailwater and storm runoff from conventional graded-furrow irrigations averaged 18 cm compared with 0.8 cm from the LID system. Water-use efficiencies were 0.9 kg m^{-3} for conventional compared with 1.4 to 1.7 kg m^{-3} with LID. Planting densities were reduced on the lower sections because of the expected reduced water supply.

Rule 7. Consider Modifying Cultural Practices

Musick (1989) presented this rule in relation to the cropping systems (largely wheat, sorghum, cotton, and corn) used in the high plains of Texas. The modifications of cultural practices pertinent to limited irrigation in that area are conservation tillage or residue management, moderate plant densities, flexible planting dates, short duration crops, and use of fallow for increased storage of precipitation between crops. Two of these, conservation tillage and flexible planting dates, merit discussion in this section.

(i) Conservation tillage : Conservation tillage, as used here, means to manage crop residues to keep them on the soil surface. A system used successfully in Texas is no-tillage with chemical weed control following deficit irrigation of wheat, followed by no-till seeding of sorghum about 11 months later, with the sorghum also grown under deficit irrigation (Musick et al., 1977). This system effectively increased precipitation storage and eliminated the need for preplant irrigation for sorghum. Two crops can be produced in three years with 11 months of fallow between each harvest and the next planting. Limited irrigation has also been combined with dryland systems in that area (Stewart et al., 1983; Unger and Wiese, 1979; Unger, 1984).

Closely related to the management of residues through conservation tillage is the application of residue mulches to the soil surface. These reduce evaporation from the soil and thus effectively increase the plant available water in the soil. The practicality of this practice depends upon the availability of crop residue and the cost of transporting and spreading it compared with the value of the water saved and increased yields that may occur.

Shekour et al. (1987) in the West Indies, applied a grass mulch to sweet corn and increased ear yields 75% over the limited irrigation control, to levels similar to those obtained with full irrigation. A mulch of cotton burrs at 22.4 Mg ha^{-1} increased precipitation storage 1.5 times and lint yields by about 40% in Texas (Koshi and Fryrear, 1971). In India, straw-mulched wheat used about 15% less water than the unmulched control (Giri and Singh, 1984). Yield increases due to mulching have been reported for sorghum (Reddy et al., 1982), lentils (Murari and Pandey, 1984), gram (Singh et al., 1982), and mustard (Damor and Vegada, 1984). Combined mulching and transpiration suppression have saved two irrigations without yield reductions of wheat (De et al., 1983), corn (De et al., 1983), and sunflower (Damor and Mane, 1982).

(ii) Flexible planting dates : Earlier than normal seeding may allow increased vegetative growth and deeper rooting prior to the onset of water stress as was observed by Musick and Rogers (1980) for sunflowers. This delayed the beginning of water stress relative to the growth stage and increased the amount of soil water extracted from the deeper profile. The yield from early planting with one irrigation at flowering was about the same as that with three irrigations and delayed planting.

Additional cultural practices, of a high application feasibility, that can be considered from the standpoint of limited irrigation are the selection of the irrigation system as well as deep tillage for increasing rooting depth especially on heavy soils.

(iii) Selection of irrigation system : As the cost of irrigation water increases, one should consider the economics of installing an irrigation system that will allow efficient application of water in the amounts and frequencies desired. Kruse et al. (1989) have reviewed the merits of various irrigation systems. Two of the options with limited water are high frequency and mist irrigation.

(a) High frequency irrigation : With a high frequency irrigation (e.g., daily irrigation by drip equal to ET rate) it is possible to maintain soil water content, at the time immediately before irrigation for the day, equal to field capacity under the emitter and to about 60 and 40% of the available water at the points 15 and 25 cm from the lateral, respectively (Singh et al., 1989).

This is especially important on sandy soils with low water holding capacities and high porosity. Because as the irrigation frequency increases, the soil water storage capacity decreases in importance and there is no risk of poor aeration. Further, an actual action between the ET demand and a small amount of water applied by daily application from the drippers at the rate equal to ET restricts the water application to the most efficient portion of the rooting volume of the soil (e.g., nearly 74% of the total root weight of the tomato plants was confined to the top 15 cm layer of the soil; Singh et al., 1989). As soon as drip irrigation at a rate less than the ET rate is practised there has been a significant decrease in yields of potatoes (Singh et al., 1978) and tomatoes (Singh et a.,1989). However if some water deficit is urgently necessary, root zone water can be managed to allow deficits staggered over various growth stages, since high frequency irrigation allows turgor recovery during at least part of every day, sufficient to allow some photosynthetic activity. This allows more efficient use of water than where water deficits are concentrated in a particular growth stage (Stegman, 1983a).

It is a known fact that sprinkler and drip systems of irrigation are best suited for high frequency applications. Both can easily be automated (as in advanced countries) or be managed manually as in developing countries and can apply water frequently and accurately in amounts equal to or less than daily crop evapotranspiration. Daily irrigation by sprinkling in trailing crops such as watermelon (*Citrullus lanatus* Thumb. Mansf.) and round gourd [*Citrullus bulgaris* var. *Fistulossus* (Stocks)] decreased yields from 20 to 32% when compared with daily drip irrigation (Singh and Singh, 1978). Hence, on loamy sand soils in hot arid region, daily irrigation is advantageous when the water is applied by drip irrigation but probably not when the water is applied by sprinkler irrigation. In daily sprinkling, much of the water was intercepted by the leaves and evaporated so that less was available to wet the soil.

Once a system is installed, the cost of applying a given amount of water is nearly independent of the number of irrigations in which it is applied. There are advantages and disadvantages to each system. Sprinkling wets the soil and canopy surfaces so that evaporation may be high and brackish water should not be used. Wind causes poor uniformity, and high canopy humidities form a niche suitable for proliferation of pest and diseases more so in coastal desert areas. (e.g., South Yemen, now Republic of Yemen).

Sprinklers can be used to cool the canopy. A drip system is generally more expensive to install and requires a higher level of management than sprinkling. Brackish water of moderate salinity can be used, with yields of some crops equal to or a little less (Singh et al.,1978; Goldberg and Shmueli, 1969), as well as weed and disease control easier than that with sprinkling. Both systems can be used to apply nutrients and pesticides along with irrigation water.

(b) Mist irrigation : During hot period a high atmospheric demand for water and the associated high transpiration can induce temporary plant water deficit each day, even though the soil water levels are high, that if added may affect plant growth adversely. Misting is the application of a fine water spray to the canopy. It effectively decreases leaf temperature (Unrath, 1972; Wright et al., 1981) through evaporative cooling and increases leaf water potential by stopping transpiration. The influence on yields has been mixed. Literature reviewed by Howell et al. (1971) showed 20 to 30% increase in dry matter yields of southern peas by misting daily between 11 a.m. and 5 p.m. The yield and quality of small fruits and many vegetables depicted similar trends. On the contrary, sugarbeets in Idaho (Kohl and Cary, 1969), corn in Alabama (Doss, 1974), and wheat in Mexico (Laing and Fisher,1975) were not benefited by misting. High initial investment as against a small

saving in water will probably limit its use to high value crops on a small scale. Sprinkler system can do the same thing as misting, therefore mist irrigation as of now is no more than a state of the art system.

(iv) Deep tillage for deeper rooting : A relatively shallow high strength soil layer, such as a plough pan, will decrease rooting depth compared with unrestricted growth thereby reducing the supply of water and nutrients available to a crop. If the crop is dependent upon stored water for an extended period, the restricted root volume can result in severe water stress. Many studies have shown that breaking a compact subsoil by deep tillage will increase rooting depth. As a result plants growing in deep tilled soil have avoided water stress during times of drought because roots reached water that was positionally unavailable without deep tillage (Doty and Reicosky, 1978; Taylor, 1983; Bennie and Botha, 1986; and others). On a sandy soil with a rigid subsoil, subsoiling resulted in enhancement of rooting depth, prevention of water stress between irrigations, and improvement in yields of dry beans (Miller, 1987), corn and potatoes (Ibrahim and Miller, 1989). The low soil strength following subsoiling may persist for at least five years thus reducing the initial expense to a reasonable cost per year of benefit. This operation may not be needed if adequate water and nutrients are available in the surface soil. Soil compaction is less likely to reduce yields of perennials than of annuals because the roots of former have opportunity to grow through old root channels or through layers at times of the year when strength is low because of high soil water contents.

IV. ROLE OF LAYERED SOILS IN PLANNING DEFICIT IRRIGATION

Knowledge about an accurate estimate of the water holding capacity of the soil is important to be able to optimally programme a deficit irrigation. This is especially important in intermittent irrigations where a given amount of water is depleted between applications. Even with high frequency irrigation in which water can be supplied daily or up to several times per day in response to evapotranspiration and where the role of storage capacity is less important, it is useful to know something about the amount of water stored in case of equipment failure.

A frequently ignored factor in estimating the water retention properties of soil in the field is the role of layers in the profile. This subject has been discussed in detail by Miller (1973).

In comparison to a uniform profile, any discontinuity in pore size distribution, such as a textural change, will increase the soil water content above the discontinuity, if enough water has been received to allow drainage to beyond the depth in question. If the discontinuity is a later of a fine texture, water can drain through the upper soil faster than through the fine layer and drain through the upper soil faster than through the fine layer and may accumulate above the layer. Positive hydraulic pressures may develop. If the discontinuity is a coarse layer, it will not conduct significant amounts of water until many of the pores are filled with water. This occurs only at water potentials much higher than those at which the soil pores above are filled. As a result, the soil water content will be greater when a coarse layer is present than in a uniform soil. As the profile drains following an irrigation, the coarse layer stops transmitting water at relatively high potentials and the water content in the soil above the layer remains higher than in a non-layered soil.

Soil profiles fall into four general types, although there are many variations and gradations among them. These types are: (1) deep soils, uniform in texture and structure; (2) profiles with slowly permeable or nonpermeable layers, including soils underlain by bedrock; (3) profiles in which soil is underlain by coarse materials; and (4) profiles in which textural gradations occur over an appreciable depth range.

The water retention properties of deep, uniform soils are dominated by texture and to some extent by structure. Since the subject of this section is layered soils, uniform profiles will not be discussed. The behaviour of the soils in the fourth group depends upon the depth, intensity, and the rate of change of texture within the profile. The soil water content above the transition zone will behave much as if the transition occurred abruptly at some depth within the zone. Accordingly only groups two and three will be discussed, with emphasis on the coarse layers.

Profiles With Less Permeable Subsoils

There are many variations of the type of profile with less permeable subsoils. Because of low permeability of the layer, water that would normally drain in a deep uniform soil is confined above the layer until it is lost through evapotranspiration or slow drainage through the layer. Inasmuch as the water would drain out of the root zone if the layer were not there, water storage in the soil above the layer is increased. Care must be exercised in irrigating these soils to prevent adverse effects of excess water and poor aeration. A slowly permeable soil will transmit water slowly, but for a long period of time after irrigation. Thus, the effective available water will be much greater with high than with low ET rates because the plant can extract water while the soil is draining. This water would eventually be lost as drainage if plants were not present.

Profiles With Layers of Coarse Materials

The effect of a coarse layer on water retention is similar to that of a slowly permeable layer except that a coarse layer will not allow formation of a positive water pressure above it. The layer will conduct water rapidly at potentials near zero and water from the overlying soil can easily escape. At relatively high potentials, the layer hydraulic conductivity will approach zero and further drainage is restricted. The major items to consider in evaluating the influence of coarse layers on water retention are (Miller, 1973) : (1) the coarseness or particle size distribution of the layer, (2) depth to the layer, and (3) desorption characteristics of the overlying soil.

An example of the large effect a coarse layer can have on water retention is given in Table 6.4.

Because of the desorption characteristics of the soil, the largest effect of the sand layer in increasing water retention is in the loamy sand and the least in the silt loam.

Table 6.4. Available water (above 1.5 M Pa %) in soils of various textures underlain at 60 cm by coarse sand. From Miller (1973).

	Available water in soil		
	Loamy sand	Loam	Silt Loam
Above the sand, cm	16.4	17.4	20.0
60 cm uniform soil, cm	6.7	11.7	1.2
Ratio : layered/uniform	2.5	1.5	1.2

V. SUMMARY

As the amount of water available for irrigation diminishes relative to demand, it becomes essential to prevent or minimize all unnecessary losses of water such as those due to deep drainage, runoff, evaporation, and weed transpiration. In addition, use of irrigation water may be extended through deficit or limited irrigation, which is a planned imposition of water stress through irrigation at rates inadequate to maintain full transpiration. The effect of water stress on yield and quality of crop product is a function of the degree and duration of stress, the marketable plant product, and the stage of growth when stress occurs.

Seven rules for deficit irrigation that serve as guides to management of irrigation water so as to minimize yield reductions and maximize water-use efficiency are : (1) Practice deficit irrigation only on relatively deep soils that have moderately high water holding capacities, (2) Use drought resistant crops, (3) Consider the contribution of precipitation to crop water requirements, (4) Consider the crop growth stage at which irrigation deficits are imposed, (5) Consider whether or not preplant irrigation is needed, (6) Manage surface irrigation systems to reduce water infiltration rates and runoff thus reducing required application amounts, and (7) Modify cultural practices.

REFERENCES

Bennie, A.T.P. and F.J.P. Botha. 1986. Soil and Tillage Res. 7:85-95.

Bielorai, H., I. Levin, and R. Assaf. 1973. *In* B. Yaron, E. Dansfors, and Y. Vaadia (eds.) Arid zone irrigation, p. 397-409. Ecological studies 5, Springer—Verlag, Berlin.

Bielorai, H., A. Mantell, and S. Moreshet. 1983. *In* T.T. Kozlowski (ed.) Water deficits and plant growth, vol. VII, p. 49-87, Academic Press, New York.

Cary, J.W. 1971. Ecology 52 : 710-715.

Chalmers, D.J., G. Burge, P.H. Jerie, and P.D. Mitchell. 1986. J. Amer. Soc. Hort. Sci. 111: 904-907.

Chalmers, D.J., P.D. Mitchell, and L. van Heek. 1981. J. Amer. Soc. Hort. Sci. 106: 307-312.

Chalmers, D.J., K.A. Olson, and T.R. Jones. 1983. *In* T.T. Kozlowski (ed.) Water deficits and plant growth, Vol. VII,p.197- 232, Adademic Press, New York.

Chauhan, D. S.,S.B. Hukkeri, and N.G. Dastane. 1970. Indian J.Agron. 15 : 46-48.

Damor, U.M. and V.D. Mane. 1982. Annals Arid Zone 21 : 25-28.

Damor, U.M. and D.A. Vegada. 1984. Trans. Indian Soc.Desert Tech. Univ. Center. Desert Studies 9: 34-35.

Day, A.D. and S. Intalap. 1970. Agron J. 62 : 27-29.

De, R., D.V.S.B. Rao, Y.Y. Rao, L.G.G. Rao, and M. Ikramullah. 1983. Irrig.Sci. 4: 215-233.

De, R., Y.Y. Rao, M. Ikramullah, and L.G.G. Rao. 1983. J.Agr. Sci. 100: 731-734.

Denmead, O.T. and R.H. Shaw. 1960. Agron. J. 52: 272-274.

Doss,B.C. 1974. Agron. J. 66:105-107.

Doty,C.W. and D.C. Reicosky. 1978.Trans. ASAE 21: 495-499.

Eck, H.V., A.C. Mathers, and J.T. Musick. 1987. Field Crop Res. 17:1-16

Ehlig, C.F. and R.D. LeMert. 1976. Soil Sci.Soc.Amer.J. 40:750- 755.

Fereres, E.,B. Amry, J. Faci, A. Kamgar, D.W. Henderson, and M. Resende. 1978. Calif. Agr. 32: 4-5.

Fischbach, P.E. and B.R. Somerhalder. 1974. Trans. ASAE 17:162- 165,171.

Fonken, D.W.,J.C. Steele, and P.E. Fischabach. 1974. Trans. ASAE 17: 889-891.

Fowler, J.L. H. Assadian and A.S. Brar. 1984. Summer Proc. Western Cotton Prod. Conf. Memphis, TN. Southwest Five State Cotton Growers Assoc. p. 68-71.

Giri,G. and R.R.Singh. 1984. Indian J.Agron. 29:173-178.

Goldberg, D. and M. Shmueli. 1969. Proceeding 1969 International Arid Lands Conference, Tucson, Arizona.

Grimes, D.W. 1977. *In* Proc. West. Cotton Prod. Conf., p. 22-25.

Grimes, D.W., R.J. Miller, and W.L. Dickens. 1970. Calif. Agr. 24: 4-6.

Hagan,R.M.,H.R. Haise, and T.W. Edminster (eds.) 1967. Irrigation of agricultural lands. Agronomy Monograph NO.2, Amer.Soc.Agron., Madison,WI.

Heermann, D.F.,H.H. Schull, and R.H. Mickelson. 1974. J.Irrigation and Drainage Div. ASCE 100: 127-141.

Hiler, E.A.,T.A. Howell, R.B. Lewis, and R.P. Boos. 1974. Trans.ASAE 17:393-398.

Hiler, E.A.,C.H.M. van Bavel, M.M. Hossain, and W.R. Jordan. 1972. Agron.J. 64:60-64.

Howell, T.A., E.A. Hiler, and C.H.M. van Bavel. 1971. Trans. ASAE 14: 906-910.

Hsiao, T.C. and K.J. Bradford, 1983. *In* H.M. Taylor, W.R. Jordan, and T.R. Sinclair (eds.) Limitations to efficient water use in crop production, p. 73-85, Amer.Soc.Agron., Madison, WI.

Ibrahim, B.A. and D.E. Miller. 1989. Soil Sci.Soc.Amer.J.53: 247- 251.

Kaspar, T.C.,H.M. Taylor, and R.M. Shibles. 1984. Crop Sci. 24: 916-920.

Kirkham, M.B. 1989. *In* Irrigation of agricultural crops, Ch.II, Amer.Soc.Agron. Monograph No 30, Madison,WI.

Kohl, R.A. and J.W. Cary. 1969. Amer. Soc. Sugarbeet Technol. 15: 416-421.

Koshi, P.T. and D.W. Fryrear. 1971. Agron. J. 63: 817-822.

Kramer, P.J. 1959. Adv. Agron. 11: 51-70.

Kriedemann, P.E. and H.D. Barrs. 1981. *In* T.T. Kozlowski (ed.). Water deficits and plant growth, Vol. VI, P. 325-417, Academic Press, New York.

Kruse, E.G., D.A. Bucks, and R.D. von Bernuth. 1989. *In* Amer. Soc. Agron. Monograph No. 30, Ch. VI, Madison, WI.

Laing, D.R. and R.A. Fischer. 1975. Agr. Meteorol. 15:285-293.

Levitt,J. 1972. Responses of plants to environmental stresses. Academic Press, New York.

Levitt, J. 1978. *In* G.A. Jung (ed.) Crop tolerance to suboptimal land conditions, p. 161-171. Amer.Soc.Agron.Spec.Publ.No.32, Madison, WI.

Loomis R.S. 1983. *In* H.M. Taylor, W.R. Jordan, and T.R. Sinclair (eds.) Limitations to efficient water use in crop production, p. 345-374, Amer. Soc. Agron., Madison,WI.

Mantell, A. and H. Bielorai. 1973. *In* B. Yarson, E. Danfors, and Y.Vaadia (eds.) Arid zone irrigation, Ecological Studies 5, p. 387-396, Springer-Verlag, Berlin.

McCree, K.J. and S.G. Rechardson. 1987. Crop Sci. 27: 539-543.

Miller, D.E. 1973. *In* R.R. Bruce, K.W. Flach, and H.M. Taylor (eds.) Field water regime, Soil Sci. Soc. Amer. Spec. Publ. No.5, p. 107-117, Madison, WI.

Miller, D.E. 1977. *In* Proc. Water Mgt. for Irrig. and Drainage, p. 269-292, ASCE, Reno, Nevada.

Miller, D.E. 1987. Soil Sci. Soc. Amer J. 51: 784-787.

Miller, D.E. and D.W. Burke. 1983. Agron. J. 75: 775-778.

Miller, D.E. and A.N. Hang. 1980. Soil Sci. Soc. Amer J. 44: 1295-1298.

Miller, D.E. and A.N. Hang. 1982. Soil Sci. Soc. Amer.J. 46: 386-389.

Miller, D.E. and M.W. Martin. 1983. Amer. Potato J. 60: 745-757.

Miller, D.E. and M.W. Martin. 1987. Amer. Potato J. 64: 109-117.

Misra, R.D., K.C. Sharma, B.C. Wright, and V.P. Singh. 1969. Indian J. agric. Sci. 39: 898-906.

Murari, K. and S.L. Pandey. 1984. Indian J. Agron. 29: 435-440.

Musick, J.T. 1987. ASAE Paper No. 87-2558, St. Joseph, MO.

Musick, J.T. 1989. Proc. Central Plains Irrigation Short Course, Colby, Kansas.

Musick. J.T. and D.A. Dusek. 1971. Trans. ASAE 14: 401-404, 410.

Musick. J.T. and D.W. Grimes. 1961. Kansas Agr. Exp. Stn. Tech. Bull. 113.

Musick. J.T. and C. Rogers. 1980. Sunflower 6: 44-48.

Musick. J.T. and J.D. Walker. 1987. Appl. Engineering in Agric. 3: 190-195.

Musick. J.T., A.F. Wiese, and R.R. Allen. 1977. Trans. ASAE 20: 666-672.

Power, J.F. 1983. *In* H.M. Taylor, W.R. Jordan, and T.R. Sinclair (eds.). Limitations to efficient water use in crop production, p. 461-470, Amer.Soc.Agron., Madison, WI.

Prihar, S.S. and P.R. Gajri. 1982. *In* Symposium on rainwater and dryland agriculture, p. 115-130, Indian Nat. Sci. Academy, New Delhi.

Proebsting, E.L., Jr. J.E. Middleton, and M.O. Mahan. 1981. J.Amer. Soc. Hort. Sci. 106: 243-246.

Proebsting, E.L.,Jr. and J.E. Middletion. 1980. J.Amer.Soc.Hort.Sci. 105: 380-385.

Proebsting, E.L.,S.R. Drake, and R.G. Evans. 1984. J.Amer.Soc.Hort.Sci.. 109: 229-232.

Proebsting, E.L., J.E. Middleton, and S.Roberts. 1977. Hort.Sci. 12: 349-350.

Radin, J.W. 1983. *In* H.M. Taylor, W.R. Jordan, and T.R. Sinclair (eds.). Limitations to efficient water use in crop production, p. 267-276, Amer.Soc.Agron.,Madison, WI.

Ram, R.S., T.C. Jain, and K.C. Jain. 1976. Indian J. agsic. Sci. 46: 81-87.

Reddy, M.G., K. Krishnamurthy, and V.C. Reddy. 1982. Mysore J.Agr. Sci. 16: 388-391.

Ritchie, J.T.,E.Burnett, and R.C. Henderson. 1972. Agron. J. 64: 168-173.

Robins, J.S. and C.E. Domingo. 1953. Agron. J. 45: 618-621.

Robins, J.S. and C.E. Domingo. 1956. Agron. J. 48: 67-70.

Robins, J.S. and C.E. Domingo. 1962. Agron. J. 54: 135-138.

Ruf, R.H., Jr. R.E. Eckert, Jr., and R.O. Gifford. 1963. Soil Sci. 96: 326-330.

Salter, P.J. and J.E. Goode. 1967. Crop response to water at different stages of growth. Commonwealth Agric. Bureaux, Farmham Royal, Bucks, England.

Schneider, A.d., J.T. Musick, and D.A. Dusek. 1969. Trans. ASAE 12: 23-26.

Shalhevet, J.,A. Mantell, A. Bielorai, and D. Shimshi. 1976. Int. Irrig. Info. Ctr. Pub. No.1 Bet Dagan, Israel.

Shekour, G.M., R.A.I. Brathwaite, and C.R. McDavid. 1987. Agron.J.79: 629-631.

Shimshi, D. 1973. *In* B. Yaron, E. Dansfors, and Y. Vaadia (eds.) Arid zone irrigation. Ecological Studies 5, p. 383-385, Springer- Verlag, Berlin.

Shimshi, D., H. Bielorai, and A. Mantell. 1973. *In* B. Yaron, E. Danfors, and Y. Vaadia (eds.) Arid zone irrigation. Ecological Studies 5, p. 369-381, Springer-Verlag, Berlin.

Singh, D.P., P. Singh, H.C. Sharma, and N.C. Turner. 1987. Field Crop Res. 16: 231-241.

Singh, A.K., A.K. Sinha, and N.M. Nimgale 1982. Trans.Ind.Soc.Desert Tech.Univ. Center. Desert Studies 7: 81-83.

Singh, S.D. 1977. *In* Desertification and its control. p. 237-255, New Delhi, Indian Council of Agric. Res.

Singh, S.D., J.P. Gupta, and P. Singh. 1978. Agron. J. 70: 948- 951.

Singh, S.D., M. Yusuf, R.C. Bhandari, and H.S. Daulay. 1976.

Indian J. agric.Sci. 46: 567-575.

Singh, S.D. and P. Singh 1978. Agron. J. 70: 945-947.

Singh, S.D. 1981. Agron.J. 73: 387-391.

Singh, S.D., Y.V.Singh, and R.C. Bhandari. 1989. Can.J.Plant Sci. 69: 991-999.

Stanhill, G. 1986. Adv.Agron. 39: 53-85.

Stegman, E.C. 1983a. *In* H.M.Taylor, W.R. Jordan, T.R. Sinclair (eds.) Limitations to efficient water use in crop production, p. 499-505, Amer. Soc. Agron., Madison, WI.

Stegman, E.C. 1983b. Adv.Irrig. 2: 1-30.

Stewart, B.A. and J.T. Musick. 1982. Adv.Irrig. 1: 1-24.

Stewart, B.A. and D.R. Neilsen (eds.) 1989. Amer.Soc.Agron. Monograph No.30, Madison, WI.

Stewart, B.A. J.T. Musick, and D.A. Dusek. 1983. Agron J. 75: 629 634.

Stone, L.R., R.E. Gwin, Jr., and M.A. Dillon. 1978. J.Soil and Water Cons. 33: 235-238.

Stringham, G.E. and J. Keller. 1979. Surge flow for automatic irrigation. *In* Proc. Irrigation and Drainage Div. Speciality Conference, Albuquerque, NM. 17-20 July, 1979, p. 132-142, ASCE, New York.

Sullivan, C.Y. and W.M. Ross. 1979. *In* H. Mussell and R.C. Staples (eds.) Stress physiology in crop plants, p. 263-281, John Wiley & Sons, New York.

Taylor, A.G., J.E. Motes and M.B. Kirkham. 1982. J. Amer Soc. Hort. Sci. 107: 387-390.

Taylor, H.M. 1983. *In* H.M. Taylor, W.R. Jordan, and T.R., Sinclair (eds.) Limitation to efficient water use in crop production, p. 76-113, Amer. Soc. Agron, Madison, WI.

Turner, N.C. 1979. *In* H. Mussell and R.C. Staples (eds.) Stress physiology in crop plants, p. 343-372, John Wiley & Sons, New York.

Turner, N.C. 1986. Adv.Agron. 39:1-51.

Unger, P.W. 1984. Soil Sci. Soc. Amer. J. 48: 885-891.

Unger, P.W. 1989. *In* B.A. Stewart and D.R. Nielsen (eds.) Irrigation of agricultural crops. Amer.Soc.Agron.Monograph No.30, Madison, WI.

Unger, P.W. and A.F. Wiese. 1979. Soil Sci.Soc.Amer J. 43: 582- 588.

Unrath, C.R. 1972. J.Amer.Soc.Hort.Sci. 97: 55-58.

Vaux, H.J., Jr. and W.O. Pruitt. 1983. Adv. *in* Irrig. 2: 62-97.

Wardlaw, I.F. 1971. Aust.J. Biol.Sci. 24: 1047-1055.

Wilson, G.D., D.G. Watts, and P.E. Fischbach. 1979. *In* Agr. Eng.Irrig.No.13, p.65 71, Nebraska Irrig Short Course, Univ, Nebraska, Lincoln.

Wright, J.L., J.L. Stevens, and M.J. Brown. 1981. Agron. J. 73: 481-490

Yaron, B.,E. Dansfors, and Y.Vaadia (eds.) 1973. Arid zone irrigation. Ecological Studies 5, Springer-Verlag, Berlin.

Ziska, L.H. and A.E. Hall. 1983. Irrig.Sci. 3: 237-245.

7

Efficient Use Of Water In Irrigated Agriculture And Limitations To Increasing Water-Use Efficiency

Claudio O. Stockle and Josep M. Villar

I. INTRODUCTION

Efficient use of water is a central issue in arid land irrigated agriculture, where water resources are usually limited and/or expensive. Efficient use of water can be evaluated from several perspectives, ranging from a plant physiology to an engineering point of view. This traditional separation of the problem into disciplines has brought confusion in terminology and definitions.

As pointed out by Sinclair et al. (1984), water-use efficiency has been used interchangeably to refer to the ratio of carbon dioxide to water vapour exchange by individual leaves observed for a few minutes to grain yield produced in response to irrigation treatments throughout an entire growing season. Irrigation engineers, on the other hand, usually refer to water-use efficiency as the fraction of water applied (at reservoir, irrigation district, turnout, farm or field levels) that is stored in the crop root zone.

Many terms and definitions are, therefore, utilized to evaluate the efficiency of water use in irrigated agriculture. In this chapter, we have chosen to use the term Irrigation Water-Use Efficiency (IWUE), defined as the ratio between the amount of crop biomass or yield produced, in kg ha^{-1}, and the amount of irrigation water used to produce such biomass or yield, in mm, which gives IWUE with units of kg ha^{-1} mm^{-1}. As it will be shown, this is a general definition that can be broken down into several components representing physiological, agronomic, and engineering factors which can be manipulated independently to improve IWUE.

II. IRRIGATION WATER-USE EFFICIENCY AND ITS COMPONENTS

The definition of IWUE given above does not indicate if the amount of irrigation water used corresponds to water transpired by the crop, to evaporation from soil and plants and crop transpiration (evapotranspiration), or to the total amount of water delivered at the field, farm or irrigation project level. Indeed, all these instances of irrigation water use are integrated in this generalized IWUE definition. By isolating them, different components of IWUE can be identified.

The ratio between biomass (B) or yield (Y) or economic yield (EY) produced and the amount of water transpired by the crop (T) to produce such biomass or yield will be referred to as Transpiration Efficiency (T_e), an important parameter from a physiological point of view.

Transpiration Efficiency can be represented as follows:

$$T_e = B / T$$

$$\text{or } T_e = Y / T \qquad [7.1]$$

To simplify the following analysis, T_e will be only presented in terms of biomass. From equation [7.1], if T_e and T are known, B can be estimated as:

$$B = (T)\ (T_e) \qquad [7.2]$$

Utilizing the conceptual definition of IWUE given above, if all inputs of water for biomass production are considered, the following expression for IWUE can be written:

$$IWUE = B / (I + P + SW) \qquad [7.3]$$

where I is irrigation amount (mm), P is precipitation amount (mm), and SW is change in soil water content in the time period of interest (mm). For irrigated systems, total water inputs are better represented if all the terms of the soil water balance are considered. Neglecting P and SW, this is expressed as:

$$I = E + T + Q + D + L \quad [7.4]$$

where E is soil evaporation (mm), Q is runoff (mm), D is percolation, and L represents other water losses (e.g., seepage from canals, droplet evaporation during sprinkler irrigation, etc.).

Substituting equation [7.2] for the numerator of equation [7.3], and equation [7.4] for the denominator (with P and SW neglected), IWUE can be expressed as follows:

$$IWUE = (T)(T_e) / (E + T + Q + D + L) \quad [7.5]$$

For irrigated lands, it is common to define an irrigation efficiency (E_{ir}), a parameter intended to measure the fraction of the total irrigation water applied that has been used beneficially, where the latter term refers to water utilized for evapotranspiration (ET), this is soil evaporation and crop transpiration combined:

$$E_{ir} = (E + T) / (E + T + Q + D + L) \quad [7.6]$$

Combining equations [7.5] and [7.6] allows incorporating the concept of irrigation efficiency into the irrigation water use efficiency definition:

$$IWUE = (T)(T_e)(E_{ir}) / (E + T) \quad [7.7]$$

Finally, equation [7.7] can be rearranged to explicitly show the IWUE components to be discussed in this chapter :

$$IWUE = T_e [T / (E + T)] [(E + T) / I] \quad [7.8]$$

In this definition, the first terms in the right hand side (transpiration efficiency) is a physiological component of IWUE, the second (transpiration fraction of ET) is a crop management related component of IWUE, and the last term (ET fraction of irrigation water applied) is an irrigation system design and operation related component of IWUE. In the sections that follow, each of these terms will be discussed.

III. TRANSPIRATION EFFICIENCY

Transpiration efficiency, T_e in Eq. 7.8, is defined as the ratio between biomass production for a given period of time, in kg ha^{-1}, and the amount of crop transpiration during the same period, in mm. As a general reference, values of T_e for a number of species are given in Table 7.1.

Biomass production is directly related to the amount of CO_2 fixed by the crop. Because both processes, CO_2 fixation and transpiration, share a common pathway (stomata) and they are largely influenced by stomatal conductance, it is common to find that biomass production and transpiration are linearly related and that T_e tends to be constant for a given cultivar, unless an effect on vapour or CO_2 gradients is involved or/and the photosynthetic capacity of the plant is altered. A few factors affecting the value of T_e deserve discussion.

Table 7.1 Transpiration efficiencies of some agricultural crops.

Species	Te (kg ha^{-1} mm^{-1})	Harvested Portion	Environment	References
Alfalfa (*Medicago sativa* L.) Harvest			Greenhouse	Walker & Richards, 1985
1	–			
2	20.4	A		
3	22.7	A		
4	20.2	A		
5	23.8	A		
6	21.1	A		
7	17.5	A		
8	14.6	A		

Table Contd....

1	2	3	4	5
Barley (*Hordeum vulgare L.*)	51.0–57.0	T	Northern Syria, rainfed	Cooper et al., 1987b.
	67.0–74.7	T	Three sites in Syria, rainfed	Shephard et al., 1987.
	62.2–66.5	T		
	55.9–62.7	T		
	30.0	A	Hot, dry summer in England	Day et al., 1978.
	43.0	A	Typical summer in England	Gregory, 1984.
Corn (*Zea mays L.*)	93.0–100.7	T	Controlled environment	King & Greer, 1986
	60.2–94.4	A	Ontario, Canada	Walker, 1986.
	55.5–85.8	A	Ontario, Canada	
Cotton (*Gossypium hirsutum L.*)	21.0	A		Howell et al., 1984.
Potato (*Solanum tuberosum*)	72.5	M		Wright & Stark, 1990.
	94.2	Total tuber yield		
Tomato (*Lycopersi-con esculentum* Mill)	16.5	M	Greenhouse	Kobbia & Ibrahim, 1986.
Wheat (*Triticum aestivum L.*)	55.0	A	Southern Autralia	French & Schultz, 1984.
	20.0	G		

(*) A = Aboveground biomass T = Total biomass
G = Grain biomass M = Marketable biomass

1. Carbon Metabolism

The main factor determining the T_e value of plant species is their carbon metabolism pathway. Most species of interest from a production standpoint may be classified as C_3 or C_4 species, depending in the number of carbon atoms of the first biochemical product during CO_2 fixation. Examples of C_4 species are maize (*Zea mays* L.), sorghum (*Sorghum bicolor* L.), sudan grass (*Sorghum vulgare sudanensis*), millet (*Setaria italica* L.), etc. and examples of C_3 species are wheat (*Triticum aestivum* L.), sunflower (*Helianthus annuus* L.), cotton (*Gossypium hirsutum* L.), alfalfa (*Medicago sativa* L.), etc.

Species with C_4 pathway are characterized by high light-saturated photosynthetic rates and the capacity to maintain these high rates at low levels of CO_2 concentration inside the air spaces of leaves (C_i). For C_4 species, the ratio of internal to external CO_2 concentration (C_i/C_a) is typically about 0.3, providing large gradients (C_a-C_i) to drive CO_2 fluxes. For C_3 species, light-saturated photosynthetic rates are lower and C_i values are higher. For C_3 species, C_i/C_a is typically about 0.7 and, therefore, the gradient to drive CO_2 fluxes is lower than for C_4 species.

The combination of high photosynthetic capacity and low internal CO_2 concentration, characteristic of C_4 species, allows these plants to fix a larger amount of CO_2 for a given amount of water transpired than C_3 species, and therefore, their transpiration efficiency is comparatively larger.

2. Vapour Pressure Deficit

Transpiration efficiency is affected by meteorological factors determining the atmospheric evaporative demand. This fact was recognized by the early transpiration efficiency research (Briggs and Shantz, 1914; Kiesselbach, 1916), as well as in later work (de Wit, 1958; Arkley, 1963). These researchers found that year to year differences in T_e values could be related to the amount of water evaporated from an evaporation pan (Briggs and Shantz, 1914; de Wit, 1958) or to the relative atmospheric humidity (Kiesselbach, 1916; Arkley, 1963). Later work showed that the variability of T_e with meteorological conditions is better related to the saturation vapour pressure deficit (VPD) of the atmosphere (Bierhuizen and Slatyer, 1965; Tanner and Sinclair, 1983).

Tanner and Sinclair (1983), using a number of simplifying assumptions, derived the following expression for T_e, similar to earlier expression by Bierhuizen and Slatyer (1965):

$$T_e = K_d / (e^*_a - e) \qquad [7.9]$$

where K_d is a crop specific coefficient, equivalent to an unadjusted transpiration efficiency, which accounts for species differences in C_i/C_a ratio and the fraction of carbohydrates, protein, and lipids in the plant biomass produced (Sinclair *et al.*, 1984), and the term $(e^*_a - e)$ corresponds to VPD, where e^*_a is the saturation vapour pressure at air temperature and e is the atmospheric vapour pressure.

As predicted by equation 7.9, T_e of crops grown in arid environments (large VPD conditions) will always be lower than that of similar crops growing in more humid environments. Illustrating this effect of VPD on T_e, Cooper et al. (1987a) reported that barley (*Hordeum vulgare* L.) grown in Rothamsted, England in an unusually hot and dry summer had a T_e value of 30 kg ha^{-1} mm^{-1} compared to 43 kg ha^{-1} mm^{-1} of a more typical summer. This decreasing trend of T_e values with increased VPD is rather unfortunate considering that arid regions are those most in need of large water-use efficiencies. Values of T_e given in Table 7.1 are valid only for environments similar (i.e., similar VPD conditions) to those prevailing at the locations where they were determined.

3. Nutrient Stress

Reports on the response of T_e to nutrient status indicate either no effect or effect at relatively low nutrient levels. Cooper et al. (1987b), in a study including measurements of root biomass, showed small but significant increases of barley T_e (from 47 to 54 kg ha^{-1} mm^{-1}) in response to fertilizer application. For wheat, results from Heitholt (1989) suggest that high nitrogen treatments would improve T_e over low nitrogen, but not over medium levels. An analysis by Arkley (1963), summarizing data for several crops from Briggs and Shantz (1913) and Kiesselbach (1916), also suggests an increase of T_e with fertility. The rate of this increase is important when nutrient levels are low, but decreases rapidly after medium levels of fertility are reached. Similar conclusions have been reported by de Wit (1958) and Tanner and Sinclair (1983). The latter authors estimated that T_e increases in response to fertilization occurred as long as production was less than half that possible when nutrients were nonlimiting.

Kallsen et al. (1984) reported an experiment with spring barley, where nitrogen rates fluctuating from 30 to 300 kg ha^{-1} did not affect T_e. Similarly, Walker and Richards (1985) concluded that there was little prospect for improving T_e through better nutrient status in general (experiment with maize) and potassium status in particular (experiment with alfalfa). These latter authors, and also Heitholt (1989), have argued that claims for increased T_e with fertility levels could be partially explained by changes in assimilate partitioning increasing the shoot-to- root ratio.

4. Water Stress

The relatively high likelihood of crop water stress in arid areas makes this type of stress an important consideration in the analysis of T_e. Because of the linkage between transpiration and carbon fixation, crops under water stress tend to decrease their biomass production and transpiration in about the same proportion, with the net result that T_e does not change. This insensitivity of T_e to water stress has been found in several studies (Briggs and Shantz, 1914; de Wit, 1958; Arkley, 1963). It must be understood that the fact that T_e remains unaffected by water stress does not mean that total biomass accumulation and yield will not be reduced as a result of water shortage. Total transpiration, and therefore total biomass production, will be reduced in response to water stress.

5. Salinity

In irrigated agriculture of arid regions, salinity is usually an important problem. Although total biomass and yield of crops can be reduced due to salinity, T_e remains largely unaffected. Salinity reduces the soil water potential by reducing the osmotic component, and therefore, makes the soil water less available to plants. In this sense, salinity is similar to water stress, reducing transpiration and photosynthesis in about the same proportion. Other stresses derived from salt toxicity also affect both processes similarly. Insensitivity of T_e to salinity has been reported by Flowers et al. (1988).

6. Atmospheric Carbon Dioxide Concentration

Significant increase of the global atmospheric CO_2 concentration is expected to occur in the next 50 to 75 years, when this concentration is expected to double that of the preindustrial era (about 270 ppm). Two responses of plants to increased atmospheric CO_2 concentration are relevant to T_e: a) total biomass and yield are expected to increase and b) stomatal conductance, and therefore transpiration per unit leaf area, is expected to decrease. These two responses would contribute to higher T_e values.

Estimates from glasshouse experiments indicate a 10 to 30% biomass production increase, for C_4 and C_3 plants, respectively, due to atmospheric CO_2 concentration doubling (Kimball, 1983). Kimball and Idso (1983), based on the analysis of 46 cases for 18 species grown under a doubled CO_2 concentration, reported an average reduction of transpiration of 34%. Based on a simulation analysis, Goudriaan and Unsworth (1990) estimated that transpiration of C_3 plants would be reduced by 10%, and that of C_4 plants by 26% in response to a doubling of the atmospheric CO_2 concentration.

7. Avenues for Improvement of Transpiration Efficiency

Possible avenues for improving T_e have been discussed by Tanner and Sinclair (1983), Sinclair et al. (1984), Richards (1987) and Cooper et al. (1987a). These include biochemical alteration, genetic alteration through selection, use of antitranspirants, and use of fertilization with CO_2.

Increasing biomass production through improvement of photosynthetic rates is one conceptual approach for improving T_e. However, opportunities for improvement of photosynthetic rates of C_4 species are very limited because the efficiency for CO_2 fixation of this biochemical pathway is already quite large. As discussed by Stanhill (1986), if the carbon metabolism responsible for the high net dry matter production rates of C_4 species could be transferred to the less efficient C_3 species, T_e of these plants would be improved in a sizeable magnitude. However, such possibility would require a major technological breakthrough and, therefore, we must assume that major improvement of T_e through biochemical alteration is unlikely for now.

Plant genetic alteration through plant breeding programmes could allow some improvement of T_e. Richards (1987) listed a number of traits that could conceivably increase T_e. They can be separated into completely inherited physiological traits, such as carbon isotope discrimination, and simpler morphological traits.

Farquhar et al. (1982) suggested that T_e and isotopic discrimination against ^{13}C during CO_2 assimilation should be correlated for C_3 species, which has been verified experimentally for various crops (Wright et al., 1988). The experimental confirmation of the relationship between T_e and isotopic discrimination supported the possibility of using this parameter as a criterion to exploit variation in T_e in breeding programmes (Wright et al., 1988).

Morphological traits include such traits as glaucousness, erect leaf postures, small leaf size, which might reduce transpiration for a given photosynthetic rate. Another approach is to increase T_e, expressed as the ratio of yield over transpiration, by selecting genotypes that increase the harvestable fraction of the total crop biomass (harvest index).

Although plant breeding technology may produce some improvement, the T_e gains attainable through these physiological and morphological traits are unlikely to be significant.

Chemicals can be used to reduce transpiration at the leaf-air interface. These materials can be grouped into three major categories (Stark and Dwelle, 1989) : a) impermeable film coatings that restrict water vapour diffusion by increasing boundary layer resistance, b) reflective coatings that reduce the absorption of radiant energy by the foliage, and c) stomatal control agents that induce stomatal closure. These materials will improve T_e only if they allow a certain degree of decoupling of transpiration from photosynthesis.

The use of water vapour impervious films and stomatal control agents, collectively referred to as antitranspirants, is based on the concept that the total diffusion pathway to water is less than that of CO_2, which has an important additional liquid-phase resistance up to the chloroplasts. Therefore, it is to be expected that any increases in the stomatal, cuticular, and boundary layer resistances will reduce transpiration more than carbon exchange (Stanhill, 1986). In the case of reflectant materials, gains on T_e are only possible if the radiation load reduction is spectrally selective and the crop canopy is light saturated. Otherwise, both transpiration and CO_2 fixation will be reduced (Stanhill, 1986).

Although successful use of antitranspirants and reflectants has been reported (e.g., Hussin and Basiouny, 1984; Kobbia and Ibrahim, 1986), results are inconsistent, particularly in the case of impermeable film coatings and reflectants (Stark and Dwelle, 1989). These materials present difficult problems in coverage, while stomatal-control agents seem more promising in that much smaller amounts of material are needed (Loomis, 1983) and often do not require complete coverage. Because massive application of antitranspirant and reflectant materials to field crops is often economically impractical, their use is generally limited to high-value crops.

In summary, significant improvement of T_e by the different avenues discussed is unlikely, unless a major technological breakthrough allowed to incorporate C_4-like carbon metabolism into C_3 species. It is possible that the impact on T_e of the expected increase of the global atmospheric CO_2 concentration could be of more significance than any attempt of improving T_e through management or genetic alteration.

IV. TRANSPIRATION FRACTION OF EVAPOTRANSPIRATION

The value of T_e determines the upper limit of IWUE, i.e., the maximum amount of biomass that can be produced per unit of irrigation water applied. To reach this limit, all irrigation water applied must be stored in the crop root zone, and all this storage must be utilized for crop transpiration. Thus, the transpiration fraction of total ET, [T/(E+T)], is an important component determining maximum IWUE.

Soil evaporation (E) is maximum when the soil surface is wet. Then, its rate is only limited by the availability of energy. As the soil dries, E declines rapidly because soil conductivity for movement of water from lower depths declines to a very low value. E is now supply limited.

Canopy cover reduces E because it reduces the amount of energy reaching the soil surface. Therefore, as the crop canopy develops, [T/(E+T)] increases and the soil evaporation fraction declines. Experimental data by Ritchie and Burnett (1971) for two row crops (cotton and grain sorghum) with soil water not limiting showed that [T/(E+T)] was close to 0.4 at a leaf area index (LAI) of 0.5, about 0.55 at LAI of 1, close to 0.8 at LAI of 2, and 0.9 to 0.95 at LAI of 3, when full crop canopy closure was estimated to occur.

The soil evaporation can be a sizeable fraction of the seasonal ET. Cooper et al. (1987a) indicated that E was about 35 and 55% of the seasonal ET for dryland crops of wheat and barley in northern Syria. In a study including 61 sites in the South Australian wheat belt, French and Schultz (1984) estimated that soil evaporation losses made up 33% of seasonal ET. Klocke et al. (1985) reported average E fraction of ET values of 18, 22, and 29% for irrigated sorghum, soybeans (*Glycine max* L.), and corn (*Zea mays* L.), respectively, in Nebraska, USA.

Management practices increasing [T/(E+T)] will increase the overall IWUE. Because soil evaporation losses are low after the canopy completely closes, improvements of [T/(E+T)] should be obtained prior to full canopy development. One approach is selecting cultivars with rapid growth of leaves early in the season. Plant population and spacings can also be manipulated to increase early ground coverage. Substantial increase of average ground coverage during early growth through genetic selection or manipulation of plant population is not likely unless, in the case of the latter, suboptimal densities for a given water supply are utilized. Certainly, the effect of these practices in improving [T/(E+T)] is difficult to evaluate experimentally. It seems, however, that the irrigation frequency during early growth may have a larger impact on [T/(E+T)] than small improvements in average ground coverage. A detailed micrometeorological study in cotton by Ham et al. (1991) showed that this ratio was 0.19 at a crop LAI of 1.7 and dry soil and increased to 0.54 at LAI for the same crop of 2.2 and wet soil.

Table 7.2 : Water-use efficeincy (WUE), based on evapotranspiration (ET)*, of some agricultural crops.

Species	ET-Based WUE (kg ha^{-1} mm^{-1})	Harvested Portion (**)	Environment	References
Alfalfa (*Medicago sativa L.*)	22.5–28.5	A	Cypress	Guitijens, 1990
	9.5–15.7	A	Nevada, USA	
	9.8–10.9	A	Imperial Valley, CA, USA	
	12.1–23.1	A	North Dakota, USA	
(Vernal)	8.0–20.0	A	dryland	Badaruddin & Meyer, 1989
Barley (*Hordeum vulgare L.*)	28.1–32.5	A	Three sites in Syria, rainfed	Shepherd et al., 1987
	19.2–31.0	A		
Beans (Pinto) (*Phaseolus vulgaris L.*)	10.0–14.0	A	Kansas, USA, irrigated	Hattendorf et al., 1988
	4.9	G		
	15.8	A		

Table contd...

Species	ET-Based WUE (kg ha^{-1} mm^{-1})	Harvested Portion (**)	Environment	References
Cabbage (*Brassica oleracea L.*)	275.0	Head	Jodhpur, India, irrigated	Singh, 1978
Cauliflower (*Brassica oleracea L.* var. *Botrytis* L.)	216.7	Head	Jodhpur, India, irrigated	Singh, 1978
Corn (*Zea mays* L.)	12.0	G	Bushland, Texas, USA	Rhoads & Bennett, 1990
	22.0	G	Israel	
	17.0	G	Davis, California, USA	
	30.0	G	Southeast USA, humid	
	35.6	A	Kansas USA, irrigated	Hattendorf et al., 1988
	12.0	G	south Africa, irrigated	Bennie and Botha, 1986
	8.4–14.3	G	Bushland, Texas, USA, irrigated	Eck, 1986
	32.0	G	Nebraska, USA irrigated	Hatlitgil et al., 1984

Table contd...

Species	ET-Based WUE (kg ha^{-1} mm^{-1})	Harvested Portion (**)	Environment	References
Corn (*Zea mays* L.)	10.7–12.8	G	Kansas, USA, irrigated	Mtui et al., 1981
	7.3–18.9	G	Kansas, USA, dryland	
	4.8–7.7	G	South Dakota, USA dryland	Olson, 1971.
	8.4–20.0	G	Minnesota, South and North Dakota, USA, dryland	Timmons et al., 1966
Cotton (*Gossypium hirsutum* L.)	8.0–9.0	Seed	Spain, irrigated	Aranda, 1979
Cowpea (*Vigna unguiculata* L.)	14.3	A	Northwestrn India, irrigated	Phogat et al., 1984
Fababean (*Fabaceae leguminous* L.)	8.0–14.0	A	North Dakota, USA, dryland	Badaruddin & Meyer, 1989
Lentil (*Lens culinaris*)	6.0–11.0	A	North Dakota, USA, dryland	Badaruddin & Meyer, 1989
	3.4–8.6	A	Syria, dryland	Cooper et al., 1987

Table contd...

Species	ET-Based WUE (kg ha^{-1} mm^{-1})	Harvested Portion (**)	Environment	References
Millet (Pearl) [*Pennisetum americanum* (L.) Leek]	5.2	G	Kansas, USA irrigated	Hattendorf et al., 1988
	31.9	A		
Mung bean (*Phaseolus aureus*)	11.9	A	North-western India, irrigated	Phogat et al., 1984
Mustard (*B. juncea* L.)	6.1	G	Hissar, India, irrigated	Singh et. al., 1985
Potatoes (*Solanum tuberosum* L.)	93.7	Tuber	Jodhpur, India, irrigated	Singh et al., 1978
Radish (*Raphanus sativas* L.)	62.0–71.0	Root	Bangalore, India, irrigated	Hedge, 1987
Red clover (*Trifolium pratense* L.)	8.0–18.0	A	North Dakota, USA, dryland	Badaruddin & Meyer, 1989
Sewan grass (*Lasiurus sindicus* Hens.)	42.0	Green biomass	Jodhpur, India, irrigated	Singh et al., 1990
Sorghum (*Sorghum bicolor* L.)	32.4	A	Kansas, USA, irrigated	Hattendorf et al., 1988
Soybeans (*Glycine max* L.)	11.0–14.0	A	North Dakota, USA, dryland	Badaruddin & Meyer, 1989

Table contd....

Species	ET-Based WUE (kg ha^{-1} mm^{-1})	Harvested Portion (**)	Environment	References
Soybeans (*Glycine max L.)*	4.6–5.8	G	Bushland Texas, USA	Singh & Malik, 1983
	5.8–7.0	G	Nebraska, USA, dryland	Clawson et al., 1986.
	5.3–7.0	G	Kansas, USA, dryland	Singh & Whitson, 1976
	5.1–6.0	G	Minnesota, USA	Timmons et al., 1967
	5.5	G	Kansas, USA, irrigated	Hattendorf et al., 1988
	16.2	A	Kansas, USA, irrigated	
Sugarcane (*Saccharum* spp. *hybrid*)	36.4–45.5	A	Florida, USA, irrigated	Shih, 1988
Sunflower (*Helianthus annuus* L.)	2.4	G	Bushland, Texas, USA, dryland	Jones, 1984
	4.2	G	Kansas, USA, irrigated	Musick et al., 1976
	20.5	A	Kansas, USA, irrigated	Hattendorf et al., 1988.
Tomatoes (*Lycopersicon lycopersicum* L. karst.)	224.0	Fruit	Jodhpur, India, irrigated	Singh, 1978

Table contd...

Species	ET-Based WUE (kg ha^{-1} mm^{-1})	Harvested Portion (**)	Environment	References
Turnip (*Brassica rapa* L.)	218.0	Root	Jodhpur, India, irrigated	Singh, 1978
Wheat (*Triticum aestivum* L.)	5.1–7.7	G	Texas, USA rainfed	Hatfield et al., 1988
	5.3–12.2	G	Texas, USA irrigated	
	8.8–9.7	G	Kansas, USA, rainfed	
	5.8–9.0	G	Kansas, USA, irrigated	
	10.8–11.9	G	Nebraska, USA, irrigated	
	10.5 10.8	G	Nebraska, USA, irrigated	
	7.2–9.0	G	North Dakota, USA, rainfed	
	8.3–11.0	G	North Dakota, USA, irrigated	
	10.2–12.6	G	Alberta, Canada, rainfed	
	10.8–14.9	G	Alberta, Canada, irrigated	

Table contd...

Species	ET-Based WUE (kg ha^{-1} mm^{-1})	Harvested Portion (**)	Environment	Refrences
Wheat (*Triticum aestivum* L.)	5.1–12.4	G	Bushland, Texas, USA	Musick et al., 1976
	10.0–12.0	G	Several sources, irrigated	Musick & Porter, 1990
	3.5		Dryland	
	8.0–10.0	G	Great Plains, Washington Utah	
	6.6–7.6	G	Jodhpur, India, irrigated	Niwas, 1975
	10.0	G	Jodhpur, India, irrigated	Singh et al., 1976

(*) ET-BASED WUE = Biomass/ET or Yield/ET
(**) A = Aboveground biomass G = Grain biomass

The use of mulches is another management practices that may reduce soil evaporation throughout the growing season. Mulches reduce the soil energy absorption and/or restrict vapour flow. Mulch contribution to E reduction is not significant if water applications are infrequent. Plastic mulches can reduce soil evaporation significantly, but due to high cost their use is limited to high-value crops.

Values of water-use efficiency based on ET, this is on the ratio of biomass or yield per unit of evapotranspired water, are presented in Table 7.2. Following the definition given by Eq. 7.8, these values correspond to T_e [T/(E+T)]. They determine the maximum attainable IWUE if all irrigation water applied was to remain in the crop root zone, available for ET.

V. EVAPOTRANSPIRATION FRACTION OF APPLIED IRRIGATION WATER

The objective of irrigation is to supply the crop root zone with enough water to avoid undesired stress that would limit productivity. This is usually done by maintaining sufficient water in the root zone to provide for plant water uptake at a rate large enough to meet atmospheric evaporative demand. In supplying the root zone, several water losses may occur. When irrigation methods are compared, the main two losses to be considered are surface runoff and deep percolation. Other losses may include evaporation during the application and drifting by wind beyond field boundaries. For the discussion hereafter, relevant water losses when irrigation systems as a whole are considered (e.g., reservoir seepage, conveyance losses, etc.) will not be addressed.

For field irrigation, the total amount of irrigation water applied (I) corresponds to ET plus the amount of water required to compensate for application losses. The main factor determining the magnitude of these losses is the irrigation method utilized to apply the water. The ratio (E+T)/I, the irrigation application efficiency, will affect the overall value of IWUE (Eq. 7.8).

The irrigation application efficiency of drip irrigation systems is high because runoff, percolation, and evaporation losses can almost be eliminated. Drip irrigation efficiencies of around 90% can be achieved. Sprinkler irrigation systems can also achieve high irrigation application efficiencies, 60 to 90%, because runoff and deep percolation can be minimized. Centre pivot systems have efficiencies of around 85%. Irrigation application efficiency of surface irrigation systems without reuse systems are about 60%, while automated gated pipe systems with reuse may attain efficiencies of 90% (Benson et al., 1981). Application efficiency of irrigation systems can be reduced drastically if they are not properly designed and/or operated.

For systems operating under conditions where salinity is a concern, a common occurrence in arid irrigated lands, leaching requirements to maintain the soil salt balance must be met. This necessary practice will however increase the amount of water percolated beyond the root system depth, will decrease irrigation application efficiency, and therefore, will reduce the attainable IWUE.

VI. IMPROVING IRRIGATION METHODS AND MANAGEMENT IS AN IMPORTANT FACTOR FOR IWUE IMPROVEMENT

As expressed by Eq. 7.8, improvement of IWUE requires improving the transpiration efficiency (T_e), maximizing the transpiration fraction of evapotranspiration, [T/(E+T)], or/and maximizing the evapotranspiration fraction of the irrigation water applied, [(E+T)/I]. As discussed, substantial improvement of T_e is not likely. The improvement of [T/(E+T)] through increased early ground coverage is not expected to be significant, unless suboptimal plant populations are utilized. However, significant progress can be achieved by improving [(E+T)/I] and [T/(E+T)] through proper selection, design, and operation of irrigation methods.

Well-designed irrigation methods differ in their application efficiency, and therefore in the magnitude of IWUE that they can achieve. When conventional furrow irrigation and drip irrigation are compared, IWUE of drip irrigation may double that of furrows (Sanders et al., (1989). Singh et al. (1978) reported that drip irrigation was capable of providing the same yield of potatoes (*Solanum Tuberosum* L.) with half the water needed for furrow irrigation, therefore doubling the IWUE. A similar results was reported by Cevik et al. (1988) in a comparison of basin and drip irrigation of a banana orchard.

Subsurface irrigation can also achieve high IWUE. Sepaskhah et al. (1976) reported more than twofold increase of IWUE when comparing subsurface and furrow irrigation of beans (*Phaseolus vulgaris* L.). Shih (1988) found slightly higher, but statistically significant, IWUE when subsurface irrigation was compared with micro irrigation in sugarcane (*Saccharum* spp. hybrid). The selection of the irrigation method is, therefore, a critical factor in determining IWUE.

The proper design and operation of the selected irrigation method is another important factor for improving IWUE. During the period 1978-1981, 278 centre-pivot and 33 side roll sprinkler systems were evaluated in the High Plains and Winter Garden areas of Texas, USA (Texas Department of Water Resources, 1983). In these evaluations, the irrigation application efficiency of each sprinkler system was defined as the average can-catch water depth of the entire area irrigated by the system divided by the equivalent depth of water pumped into the system. As such, these efficiencies did not include runoff from the irrigated fields and percolation losses which, if considered, would reduce the reported values.

Results from these evaluations in Texas showed that the average application efficiency measured in centre-pivot system was 83%, with a range of 45 to 100 %, and with 16% of the systems operating at less than 70% efficiency. Efficiencies measured in side-roll systems averaged 74%, with a range of 35 to 100%, and with 21% of the systems operating at less than 60% efficiency. Considering that these systems may achieve efficiencies of 85 to 90% when properly designed and operated, it seems clear that large improvements in application efficiency and, therefore, in IWUE (Eq. 7.8) are possible through proper design and operation. Simple modifications in the operation of 17 of the evaluated systems, consisting of matching operating and systems requirement pressures and of replacing nozzles with the correct size, improved application efficiencies by 10%.

In another example, traditional and improved surface irrigation methods and irrigation scheduling were compared at field scale in central Chile (Gurovich, 1979). The study included three sites and years and 11 crops. As the result of simple improvements, average yields were increased by 17%, the average amount of water applied was reduced from 14,535 to 9,995 $m^3 ha^{-1}$, and the average IWUE was increased 70%.

It is interesting to note, however, that the irrigation method in itself, when runoff losses are disregarded and irrigation frequencies of compared methods are not too different, has little influence on IWUE. For example, Hedge (1987) reported no difference in IWUE between basin and furrow irrigation of radish (*Raphanus sativas* L.), Ellies et al. (1986) found no IWUE difference between graded furrow, level furrow, sprinkler, and trickle irrigation of onions, and Ravelo et al. (1977) measured similar IWUE between trickle and sprinkler irrigation of grain sorghum.

Irrigation scheduling, this is the decision of when and how much water to apply, may also have an influence on IWUE. For example, irrigation scheduling providing excessive irrigation will result in increased water losses, decreased [(E+T)/I], and decreased IWUE.

Schedules aimed to replenish crop ET with frequent irrigations seem to increase IWUE. Frequent irrigation keeps a high soil water potential, which helps to assure that plant water uptake will meet the atmospheric evaporative demand, avoiding temporary stress during the high-demand hours of the day. This contributes to increased transpiration and biomass production, particularly in hot and dry environments, provided that frequent irrigation does not limit soil aeration. In addition, reduced volumes of water applied in each irrigation, reduce the likelihood of runoff and deep percolation losses. These losses become non-existent with properly managed drip irrigation.

Although frequent irrigation increases the opportunity of soil evaporation before canopy closure, all together, the experimental evidence seems to suggest that the net result of frequent irrigations is toward increased IWUE, a conclusion that is independent of the irrigation method utilized provided that soil aeration is not limited. Freeman et al. (1976), experimenting with tomatoes (*Lycopersicon esculentum* Mill) in Griffith, Australia found a consistent trend towards increasing yields and IWUE with reduced irrigation intervals (more frequent irrigations) for both, drip and furrow irrigation. They cited the work of Jobling (1974), who summarized the results of 15 trials in which various crops were studied and found a similar positive trend with decreased irrigation interval. Srinivas et al. (1989) also reported increased yields and IWUE of furrow and drip irrigated watermelons when irrigation frequency was increased.

It also seems that deficit irrigation, this is the practice of replenishing less than 100% crop ET, may also increase IWUE.

Although deficit irrigation may lead to decreased transpiration and yield, the increase of [(E+T)/I] due to the reduction of runoff and percolation losses may counteract this effect and increase IWUE. Bielorai (1982) compared drip and sprinkler irrigation of a mature grapefruit grove applying two amounts of water, 80% and 100% of seasonal ET. He found that, despite yield reductions of 11 and 13% for drip and sprinkler irrigation, respectively, IWUE was increased due to the 20% reduction in applied water. Similarly, Srinivas et al. (1989) reported yield reductions of drip irrigated watermelons but increase of IWUE as irrigation amounts were reduced to replenish 100, 75, 50 and 25% of pan evaporation.

VII. CONCLUDING REMARKS

The overall irrigation water-use efficiency (IWUE) corresponds to the ratio of biomass (or marketable yield) produced per unit of total irrigation water applied. The transpiration efficiency (T_e), this is the ratio of biomass (or marketable yield) produced per unit of water transpired by the crop, set the physiological upper limit of IWUE. This limit will be reduced by various sources of water-losses in the chain of events between irrigation water deliver and crop transpiration. Significant improvement of T_e is unlikely with the exception, perhaps, of the possible increase of plant photosynthetic rates and reduction of transpiration that may result from the increasing trend of the global atmospheric CO_2 concentration. The magnitude of the transpiration fraction of the water evaporated from irrigated fields modifies the potential IWUE determined by T_e. Management aimed to maximize crop transpiration and minimize soil evaporation would increase IWUE. However, again, not significant improvement of ET partitioning into transpiration seems likely unless suboptimal plant populations with respect to applied water are optimized. The fraction of irrigation water applied that is made available for evapotranspiration (i.e., the irrigation application efficiency) also modifies the upper limit of IWUE set by T_e. Because the application efficiency fluctuates greatly with different irrigation methods, and with their design and operating conditions, improvement of this efficiency seems to offer large potential for improvement of IWUE.

REFERENCES

Aranda, J.M. 1979. *In* E. Welte (ed.). Proc. water and fertilizer use for food production in arid and semiarid zones. p. 177-183. Nov. 26-Dec. 1, Garyounis Univ., Benghazi, Libya.

Arkley, R.J. 1963. Hilgardia 34:559-584.

Badaruddin, M. and D.W. Meyer. 1989. Crop Sci. 29:1212-1216.

Bennie, A.T.P. and F.J.P. Botha. 1986. Soil & Till. Res. 7:85-95.

Benson, V.W., C.A. Everson, and R.L. Sharp. 1981. U.S. Dept. of Agric., Econ. Res. Serv. ERS-670. 73 p.

Bielorai, H. 1982. Irrig. Sci. 3:89-100.

Bierhuizen, J.F. and R.O. Slatyer. 1965. Agric. Meteorol. 2:259- 270.

Briggs, L.J. and H.L. Shantz. 1914. J. of Agric. Res. (Washington, D.C.) 3:1-63.

Briggs, L.J. and H.L. Shantz. 1913. U.S. Bur. Plant Indus. Bull. 284, 49 p.

Bristow, K.L. 1983. Ph.D. dissertation, Washington State University. 74 p.

Brun, L.J., J.W. Enz, J.K. Larsen, and C. Fanning. 1986. J. Soil Water Cons. 41:120-122.

Cevik, B., N. Kaska, O. Tekinel, M. Pekmezci, N. Yaylali, and S. Paydas. 1988. Acta Hort. 228:213-218.

Clawson, K.L., J.E. Speiht, B.L. Blad, and A.F. Garay. 1986. Agron. J. 78:483-487.

Cooper, P.J.M., P.J. Gregory, D. Tully, and H.C. Harris. 1987a. Exp. Agric. 23:113-158.

Cooper, P.J.M., P.J. Gregory, J.D.H. Keatinge, and S.C. Brown. 1987b. Field Crops Res. 16:67-84.

Day, W., B.J. Legg, B.K. French, A.E. Johnstone, D.V. Lawlor, and W. de Jeffers. 1978. J. of Agric. Sci. (Cambridge) 91:599-623.

de Wit, C.T. 1958. Versl. Landbouwk. Onderz. 64.6 Inst. of Biol. and Chem. Res. on Field Crops and Herbage, Wageningen, The Netherlands.

Eck, H.V. 1986. Agron. J. 78:1035-1040.

Ellis, J.E., E.G. Kruse, A.E. McSay, C.M.U. Neale, and R.A. Horn. 1986. Hort. Sci. 21(6):1349-1351.

Farquhar, G.D., M.H. O'Leary, and J.A. Berry. 1982. Austr. J. Plant Phys. 9:121-137.

Flowers, T.J., F.M. Salama, and A.R. Yeo. 1988. Plant, Cell and Env. 11:453-459.

Freeman, B.M., J. Blackwell, and K.V. Garzoli. 1976. Agric. Water Manage. 1:21-31.

French R.J. and J.E. Schultz. 1984. Austr. J. Agric. Res. 35:743- 764.

Goudriaan, J. and M.H. Unsworth. 1990. Pages 111-130. *In* Impact of Carbon Dioxide, Trace Bases, and Climate Change on Global Agriculture. ASA Special Publication No. 53.

Gregory, P.J. 1984. Outlook in Agric. 13:208-215.

Guitjens, J.C. 1990 *In* B.A. Stewart and D.R. Nielsen (eds.). Pages 537-568. Agronomy Monograph No. 30. ASA-CSSA-SSSA, Madison, WI.

Gurovich, L.A. 1979. Irrig. Sci. 1:97-105.

Ham, J.M., J.L. Heilman, and R.J. Lascano. 1991. Agron. J. 83:744-753.

Hatfield, J.L., A. Bauer, E.T. Kanemasu, D. Major, B.L. Blad, R.J. Reginato, and K.G. Hubbard. 1988. Agric. and For. Meteorol. 44:187-195.

Hatlitgil, M.B., R.A. Olson, and W.A. Compton. 1984. Fertilizer Res. 5:321-333.

Hattendorf, M.J., M.S. Redelfs, B. Amos, L.R. Stone, and R.F. Gwln. 1988. Agron. J. 80:80-85.

Hedge, D.M. 1987. Irrig. Sci. 8:13-22.

Heitholt, J.J. 1989. Agron. J. 81:464-469.

Howell, T.A., K.R. Davis, R.L. McCormick, H. Yamada, V.T. Walhood, and D.W. Meek. 1984. Irrig. Sci. 5:195-214.

Hussin, M.M. and F.M. Basiouny. 1984. Proc. of Florida State Hort. Soc. 97:348-350.

Idso, S.B. 1990. *In* Impact of Carbon Dioxide, Trace Bases, and Climate Change on Global Agriculture. ASA Special Publication No. 53. Pages 61-69.

Jobling, G.A. 1974. Publ. New Zealand Agric. Eng. Inst., Lincoln College, Canterbury, N.Z. 68 p.

Jones, O.R. 1984. Agron. J. 76:229-235.

Kallsen, C.E., T.W. Sammis, and E.J. Gregory. 1984. Agron. J. 76:59-64.

Kiesselbach, T.A. 1916. Nebraska Agric. Exp. Stn. Res. Bull. No. 6. Lincoln, NE.

Kimball, B.A. 1983. Water Cons. Lab. Rept. 14. U.S. Dept. of Agric./ARS. Phoenix, AZ.

Kimball, B.A. and S.B. Idso. 1983. Agric. Water Manage. 7:55 72.

King, K.M. and D.H. Greer. 1986. Agron. J. 78:515-521.

Klocke, N.L., D.L. Martin, and D.F. Heermann. 1985. *In* Amer. Soc. of Agric. Engr. Pages 335-341. Proc. National Conf. on Advances in Evapotranspiration, Dec. 16-17, Chicago, IL.

Kobbia, T.E. and A. Ibrahim. 1986. Acta Hort. 190:281-289.

Loomis, R.S. 1983. *In* H. Taylor et al. (eds.). Pages 345- 374. ASA-CSSA-SSSA, Madison, WI.

Mtui, T.A., E.T. Kanemasu, and C. Wassom. 1981. Agron. J. 73:639- 643.

Musick, J.T., L.L. New, and D.A. Dusek. 1976. Trans. ASAE 19:489- 493.

Musick, J.T. and K.B. Porter. 1990. *In* B.A. Stewart and D.R. Nielsen (eds.). Pages 597-638. Agronomy Monograph No. 30. ASA-CSSA-SSSA, Madison, WI.

Niwas, Ram. 1976. Ph.D. Thesis, University of Jodhpur (India).

Olsdon, T.C. 1971. Agron. J. 63:104-106.

Phogat, B.S., D.P. Singh, and Phool Singh. 1984. Irrig. Sci. 5:47-60.

Ravelo, C.J., E.A. Hiler, and T.A. Howell. 1977. Trans. ASAE 20:96-99/104.

Rhoads, F.M. and J.M. Bennett. 1990. *In* B.A. Stewart and D.R. Nielsen (eds.). Pages 569-596. Agronomy Monograph No. 30. ASA-CSSA-SSSA, Madison, WI.

Richards, R.A. 1987. Pages 133-149. *In* Srivastava et al. (eds.). ICARDA. John Wiley & Sons.

Ritchie, J.T. and E. Burnett. 1971. Agron. J. 63:56-62.

Sanders, D.C., T.A. Howell, M.M.S. Hile, L. Hodges, D. Meek, and C.J. Phene. 1989. J. Amer. Soc. of Hort. Sci. 114:904-908.

Sepaskhah, S.A., S.A. Sichani, and B. Bahrani. 1976. Trans. ASAE 19:1089-1092/97.

Shepherd, K.D., P.J.M. Cooper, A.Y. Allan, D.S.H. Drennan, and J.D. H. Keatinge. 1987. J. Agric. Sci. (Cambridge) 108:365-378.

Shih, S.F. 1988. Trans. ASAE 31:142-148.

Sinclair, T.R., C.B. Tanner, and J.M. Bennett. 1984. Biol. Sci. 34:36-40.

Singh, B.P. and E.N. Whitson. 1976. Agron. J. 68:834-835.

Singh, D.P., Phool Singh, Pratap Singh, Maharaj Singh, and H.C. Sharma. 1985. Indian J. Agron. 30:59-63.

Singh, S.D. 1978. Agron. J. 70:951-954.

Singh, S.D., J.P. Gupta, and P. Singh. 1978. Agron. J. 70:948- 951.

Singh, S.D., M. Yusuf, R.C. Bhandari, and H.S. Dauley. 1976. Indian J. agric. Sci. 46:567-575.

Singh, S.D., Y.V. Singh, and K.C. Singh. 1990. Indian J. agric. Sci. 60:23-28.

Singh, T. and D.S. Malik. 1983. Irrig. Sci. 4:239-245.

Srinivas, K., D.M. Hedge, and G.V. Havanagi. 1989. J. Hort. Sci. 64:115-124.

Stanhill, G. 1986. Adv. in Agron. 39:53-85.

Stark, J.C. and R.B. Dwelle. 1989. Amer. Potato J. 66:563-574.

Tanner, C.B. and T.R. Sinclair. 1983. *In* H. Taylor et al. (eds.). Pages 1-27. ASA-CSSA-SSSA, Madison, WI.

Texas Department of Water Resources. 1983. LP-191. 107 p.

Timmons, D.R., R.F. Holt, and J.T. Moraghan. 1966 Agron. J. 58:429-432.

Timmons, D.R., R.F. Holt, and R.L. Thompson. 1967. Agron. J. 59:262-265.

Walker, G.K. 1986. Fields Crops Res. 14:29-38.

Walker, G.K. and J.E. Richards. 1985. Agron. J. 77:263-269.

Wright, J.L. and J.C. Stark. 1990. *In* B.A. Stewart and D.R. Nielsen (eds.). Pages 859-888. Agronomy Monograph No. 30. ASA-CSSA-SSSA, Madison, WI.

Wright, G.C., K.T. Hubick, and G.D. Farquhar. 1988. Austr. J. Plant Physiol. 15:815-825.

8

Trickle Irrigation Design And Evaluation Based On The Statistical Uniformity

Vincent F. Bralts and I-Pai Wu

I. INTRODUCTION

Irrigation as a whole, and drip irrigation in particular, has experienced unprecedented growth over the past decade in the United State, India, and around the world. In 1985 total area under drip irrigation in the United States was over 300,000 hectares (750,000 acres). This compares with only 50,000 hectares (125,000 acres) in 1970.

A drip irrigation system, Figure 8.1, consists of a water supply and pump followed by a network of mainlines and submains, laterals, and emitters. The mainline is the primary artery for delivery of water to the various irrigation zones. Within each zone there is usually a number of submain units. Submain units can consist of 1 to 5 hectares (2.5 - 12.5 acres), while a zone consists of 20 to 50 hectares (50 - 125 acres).

Hydraulic design is probably the most important factor in the ultimate success or failure of a drip irrigation system. Over the past decade a significant amount of research has been done in this area. To assist in the improved design of drip irrigation systems, Keller and Karmeli (1974) developed the Emission Uniformity Concept, Wu and Gitlin (1974) developed the Emitter Flow Variation Concept, and Bralts et al. (1981a and 1981b) developed the Statistical Uniformity Concept. The design of a drip irrigation submain unit for optimum emitter uniformity is very important, because once the emitters, lateral, and submain components have been chosen, very little additional flow control is possible. Thus, the engineer making the design decisions regarding pipe size and emitter selection must have a method of determining submain unit emitter flow uniformity at the design stage.

In addition to the delivery of water, a drip irrigation system can be an effective system for the application of fertilizer and other chemicals to plants. It has been shown that a savings of fertilizer can be attained using drip irrigation systems (Kesner et al., 1985). In addition, drip irrigation systems can reduce labour and the energy cost of fertilizer application. Despite these advantages, the use of fertilizer injection through drip irrigation systems has not been fully realized. An important reason for the lack of development has been that irrigators were not certain that their drip irrigation systems were performing at an acceptable level of uniformity for fertilizer injection. The problem had been a lack of the simple user oriented field evaluation tools.

The field evaluation of drip irrigation submain units is important to the design engineer to confirm the successful design of a submain unit to the desired emitter flow uniformity specifications. Secondly, field evaluation is important to an irrigator in deciding if the submain unit performance is acceptable for fertilizer injection. Thirdly, field evaluation is important as a diagnostic tool for problem submain units.

In this chapter drip irrigation design and evaluation procedures based on the statistical uniformity concept will be developed. The coefficient of variation and the constant odds uncertainty formulation will be used to relate hydraulic design and field evaluation procedures for drip irrigation submain units. In particular hydraulic variation, manufacturer's variation, emitter plugging variation, and the number of emitters per plant will be related to submain unit performance. In addition the relationships of the statistical uniformity to irrigation and application efficiency and schedules will be discussed. Based on the above simplified

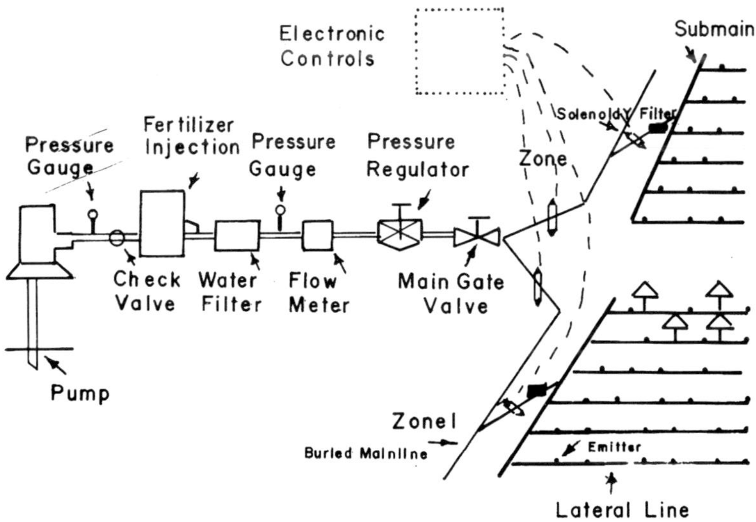

Fig 8.1. Drip irrigation system components

graphical techniques for hydraulic design and field evaluation of drip irrigation systems will be presented.

II. BASIC HYDRAULICS

The hydraulic principles of drip irrigation, based upon the classical equations of continuity and energy, have been developed by various researchers, (Wu and Gitlin, 1974; Howell and Huler, 1974a, 1974b; and Keller and Karmeli, 1974). The following development will closely follow the theory and nomenclature used by Wu et al. (1979).

A. Pipe Flow Equations

The flow in drip irrigation lateral or manifold pipe can be considered to be hydraulically steady, spatially varied pipe flow. This means that the total flow through the pipe is changing, usually decreasing, with respect to length. The pressure distribution or energy gradient line along the pipe is also changing, usually decreasing, as it is affected by friction and elevation. Figure 8.2

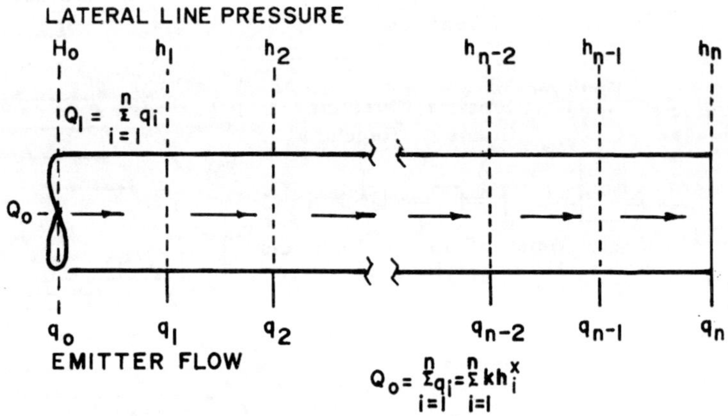

Fig 8.2. Water and pressure distribution along a drip irrigation lateral line (Wu and Gitlin, 1974).

represents the flow and pressure distribution along a drip irrigation lateral line.

By considering drip irrigation pipes as hydraulically smooth, any one of several empirical equations can be used to calculate head loss due to friction. In this chapter only two such equations will be discussed. The first equation which is based on the Darcy-Weisbach equation begins with

$$h_f = f\frac{L}{D}\frac{V^2}{2g} \tag{8.1}$$

where h_f = head loss due to friction, f = dimensionless friction factor, L = length of pipe, V = velocity of water in the pipe, D = diameter of the pipe, and g = acceleration of gravity.

Since drip irrigation lateral lines can be assumed to be hydraulically smooth and their flow fully turbulent, the Blasius empirical formula for turbulent flow in a smooth pipe can be substituted for f (Wu and Gitlin, 1974; Howell et al., 1981). The Blasius formula is,

$$f = \frac{0.3164}{R_e^{0.25}} \qquad (4{,}000 \le R_e \le 100{,}000) \qquad [8.2]$$

where f = friction coefficient and R_e = Reynold's number.

Walters and Keller (1978) combined equations [8.1] and [8.2] at 20 degrees Celsius and found

$$h_f = \frac{7.89 \times 10^5 Q^{1.75} L}{D^{4.75}} \qquad [8.3]$$

where h_f = head loss in metres, Q = flow rate in litres per second, D = pipe diameter in millimeters, and L – pipe length in metres.

The second empirical equation which is commonly used in hydraulic design is the Hazen-William's formula, (Keller and Karmeli, 1975; Jeppson, 1982). In equation form,

$$h_f = \frac{1.22 \times 10^{10}\ Q^{1.852}\ L}{C^{1.852}\ D^{4.871}} \qquad [8.4]$$

where C = pipe roughness coefficient, and all other variables previously defined. If a C value of 150 for smooth pipe is substituted into equation [8.4], we obtain the following empirical equation

$$h_f = \frac{11.38 \times 10^5\ Q^{1.852}\ L}{D^{4.871}} \qquad [8.5]$$

Both equations [8.3] and [8.5] are in the same units and result in very similar solutions to drip irrigation hydraulic problems. The major difference between the two equations is that the Darcy-Weisbach equation can be corrected for viscosity through the Reynold's number term in the Blasius equation while the Hazen-William's equation cannot. Figure 8.3 is a comparison of the two equations for 27 mm (1 inch pipe) and taken from Hughes and Jeppson (1978).

Howell et al. (1981) reviewed the work done by Hughes and Jeppson (1978) and commented that C values of 130 and 150 in the

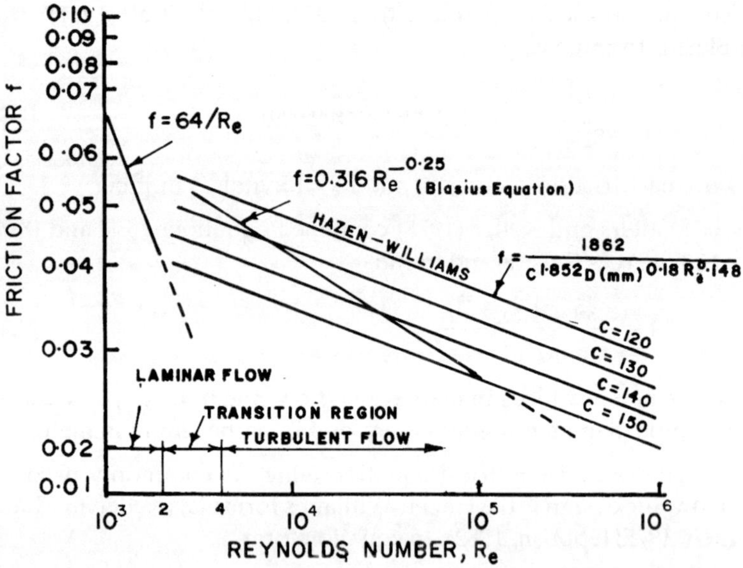

Fig 8.3. Moody diagram with fraction factors (Hughes and Jeppson, 1978).

Hazen- William's equation are clearly dependent upon the R_e. Furthermore, when comparing the two equations at the same velocity, the C value of the Hazen-William's equation seemed dependent upon pipe diameter. Howell et al. (1981) found C equal to 130 for 14-15 mm (0.58 in) plastic pipe, C equal to 140 for 18-19 mm (0.75 in) plastic pipe and C equal to 150 for 25-27 mm (1 in) plastic pipe. Generally, underestimating the C value results in a more conservative design.

Both equations [8.3] and [8.5] are generalizable into the form used by Wu and Gitlin (1975) and Wu et al. (1979), as

$$\Delta H = -a\, Q^{m} L \qquad [8.6]$$

where ΔH = head loss due to friction ($-h_f$), Q = total lateral line flow, a = a pipe constant, L = lateral line length, and m = the pipe flow exponent. Equation [8.6] will be used to approximate the energy grade line due to friction required in the next section.

B. Emitter Flow Equation

Drip irrigation emitters vary in their design from elaborate pressure compensating devices to long flow path and simple orifice type emitters. In general, the flow characteristics of emitters been shown by Karmeli (1977) and Wu et al. (1979) to be

$$q = kh^x \qquad [8.7]$$

where q = emitter flow rate, k = constant of proportionality, h = pressure head at the emitter, and x = emitter discharge exponent.

Equation [8.7] can be derived from a combination of the Bernoulli energy and continuity equations. The constant of proportionality, k, in the emitter flow equation contains the variables such as the coefficient of discharge, emitter geometry, and the acceleration of gravity.

The value of x in equation [8.7] characterized the type of emitter and/or the flow regime in a long flow path emitter. For example, orifice type emitters have an emitter discharge exponent of 0.5. In long flow path emitters x = 0.5 for fully turbulent flow and x = 1.0 for laminar flow (Karmeli, 1977). An emitter with an x value of less than 0.5 would be pressure compensating in nature (Wu et al., 1979). The ideal value of x is zero which would make the emitter fully pressure compensating.

A special form of the emitter flow equation when considering emitter plugging is

$$q = (1 - \alpha)\, kh^x \qquad [8.8]$$

where α = the degree of plugging and all other variables are as previously defined. Equation [8.8] will be used when considering design and evaluation procedures including emitter plugging.

C. Hydraulic Analysis Technique

Since the flow in the lateral line is spatially varied with decreasing discharge, the energy grade line is an exponential curve rather than a straight line. The solution of the energy gradient in drip irrigation submain units can be determined by the approximation of by iterative procedures. In this section the approximation procedure developed by Wu and Gitlin (1975) will be presented.

The dimensionless energy gradient procedure developed by Wu and Gitlin (1975) assumes that all emitters along a lateral line discharge the same flow. This approximation results in a curve

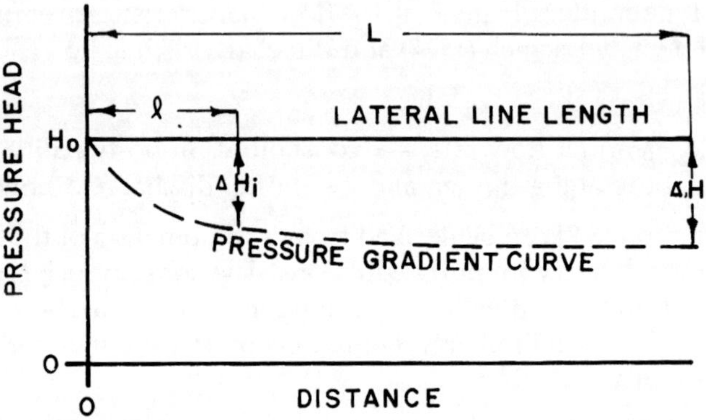

Fig 8.4. Lateral line notation.

which can be used to directly calculate the head loss at any point along any lateral line. When assuming constant emitter flow, along the lateral line, the shape of the energy grade line can be expressed dimensionlessly to be the energy drop ratio (R_i) as shown by Wu and Gitlin (1975), as

$$R_i = \frac{\Delta H_i}{\Delta H} = 1-(1-i)^{m+1} \qquad [8.9]$$

where i = ratio l/L. l, L, ΔHi and ΔH are defined in Figure 8.4 and m is as defined in equation [8.6]. If the Hazen-William's equation is used, then, equation [8.9] becomes

$$R_i = 1 - (1-i)^{2.852} \qquad [8.10]$$

Figure 8.5 illustrated the theoretical shape of the dimensionless energy gradient curves for various flow conditions (Wu and Gitlin, 1975).

The total pressure variation along a lateral line can be expressed as a combination of the original pressure and the variation due to energy slope and terrain. Expressed summarily, using the dimensionless energy gradient line concept and uniform slope, Wu et al. (1979) developed the equation

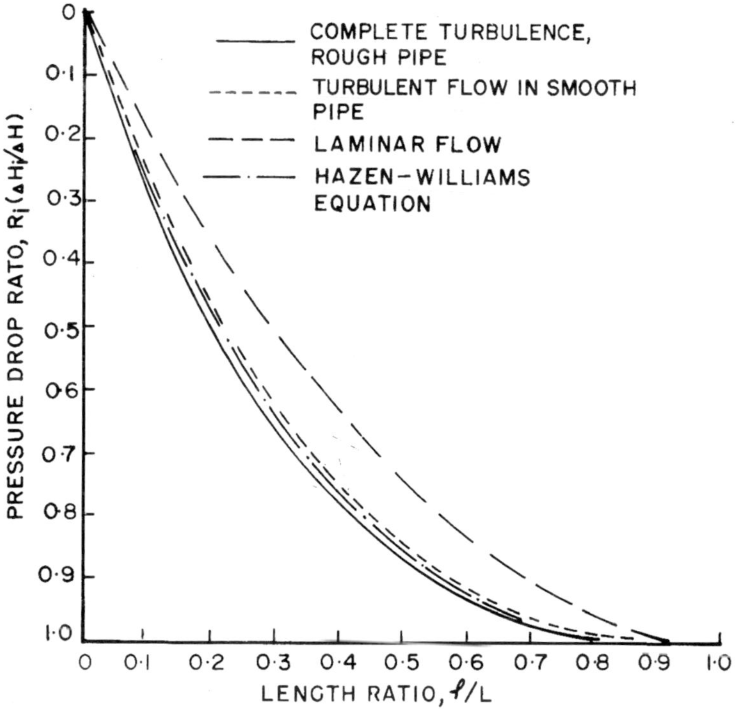

Fig 8.5. Dimensionless energy gradient curve (Wu and Gitlin, 1974).

$$h_i = H_o - R_i \Delta H \pm R_i' \Delta H' \qquad [8.11]$$

where h_i = pressure head for a given length ratio, H_o = pressure head at the origin, $R_i\Delta H$ = pressure head loss due to friction, and $R_i\ \Delta H$ = pressure head loss or gain due to elevation.

Figure 8.6a and 8.6b show schematically the pressure distributions due to various situations and the resulting energy gradient line.

Since the emitter flow is related to pressure (h), as shown in equation [8.7], the emitter flow rate at any point along the lateral line will be equal to a combination of equations [8.7] and [8.11], or combining we obtain

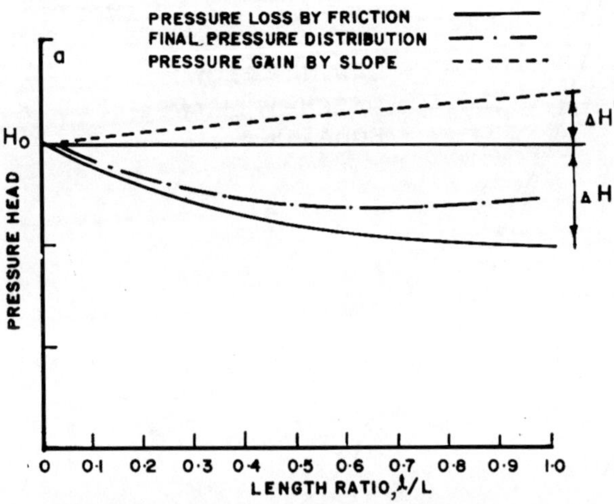

Fig 8.6a. Pressure distribution for down-slope conditions (Wu et al., 1979).

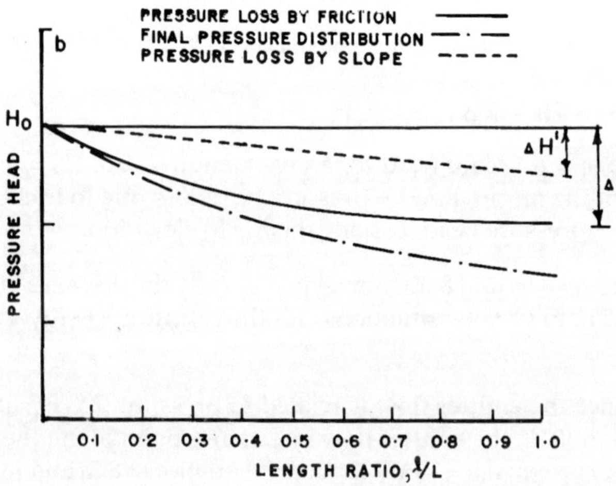

Fig. 8.6b. Pressure distribution for up-slope conditions (Wu et al., 1979).

$$q_i = kh_i^x = k(H_o - R_i\Delta H \pm R_i'\Delta H')^x \quad [8.12]$$

Furthermore, if equation [8.12] is divided by the emitter flow equation for the first emitter $q_o, q_o = kH_o^x$ then the resulting equation becomes independent of the coefficient k.

In simplified form equation [8.12] becomes

$$q_i = q_o\left(1 - R_i\frac{\Delta H}{H_o} \pm R_i'\frac{\Delta H'}{H_o}\right)^x \quad [8.13]$$

where all variables are as previously defined. With equation [8.13] the emitter flow can be calculated at various points along a lateral line once the emitter flow rate at the original pressure is known.

The dimensionless energy gradient line concept can also be used for drip irrigation submain manifolds where lateral lines are considered similar to uniformity spaced emitters. Using this concept, equation [8.11] can be rewritten as

$$h_i = H_s - R_j\Delta H \pm R_j'\Delta H' \quad [8.14]$$

where h_j = pressure head for a given submain manifold length ratio, H_s = original pressure at the head of the submain unit, $R_i\,\Delta H$ = submain manifold pressure head loss due to friction, and $R_j\,\Delta H$ = submain manifold pressure head loss or gain due to elevation.

Combining equations [8.11] and [8.14] to determine the pressure at any point in a submain unit results in

$$h_{ji} = H_s - R_j\,\Delta H_s - R_i\,\Delta H_i \pm R_j'\,\Delta H_s' \pm R_i'\,\Delta H_i' \quad [8.15]$$

where h_j = pressure head at the submain manifold length ratio (j) and the lateral line length ratio (i), $R_j\,\Delta H_s$ = submain manifold head loss due to friction, $R_i\,\Delta H_l$ = lateral line head loss due to friction, $R_j\,'\Delta H's$ = submain head loss or gain due to elevation, and $R_j\,'\Delta H'l$ = lateral line head loss or gain due to elevation.

Equation [8.15] can be modified to calculate the emitter flow relative to the emitter flow at the original pressure by the equation

$$q_{ji} = q_s\left(1 - R_j\frac{\Delta H_s}{H_s} - R_i\frac{\Delta Hl}{H_s} \pm R_j'\frac{\Delta H's}{H_s} \pm R_i'\frac{\Delta Hl'}{H_s}\right)^x \quad [8.16]$$

where all variables are as previously defined. Thus using equation [8.16] the approximate emitter flow at any point in the submain unit can be determined once the basic parameters of submain unit hydraulic system are known.

III. THE STATISTICAL UNIFORMITY CONCEPT

The statistical uniformity concept consists of a statistical approach to the uniformity of emitter flow and irrigation application efficiency based on the coefficient of variation. Statistically speaking the coefficient of variation is a term used to define the normalized dispersion of a random variable. The normalization process is much like the unit vector concept in engineering or physics and consists of dividing the standard deviation by the mean. The result is a normalized measure of dispersion which can be used to compare or combine the relative variation of different systems or subgroups. When used in conjunction with the constant odds formulation the coefficient of variation can be considered a measure of the relative magnitudes of various contributors to the variance.

In this section the statistical uniformity concept will be presented and expanded to incorporate the variation due to submain unit hydraulics, emitter manufacturing, emitter plugging, and the number of emitters per plant into the design procedures for drip irrigation systems.

A. Statistical Uniformity

The statistical uniformity was first presented by Wilcox and Swailes (1947) for the evaluation of sprinkler irrigation systems. The so called "statistical uniformity coefficient" wad based on the coefficient of variation and can be defined by the equation.

$$U_s = 100(1 - V_y) = 100\left(1 - \frac{S_y}{\bar{y}}\right) \qquad [8.17]$$

where U_s = statistical uniformity coefficient as a percentage, V_y = coefficient of variation of the depth of irrigation water, S_y is the standard deviation of irrigation depth, and y is the mean depth of water applied.

A similar statistical approach can be developed for drip irrigation systems where the random variable y, the depth of water in sprinkler irrigation is replaced by q (the emitter flow rate or volume from equation [8.7] and results in

$$U_s = 100\,(1 - V_q) = 100\left(1 - \frac{S_q}{\bar{q}}\right) \qquad [8.18]$$

where U_s = statistical uniformity coefficient as a percentage, V_q is the coefficient of variation of emitter flow, S_q is the standard deviation of emitter flow, and $\bar{q}$ is the mean emitter flow rate or volume.

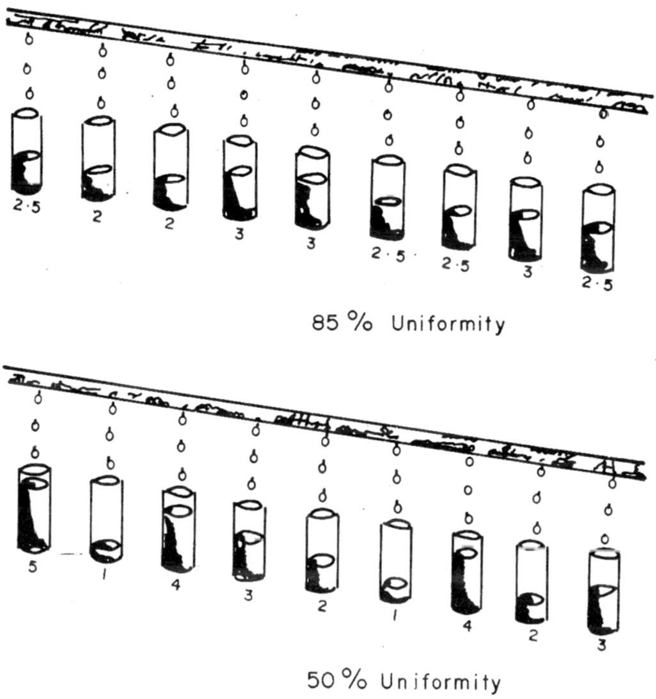

Fig 8.7. Illustration of the statistical uniformity.

The general criteria for an acceptable statistical uniformity U_s, are: 90% or greater, excellent; 80-90%, very good; 70-80%, fair; 60-70%, poor; and, less than 60%, unacceptable. The irrigation catch cans in Figure 8.7 illustrate the basis for the statistical uniformity criteria. From the volumes in the catch cans it should be obvious that 85% is a very acceptable uniformity whereas 50% is unacceptable.

Bralts et al. (1981a, b) recommended the use of the statistical uniformity coefficient for drip irrigation lateral lines. The justification for their recommendation was that all of the factors such as lateral line friction, elevation differences, emitter manufacturing, and emitter plugging could be included in the final uniformity estimate. Bralts et al. (1981a, b) illustrated that for single chamber tubing with orifice type emitters, the coefficients of hydraulic and manufacturer's variation were independent and could be combined into the total coefficient of variation for the lateral line by the equation.

$$V_{qt} = [V_{hl}^2 + V_{me}^2]^{0.5} \qquad [8.19]$$

where V_{ql} = total coefficient of variation, V_{hl} = hydraulic coefficient of variation for the lateral line, and V_{me} = manufacturing coefficient of variation for the emitters along the lateral line. The coefficient of total variation can then be used in equation [8.18] to determine the statistical uniformity of a drip irrigation lateral line.

B. Uncertainty Formulation

The variation of emitter flow in a drip irrigation submain unit is the result of a variety of factors. The primary factor is hydraulic design. Other important factors are emitter type, emitter manufacturing, emitter plugging, and the number of emitters per plant. For this development the variation of emitter flow in the submain unit will be determined as a probability combination of the relevant factors.

Uncertainty prediction for any quantity R which is a function of x_i's, a list of real variables ($R = f(x_1, x_2, \ldots, x_n)$), can be estimated in a worst case combination or a constant odds combination as by (Moffat, 1982) to be

Worst Case :

$$\delta R = \left[\frac{\delta R}{\delta x_1}\delta x_1\right] + \left[\frac{\delta R}{\delta x_2}\delta x_2\right] + \ldots + \left[\frac{\delta R}{\delta x_n}\delta x_n\right] \qquad [8.20]$$

Constant Odds:

$$\delta R = \left(\left[\frac{\delta R}{\delta x_1}\delta x_1\right]^2 + \left[\frac{\delta R}{\delta x_2}\delta x_2\right]^2 + \ldots + \left[\frac{\delta R}{\delta x_n}\delta x_n\right]^2\right)^{0.5} \qquad [8.21]$$

Both equations [8.20] and [8.21] were obtained by applying the Taylor series expansion to the function of R and retaining the linear terms.

Kline and McClintock (1953) reported that the constant odds formulation, assessed the uncertainty of R accurately for most engineering applications. When using the constant odds equation there are three prerequisites which must be met:

1. Each x_i is an independent variable.
2. Each x_i be from a Gaussian distribution.
3. The odds must be the same for each input.

If the variance of x_i is substituted for δx_i^2 in equation [8.21] then only the first and last prerequisites must be met.

In the case of a quantity R which is a known function of a product string $(R = f(x_1^a,\ x_2^b, \ldots, x_n^z))$, the constant odds combination becomes

$$\frac{\partial R}{R} = \left(\left[a\ \frac{\delta x_1}{x_1}\right]^2 + \left[b\ \frac{\delta x_2}{x_2}\right]^2 + \ldots + \left[z\ \frac{\delta x_n}{x_n}\right]^2\right)^{0.5} \qquad [8.22]$$

The statistical analogy to this relationship would be the substitution of the coefficient of variation for $\delta R/R$ and $\delta x_i/x_i$, respectively.

For a drip irrigation submain unit, the statistical equation for an unbiased estimate of the variance using emitter flow (q) as the random variable is,

$$S_q^2 = \sum_{u=1}^{n} \frac{(q_u - \bar{q})^2}{(n-1)} \qquad [8.23]$$

where S_q^2 = sample variance of the random variable q, q_u = total number of observations, u = subscript identifying a particular emitter or observation, q_u = random variable, and $\bar{q}$ = mean emitter discharge or flow.

The coefficient of variation (V_q) is a useful measure of emitter flow variation and can be used in conjunction with the statistical uniformity equations. The coefficient of variation is the standard deviation divided by the mean; or, in equation form

$$V_q = \frac{S_q}{\bar{q}} = \sqrt{\frac{S_q^2}{\bar{q}^2}} \qquad [8.24]$$

Application of the constant odds formulation to the emitter flow equation [8.7] results in

$$V_q = [V_k^2 + x^2 V_h^2]^{0.5} \qquad [8.25]$$

where V_q = coefficient of variation of emitter flow, V_k = coefficient of variation of the emitter constant k, x = emitter discharge exponent, and V_h = coefficient of variation of the hydraulic pressure h.

The terms V_k and V_h in equation [8.25] are the same as the emitter coefficient of manufacturer's variation (V_{me}) and the hydraulic variation (V_{hl}) as defined in equation [8.19]. The primary difference is that Bralts et al. (1981a) treated h^x as a single variable, whereas equation [8.25] does not. The major advantage of the above form is that V_h can be calculated directly from the hydraulic pressure heads, h_i, without first calculating emitter flows, q_i.

Similarly if we apply the constant odds formulation to the emitter flow equation including plugging (Equation [8.8]) the result is

$$V_q = [V_p^2 + V_k^2 + x^2 V_h^2]^{0.5} \qquad [8.26]$$

where V_p = coefficient of variation of emitter plugging, and all other variables are as previously defined. In equation [8.26] the subscript p has been substituted for (1 α) in equation [8.8]. The substitution was made for clarity only.

Finally, it has been shown by numerous researchers that the coefficient of variation must be adjusted for the number of emitters per plant (Keller, Karmeli, 1975 and Solomon, 1977). This is done by dividing the coefficient of variation by the square root of the number of emitters per plant (e) and results in

$$V_q = e^{-1/2}[V_p^2 + V_k^2 + x^2 V_h^2]^{0.5} \qquad [8.27]$$

Equation [8.27] thus combines the variation caused by submain hydraulics, emitter manufacturing, emitter plugging, and the number of emitters per plant. When combined with the statistical uniformity (equation [8.18] the result is a statistically based measure of emitter flow variation.

IV. ENGINEERING APPLICATIONS

Practically speaking, theoretical equations are of little use without some simplified method of application. Given the hydraulic design and statistical uniformity equations developed in the preceding sections a computer programme can be written to determine the various contributors to the variance. For this chapter, however, design charts have been written to illustrate the utility of the statistical uniformity with respect to hydraulic design and field evaluation of drip irrigation submain units. The theoretical development of individual design charts can be found in papers by Bralts et al. (1981a and 1981b) and Bralts and Edwards (1983).

A. Applications to Design

As mentioned earlier flow in a drip irrigation system is the result of a number of factors. The most important of these are hydraulic variation, manufacturer's variation, emitter plugging variation, and the number of emitters per plant. To the design engineer it is important that emitter flow variation or the uniformity of water distribution be known so that the acceptability of a particular design can be determined at the design stage.

Design Objective ; To determine the Statistical Uniformity (Us) as defined by the equation

$$U_s = 100\,(1 - V_q) \quad \text{Statistical Uniformity}$$

Where

$$v_q = e^{-1/2}\,[V_p^2 + V_k^2 + x^2 V_h^2]^{1/2} \quad \text{Emitter Flow Variation}$$

EXAMPLE PROCEDUE :

Step 1 : $V_q = [x^2 V_h^2]^{1/2}$ Hydraulic Variation

Step 2 : $V_q = [V_k^2 + x^2 V_h^2]^{1/2}$ Hydraulic plus Emitter Manufacturer's Vaiation

Step 3 : $V_q = [V_p^2 + V_k^2 + x^2 V_h^2]^{1/2}$ Including emitter plugging variation

Step 4 : $V_q = e^{-1/2}\,[V_p^2 + V_k^2 + x^2 V_h^2]^{1/2}$ Considering the Number of Emitters per Plant

Fig 8.8 Flow chart of design objective and example procedure.

In this section a series of nomographs based on the dimensionless energy, gradient line the constant odds uncertainty formulation, and the statistical uniformity will be presented. Along with the monographs a series of examples have been developed to illustrate the use and applicability of the design and procedures. Figure 8.8 is a flow chart of the design procedure and how it relates to the evaluation of equation [8.27]. In addition a step by step design example is presented as part of this discussion.

1. Hydraulic Variation

The term hydraulic variation has been used rather loosely by numerous researchers, to describe the variation in hydraulic pressures found along a drip irrigation lateral line. In this section the coefficient of variation of emitter pressures found along a drip irrigation lateral line or in drip irrigation submain unit.

From an engineering standpoint the coefficient of variation due to hydraulics (V_h) can be determined using the dimensionless energy grade line (Equation [8.15] or a network analysis solution. Statistically, the coefficient of variation due to hydraulics can be determined using the standard statistical equations for the variance and the mean pressure head (h). In equation form

$$V_h = \frac{S_h}{\bar{h}} = \sqrt{\frac{s_h^2}{\bar{h}^2}} \qquad [8.28]$$

where V_h = coefficient of variation due to hydraulics, S_h = standard deviation of the hydraulic pressures, $\bar{h}$ = mean hydraulic pressure, and S_h^2 = variance of the hydraulic pressure.

When considering the components of the constant odds uncertainty formulation it can be shown that when plugging and emitter manufacturer's variation are held constant, emitter flow variation due to hydraulics can be formed using the equation

$$V_q = x\,V_h = [x^2\,V_h^2]^{0.5} \qquad [8.29]$$

where V_q = coefficient of emitter flow variation, x = the emitter discharge exponent, and V_h = the coefficient of variation due to hydraulics. Thus once the coefficient of variation due to hydraulics

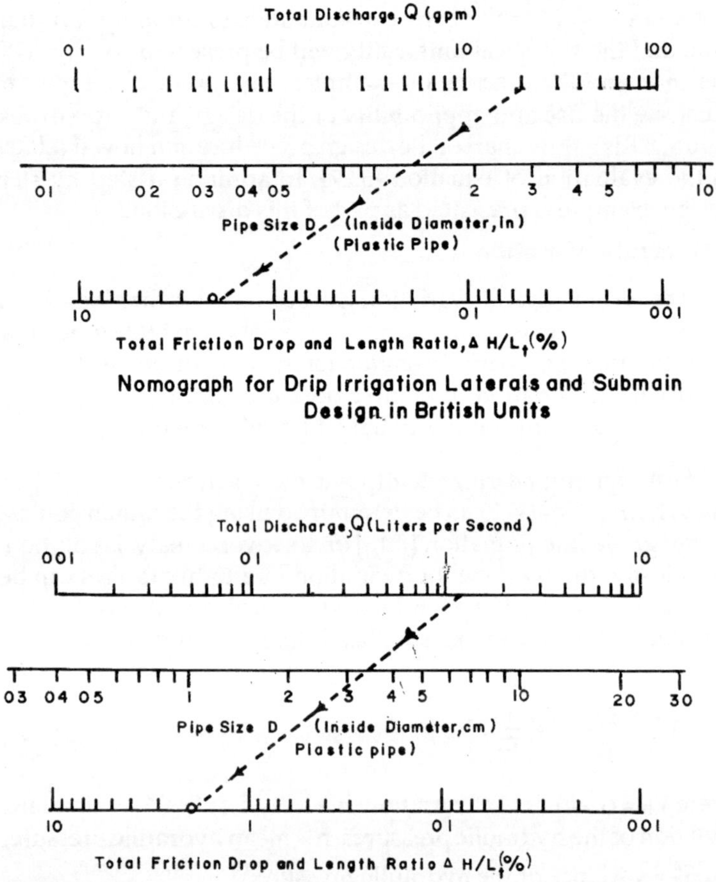

Fig 8.9. Nomograph for total friction drop (ΔH) and lenght (L_t) ratio $\Delta H/L_t$ (%)

is known variation of emitter flow due to hydraulic and the statistical uniformity can be easily determined.

Equation [8.29] also shows that the emitter flow variation is directly proportional to the emitter discharge exponent x when the total system flow rate is held constant. If the variation of emitter flow is too high for an acceptable uniformity then a new emitter

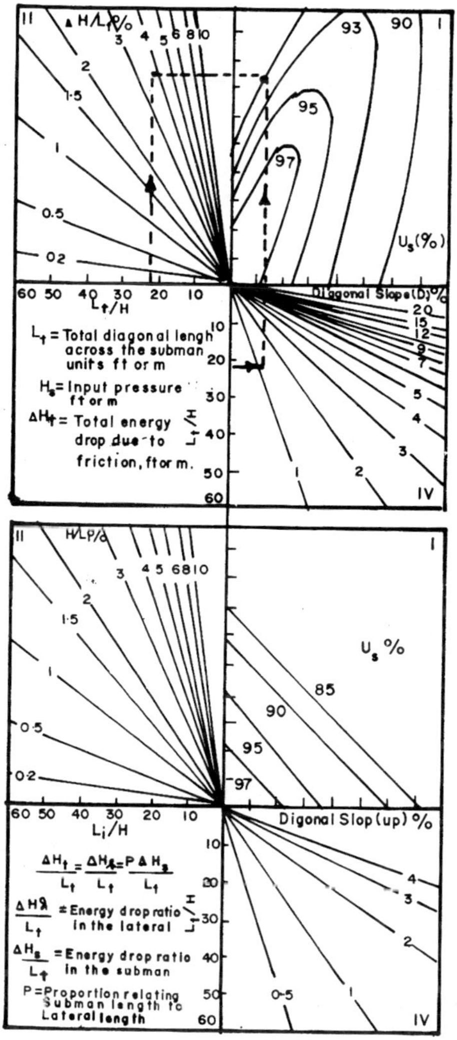

Fig 8.10. Drip irrigation design calculator using statistical unformity.

with a lower emitter discharge exponent, but a similar flow rate, should be installed. The expected reduction in the coefficient of variation would be equal to the proportion of the respective emitter discharge exponents. In the extreme case where an emitter is chosen with a discharge exponent of zero, the coefficient of variation due to hydraulics is also zero. In other words, the emitter is completely pressure compensating.

For clarity sake the hydraulic design charts and monographs represented next are based on the coefficient of variation due to hydraulics and an emitter discharge exponent of 0.5. This was done so that the complete hydraulics variation term in the constant odds formulation (Equation [8.27]) could be determined from one nomograph and so that subsequent design nomograph relating to emitter manufacturing and plugging could be more easily used.

The hydraulic design charts and nomographs shown in Fig. 8.9 and 8.10 are based on the dimensionless energy gradient line concept equation [8.16] and are simple modifications of the four quadrant statistical uniformity design charts presented by Bralts (1981a). The major changes are the use of: L_t, the diagonal distance across the submain unit, in the place of L, the length of lateral; S_t, the diagonal slope, in place of S, the lateral line slope; and, the need to add the submain portion of the $\Delta H/L_t$ to the lateral $\Delta H/L_t$. By using L_t and S_t, it is possible to combine the submain and lateral line variables into a single variable. The addition of a portion of the $\Delta H/L_t$ is necessary so that the submain and lateral line hydraulic variables can be combined (Bralts and Edwards, 1983). The result is a two dimensional nomograph for submain unit design.

The procedure for using the submain unit statistical uniformity design charts to check the acceptability of a given submain design is as follows :

a. Establish a trial L_t/H_s and total submain unit flow.
b. From the proper nomograph, Fig. 8.9, use the total submain unit flow and the pipe size for the submain manifold and the lateral line to determine the $\Delta H/L$ for each.

c. Combine the $P\Delta H/L$ of the submain manifold with $\Delta H/L_t$ of the lateral, where P is the proportion of the submain manifold length to that of the lateral.

d. Find the L_t/H_s on the x axis and move vertically to the combined $\Delta H/L_t$ in quadrant II; then establish a horizontal line into quadrant I.

e. Find the L_t/H_s on the y axis and move horizontally in quadrant IV to the per cent slope, S_t; then establish a vertical line into quadrant I.

f. The intersection point of the lines in quadrant I determines the statistical uniformity of the design.

A limitation of the above design chart is the assumption that the emitter discharge exponent x = 0.5. This limitation can be overcome by simply multiplying the V_q related to the statistical uniformity found on the nomograph by 2x′ where x′ is the actual emitter discharge exponent.

A design example has been developed to illustrate the combined nature of the variation and the resulting statistical uniformity. The steps are sequential and the information obtained in the first step will be used in subsequent procedures. The design example Step 1 has been written to illustrate the use of Figures 8.9 to 8.10 in the determination of hydraulic variation and statistical uniformity.

Design Example : Step 1

Given : A submain unit consisting of a 30-56 m (100 ft.) long, 31.75 mm (1.25 in.) diameter, manifold pipe on a 2% downhill slope and 10 equally spaced 152.5 m (500 ft.) long, 12.5 mm (1/2 in.) diameter lateral lines at a 1% downhill slope. The input pressure is 69 kPa (23.1 ft.) and the total submain discharge has been estimated at 1.26 1/sec (20 gpm). The emitter discharge exponent is x = 0.4 as given by the manufacturer. See. Fig. 8.11 for a layout of the design example submain unit.

Required : Determine the statistical uniformity for the hydraulic variation or the example submain unit.

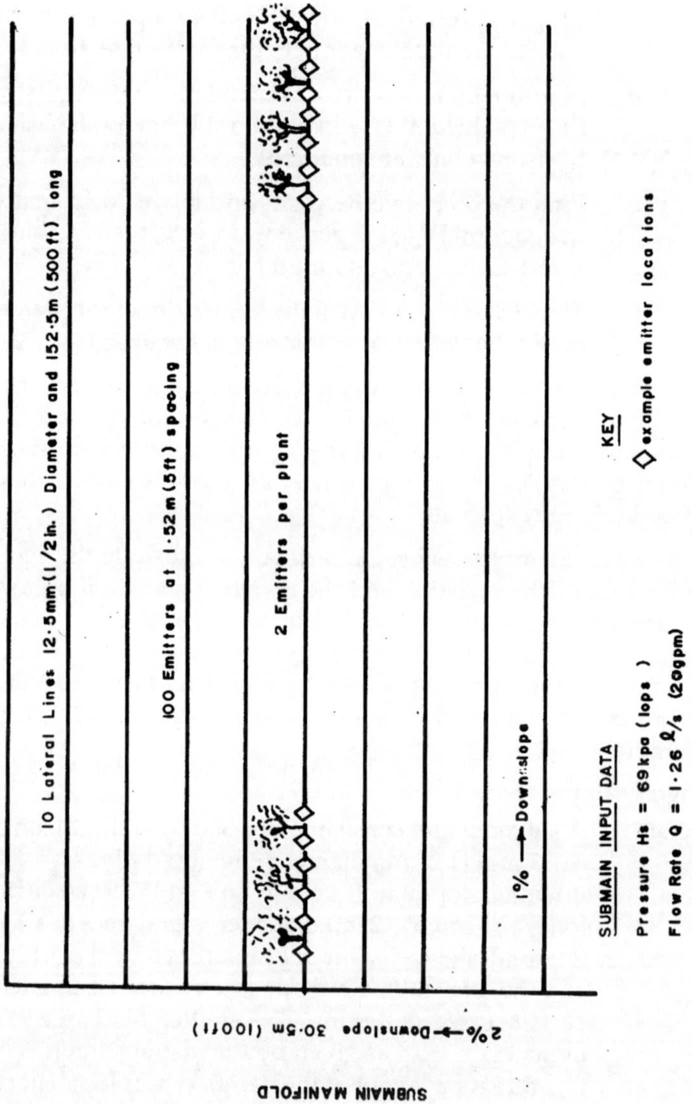

Fig 8.11. Submain unit design example layout

Solution :

1. Calculate L_t/H_s.

$$\left(\text{Answer}: \frac{L_i}{H_s} = \frac{(100^2 + 55^2)^{1/2}}{23.1} = \frac{510}{23.1} = 22\right)$$

2. Determine $\Delta H/L_t$ for the manifold and lateral line using Figure 8.9.

$$\left(\text{Answer}: \text{ Manifold } \frac{\Delta H}{L_t} = 2.0 \text{ and Lateral } \frac{\Delta H}{L_t} = 3.0\right)$$

3. Calculate the combined $\Delta H/L_t$

$$\left(\text{Answer}: \frac{\Delta H}{L_t} = 0.2(2.0) + 3.0 = 3.4\right)$$

4. Calculate S_t.

$$\left(\text{Answer}: S_t = \frac{(100 \times 0.02 + 500 \times 0.01)}{(100^2 + 500^2)^{1/2}} = \frac{7}{510} = 0.0137\right)$$

5. Using the nomograph in Figure 8.10 determine the statistical uniformity.

(Answer : $U_s = 90\%$)

6. Correcting for the emitter discharge exponent.

(Answer : $V_q = 2x' \;\; V_h = 2(0.4)(0.1) = 0.08$ or $U_s = 92\%$)

Comments : The hydraulic design is accepted based on the statistical uniformity of 92%.

2. Manufacturer's Variation

The concept of manufacturer's variation of emitters was first developed by Keller and Karmeli (1974) and later researched by Solomon (1977). The technique used to find the manufacturer's variation consisted of measuring the variation of emitter flow under a constant head or pressure. This experimental procedure is the same as applying the constant odds uncertainty formulation to the emitter flow equation without emitter plugging while at a constant pressure. The result is the coefficient of emitter flow

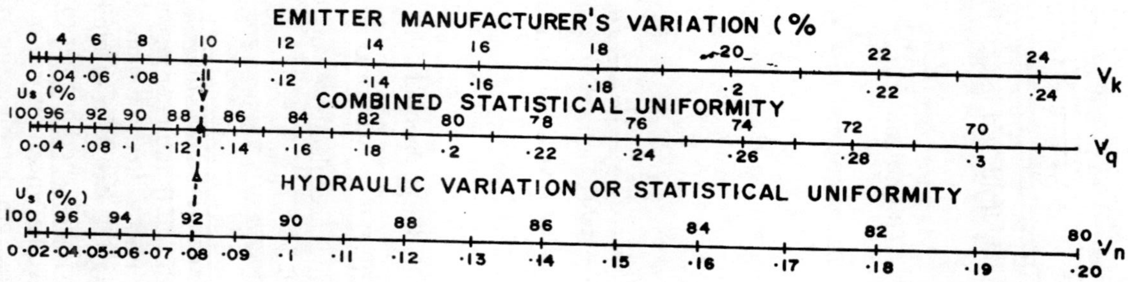

Fig 8.12. Nomograph to determine the statistical unfcrmity given the hydraulic and manufacturer's coefficient of variation.

variation due to the proportionality constant k or manufacturer's variation as described by Keller and Karmeli (1974). The result in equation form is

$$V_q = V_k = V_{me} = \frac{S_q}{\bar{q}} = \sqrt{\frac{S_q^2}{\bar{q}^2}} \qquad [8.30]$$

where all variables are as previously defined.

The combination of the manufacturer's variation and the hydraulic variation can be accomplished by a simple application of the constant odds uncertainty formulation to the emitter flow equation without emitter plugging. The result is

$$V_q = [V_k^2 + x^2 V_h^2]^{1/2} \qquad [8.31]$$

where all variables are as previously defined. Thus the emitter flow variation and statistical uniformity can be determined once the hydraulic and manufacturer's coefficients of variation are known.

A nomograph presented by Bralts et al. (1981a) Fig. 8.12, for combining lateral line hydraulic and emitter manufacturer's variation, can also be used to combine the submain unit hydraulic and manufacturer's variation for drip irrigation submain units. The procedure is simply to find the hydraulic variation or statistical uniformity from the design example Step 1 on the top line and the manufacturer's variation on the bottom line of the nomograph in Fig. 8.12. The line connecting the points will intersect the total or combined statistical uniformity line in the centre of the nomograph.

Design example Step 2 is a continuation of the design example begun in the previous section and has been written to illustrate the effect of manufacturer's variation or system uniformity.

3. Emitter Plugging Variation

The variation of emitter flow due to emitter plugging (emitter plugging variation, V_p) has been researched by numerous individuals. The approach presented by Bralts et al. (1981b) will be used here. From a design stand point emitter plugging is presumed to occur at some point after installation of the system and its effects should be estimated as part of the design procedure.

Design Example : Step 2

Given : The hydraulic variation or statistical uniformity from Step 1 is 92%. The emitter manufacturers variation as reported by the manufacturer is 10%.

Required : Determine the statistical uniformity including manufacturer's variation.

Solution : Using nomograph in Fig. 8.12, the combined statistical uniformity can be determined.
(Answer : $U_s = 87\%$)

Comments : The manufacturer's variation reduced the statistical uniformity from 92% to 87%. This is still an acceptable uniformity for the drip submain unit based on the criteria given in Section II.

There are at least five possible cases of emitter plugging. The specific cases are as follows :

Case I - All plugging is partial and affects all emitters equally.

Case II - All plugging is partial and affects a specific proportion of the emitters equally.

Case III - All plugging is complete and affects all emitters equally.

Case IV - All plugging is complete and affects a specific proportion of the emitters equally.

Case V - All plugging is a combination of complete plugging and partial plugging or cases I, II, and IV above.

Considering Case I through IV above, it can be shown that each of these conditions is a special case deriving from the general Case V. For this reason Case V will be developed further. The authors acknowledge the case of varying degrees of partial plugging with in submain units exists, however, the solution for that case is beyond the scope of this presentation.

Bralts et al. (1981b) developed an equation for Case B where all plugging is a combination of complete and partial plugging. The equation for Case V is

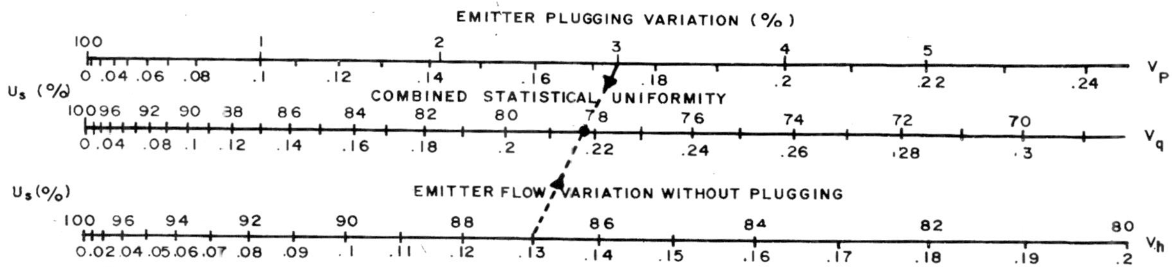

Fig 8.13. Nomog-aph for calculating the statistical uniformity given the percentage of emitter plugging

$$V_p = \left[\frac{n\,[\theta + p'(1-\alpha)^2]}{[\theta + p'(1-\alpha)]^2}\right]^{1/2} \qquad [8.32]$$

where V_p = the coefficient of variation of emitter flow due to emitter plugging, n is the total number of emitters, θ is the number of openly flowing emitters, p' is the percentage of emitters partially plugged, and α is the degree of partial plugging. Thus given the varying degrees of complete and partial plugging, equation [8.32] can be used to determine the coefficient of emitter flow variation due to emitter plugging. Once the coefficient of variation due to emitter plugging is determined it can be combined with hydraulic and manufacturer's variation using the equation

$$V_q = [V_p^2 + V_k^2 + x^2 V_h^2]^{1/2} \qquad [8.33]$$

where all variables are as previously defined.

To simplify the design procedure being presented in this chapter only the complete emitter plugging case will be considered. The nomograph presented in Fig. 8.13 will combine the emitter flow variation or the statistical uniformity due to hydraulics and manufacturing with the coefficient of variation due to complete emitter plugging. To use the nomograph identify the percentage of complete emitter plugging on the top line and the statistical uniformity for hydraulic plus manufacturer's variation on the lower line. Connect the two points and the combined statistical uniformity including emitter plugging can be read on the center line.

Design example Step 3 is a continuation of the design example begun in the previous section and has been written to illustrate the efforts of emitter plugging on the statistical uniformity.

Design Example : Step 3

Given : The statistical uniformity due to hydraulics and manufacturer's variation as determined from the previous design step is 87%. The emitter plugging rate is 3%.

Required : Determine the statistical uniformity including emitter plugging.

Solution : Using the nomograph in Figure 8.13 the statistical uniformity including emitter plugging can be determined.
(Answer : $U_s = 78\%$)

Comment : The statistical uniformity including 3% emitter plugging is less than 80% and thus would not be acceptable if the emitter plugging is allowed to occur. Either a strict water quality control programme should be considered or additional emitters per plant should be installed.

4. Number of Emitters per Plant

Statistically speaking, if there are several emitters per plant, the variations of emitter flows tend to cancel one another. The result is that the total variation of emitter flow to any one plant is less than the variation of emitter flow from the individual emitters (Solomon, 1979).

If we assume that each emitter is independent then the variations of flow is additive across emitters. This condition of orthogonality allows the emitter flow variation to be combined with the coefficient of variation as

$$V_q = e'^{-1/2} [V_p^2 + V_k^2 + x^2 V_h^2]^{1/2} \qquad [8.34]$$

where all variables are as previously defined. Equation [8.34] illustrates the effect of emitter spacing or grouping on the coefficient of variation. If more than one emitter can affect the amount of moisture available to the plant due to such factors as plant spacing, emitter spacing, and/or translocation of water in the soil, the coefficient of variation of emitter flow should be adjusted in this fashion at the design phase. Figure 8.14 consists of a nomograph to solve Equation [8.34].

Design example Step 4 is a continuation of the design example begun in the previous section and has been written to illustrate the effects of emitter number per plant or grouping on the statistical uniformity.

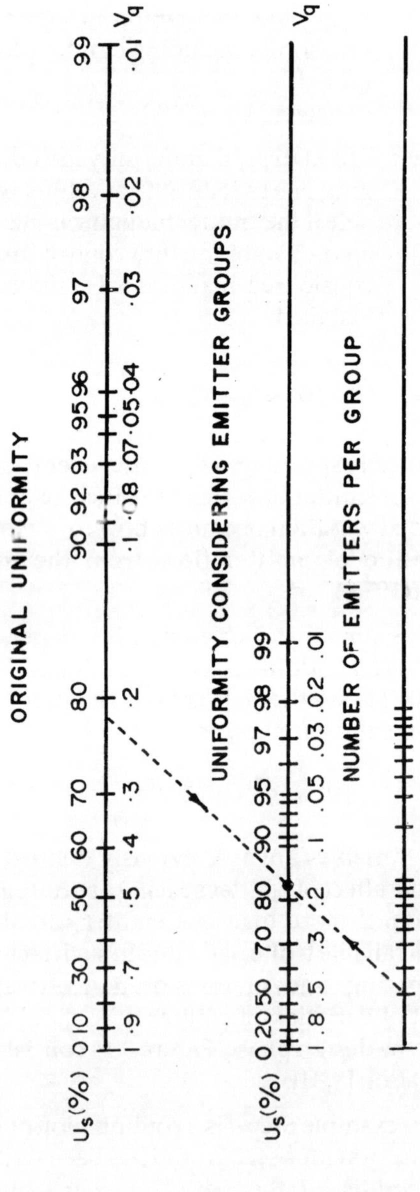

Fig 8.14. Nomograph to determine the statistical uniformity given the number of emitters per plant.

Design Example : Step 4

Given : The statistical uniformity including emitter plugging is 78% as determined in the previous design step and the fact that there are 2 emitters per plant.

Required : Determine the statistical uniformity considering 2 emitters per plant.

Solution : Using the nomograph presented in Fig. 8.14 the statistical uniformity considering the number of emitters per plant can be determined.
(Answer : U_s = 84%)

Comments : The resulting design uniformity is accepted based on the criteria given in Section II.

B. Applications to Evaluation

The evaluation of a drip irrigation system encompasses significantly more than the uniformity of emitter flow. However, the accurate estimation of system uniformity is probably the single most important indicator of system performance. Field uniformity estimation of drip irrigation systems is important from several perspectives. First, from the engineer's perspective, field uniformity estimation is important in confirming whether a design was satisfactory. Second, from the purchaser's perspective, field uniformity estimation is important in confirming product performance. Third, from the irrigator's perspective, field uniformity estimation is important when considering irrigation application efficiency and schedules.

To date several methods have been proposed for field uniformity estimation. Keller and Karmeli (1974) proposed a modified form of the Soil Conservation Service irrigation system evaluation equation as the absolute emission uniformity (EU_a). The field evaluation procedure of this method consisted of collecting emitter flows from each of 16 areas in a submain unit and calculating the emission uniformity. The emission uniformity concept has been presented in numerous publications including Keller and Karmeli (1975), Merriam and Keller (1978) and the Soil Conservation Service National Engineering Handbook (1980).

General Criteria for EU_a values are: 90% or greater, excellent; 80-90%, good; 70-80%, fair; and, less than 70%, poor. The primary disadvantage of this method is its non statistical basis. For this reason, confidence limits and a further break down of the components of emitter flow variation is not possible.

A second method of field evaluation based upon the design procedures developed by Wu and Gitlin (1974, 1975), would be an estimation of emitter flow variation based on submain pressures. In this case, the field evaluation procedure consist of finding the minimum and maximum pressures in the submain unit and then calculating the emitter flow variation (q_{var}). General criteria for q_{var} values are : 10% or less, desirable; 10-20%, acceptable; greater than 20%, not acceptable. A major limitation of this procedure is that it does not include emitter flow variation due to manufacturing or plugging, factors which could be significant in the submain uniformity.

Bralts and Kesner (1983) proposed a method of submain uniformity estimation based upon the statistical uniformity. The advantages of this method include simplicity of field determinations and the ability to set confidence limits on uniformity estimates.

The general criteria for an acceptable U_s as mentioned earlier are : 90% or greater, excellent; 80-90%, very good; 70-80%, fair; 60-70%, poor; and less than 60%, unacceptable. An important advantage of this method is that the determination of the system uniformity will lend information to assist in diagnosing the problem areas such as emitter performance and hydraulic variation. Another advantage of this method is that once the statistical uniformity is known, the effects of uniformity on irrigation efficiencies and schedules can be determined.

From an evaluation perspective the constant odds formulation can also be employed. When an emitter is initially installed, the major factors affecting uniformity of emitter flow are still hydraulic and manufacturer's variation. Over time, however, other factors such as emitter wear and partial emitter plugging will affect the performance of the emitters in the field. For this reason a new variable which describes the variation of emitter flow due to the

Evaluation objective : To determine the Statistical Uniformity of emitter flow using the equation:

$$U_s = 100\,(1 - V_q)$$

Where

$$V_q = [V_{pf}^2 + x^2 V_h^2]$$

EVALUATION PROCEDURE : determine the components of the emitter performance variation.

Step 1 :	$\hat{V}_q = [V_q^2]^{1/2}$	Emitter flow variation
Step 2 :	$\hat{V}_q = [x^2 V_h^2]^{1/2}$	Hydraulic variation
Step 3 :	$V_{pf} = [\hat{V}_q^2 - x^2 V_h^2]^{1/2}$	Emitter performance variation

Step 4 : Considering the plugging variation

Step 5 : Considering the number of emitters per plant

Step 6 : Application efficiency and schedules

Fig 8.15. Flow diagram of evaluation example procedure.

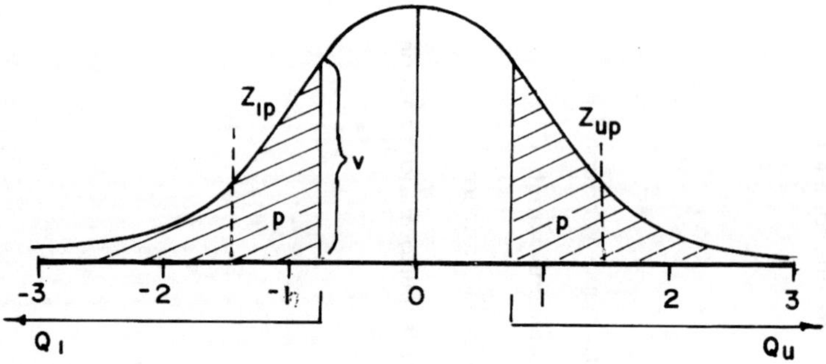

$$y = \frac{1}{\sqrt{2\pi}} e^{-\frac{x^2}{2}}$$

Fig 8.16. Standard normal distribution curve.

field situation is desirable. If the coefficient of manufacturer's variation in equation [8.31] is redefined to include emitter field performance the result is

$$V_q = [V_{pf}^2 + x^2 V_h^2]^{1/2} \quad [8.35]$$

where V_{pf} – coefficient of variation for emitter performance, and all other variables are as previously defined.

In this section a series of nomographs based on the constant odds uncertainty formulation and the statistical uniformity will be presented. The series of nomographs have been developed to assist the irrigation specialist in field evaluations of a drip irrigation system. Figure 8.15 is a flow chart of how the evaluation procedure related to equation [8.35] and the evaluation procedure.

1. Emitter Flow Variation

To the drip irrigation system designer, purchaser or irrigator, it is essential that the constant odds theory be translated into practical form. One approach would consist of programming a small, hand-held calculator to compute S_q, $\bar{q}$, and U_s based upon standard statistical methods and the statistical uniformity, equation [8.18]. Another possibility would be to make some specific assumptions regarding normality of emitter flows, sample size and the proportion of observations to be used in the estimate and develop a nomograph to solve the simplified equations. The latter approach will be shown here.

Bralts and Kesner (1983) developed an equation for determining the coefficient of variation of emitter flow based on the location of the tails in a normal probability density function (see Fig. 8.16). For the special case where the probability in each tail is 1/6 the equation for estimating the coefficient of emitter flow variation is

$$\hat{V}_q = 0.667\left(\frac{Q_{us} - Q_{ls}}{Q_{us} + Q_{ls}}\right) \qquad [8.36]$$

where $\hat{V}_q$ = the estimated coefficient of emitter flow variation, Q_{us} is the sum of the observations in the upper 1/6 of the distribution, and Q_{ls} is the sum of the observations in the lower 1/6 of the distribution.

The simplicity of the above equation for estimating the coefficient of variation is evident. If 18 random measurements of emitter flow rate were made, it would only be necessary to sum the three highest and the three lowest values to estimate the coefficient of variation.

The confidence limits for the coefficient of variation (V_q) on samples from a normal population (Sokal and Rohlf, 1969) can be expressed as

$$P\,(V_q - T_{\alpha/2}\, Sv_q \,|\, V_q^* \,|\, V_q + Sv_q) = 1 - \alpha \qquad [8.37]$$

where V_q = sample coefficient of variation, $t_{\alpha/2}$ = student t value for given α, α = confidence level desired, V^*_q = actual coefficient of variation for the full submain unit, Sv_q = standard deviation of the coefficient of variation, calculated from the equation

$$Sv_q = \frac{V_q}{\sqrt{n^2}}\sqrt{1 + 2V_q^2} \qquad [8.38]$$

Using equations [8.37] and [8.38] the confidence limits for the sample coefficient of variation ($\hat{V}_q$) can be found. Since the confidence limits of the estimated coefficient of variation are dependent on the assumption of a normal distribution, the above confidence limits can only be used as the approximate confidence limits of $\hat{V}_q$.

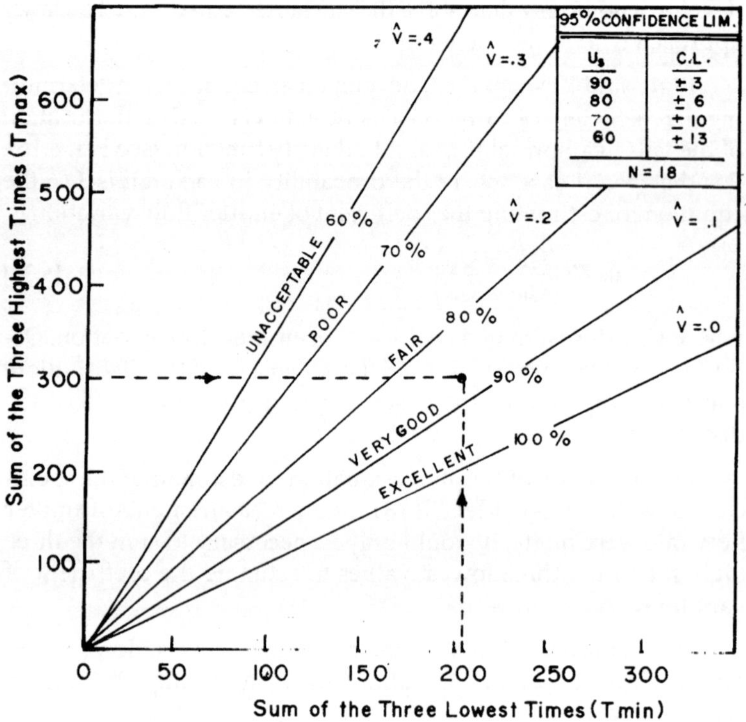

Fig 8.17. Nomograph for estimating statistical uniformity based on emitter discharge rates.

Equation [8.36] can also be rearranged to demonstrate the linear nature of the terms Q_{us} and Q_{ls}

$$Q_{us} = \left(\frac{0.667 + \hat{V}_q}{0.667 - \hat{V}_q}\right) Q_{ls} \qquad [8.39]$$

where all variables are as previously defined. Thus, for any given coefficient of variation $\hat{V}_q$, the Q_{us} varies linearly with Q_{ls}.

Using the inverse relationship of minimum time to maximum emitter flow rate, equation [8.38] can be rewritten as

$$T_{max} = \left(\frac{0.667 + \hat{V}_q}{0.667 - \hat{V}_q}\right) T_{min} \qquad [8.40]$$

where T_{max} = the sum of the top one sixth of the emitter flow times required to fill a specific container with water, T_{min} = the sum of the bottom one sixth of the emitter flow times required to fill a specific container with water, and V_q is as previously defined. This relationship can now be translated into a nomograph. Figure 8.17 represents an example of a nomograph to determine the statistical uniformity designed to use the highest one sixth and lowest one sixth of the times required to fill a given container. The procedure for using the nomograph in Fig. 8.17 is as follows :

a. Select a container for making the flow rate determination (100 or 200 ml).

b. Choose 18 emitters at random in the submain unit and measure the time it takes to fill the container.

c. T_{max} is the sum of the three highest times it takes to fill the container.

d. T_{min} is the sum of the three lowest times it takes to fill the container.

e. Find T_{max} on the vertical axis and draw a horizontal line.

f. Find T_{min} on the horizontal axis and draw a vertical line.

g. The intersection of the two lines gives the statistical uniformity.

The following evaluation example has been developed to illustrate the diagnostic nature of the constant odds uncertainty formulation (equation [8.35]). The steps in the evaluation are sequential and the information obtained in the first step will be needed in the second. To further illustrate the related nature of the design and evaluation procedures the filed evaluation example has been carried out on the exact same submain as the design example. Thus the evaluation procedure should yield the coefficients of hydraulic and manufacturer's, used in the design of the submain unit.

Evaluation Example : Step 1

Given : The following field data was collected from the example submain unit (see Fig. 8.18); Time (sec) to fill a

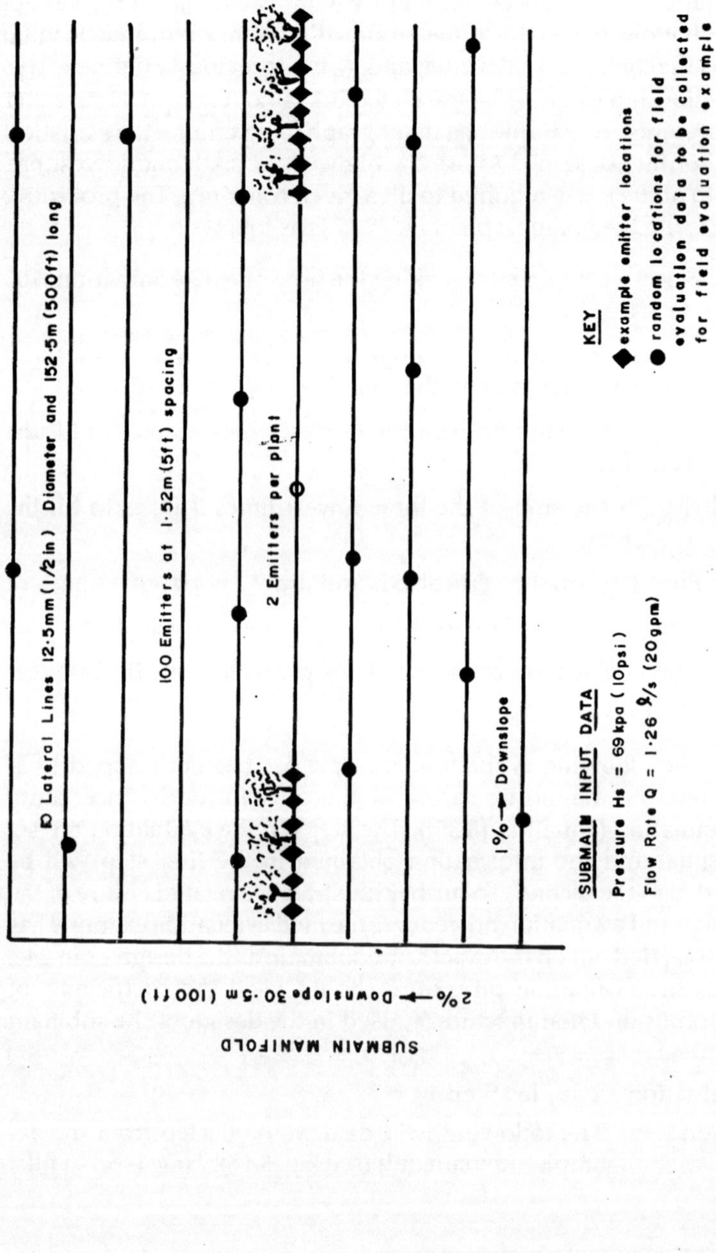

Fig 8.18. Evaluation example submain unit.

200 ml container : 69, 84, 72, 76, 80, 86, 73, 97, 80, 74, 90, 82, 101, 73, 86, 102, 70, and 66.

Required : Estimate the statistical uniformity of the submain unit.

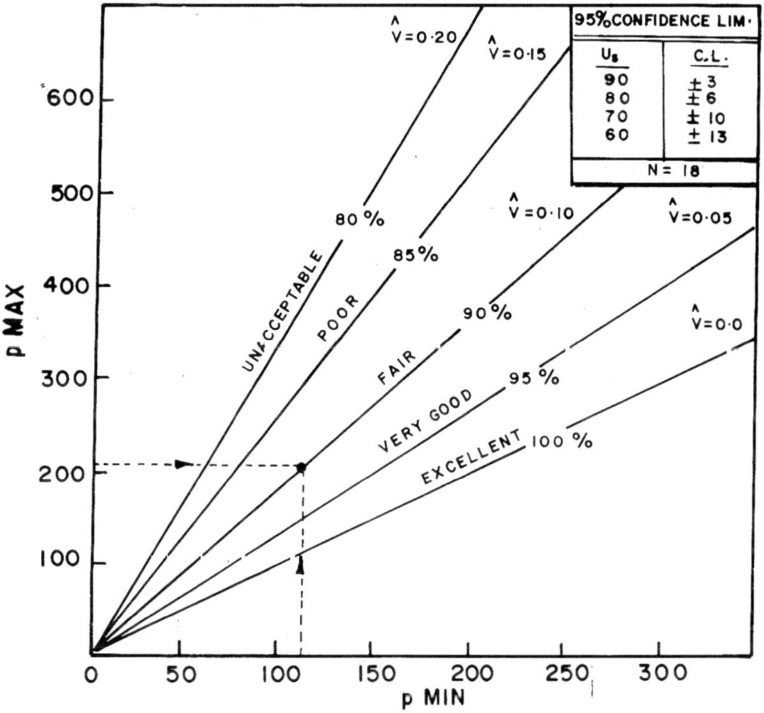

Fig 8.19. Nomograph for estimating hydraulic uniformity based on pressure distribution measurements (sum of three highest and lowest).

Solution : 1. Calculate T_{max} and T_{min} using the three highest and three lowest times.

(Answer : T_{max} = 97+101+102 = 300, and T_{min} = 69+70+66 = 205)

2. Use Fig. 8.17 to determine the statistical uniformity of the given submain unit.

(Answer : U_s = 87% ± 3%)

Comments : The statistical uniformity is satisfactory and matches that found in the design example Step 2.

2. Hydraulic Variation.

Field evaluation of the hydraulic variation is important to the drip irrigation system designed and purchaser to confirm the proper design and operation of a drip irrigation submain unit. The coefficient of variation due to hydraulics V_h can be determined by taking a random sample of the pressures at various emitters similar to the emitter lows in the previous section. The coefficient of variation of emitter flow due to hydraulics xV_h can also be determined once the emitter discharge exponent is known. If the equation

$$\hat{V}_q = [x^2 \hat{V}_h^2]^{1/2} \qquad [8.41]$$

is used a graphical technique similar to that used to develop the emitter flow uniformity nomograph can be used to determine the coefficient of emitter flow variation due to hydraulics. Fig. 8.19 consists of a nomograph to determine the statistical uniformity of emitter flow due to hydraulics where x = 0.5. For instances where x is known to be something other than 0.5 the coefficient of variation can be adjusted by multiplying the coefficient of variation of emitter flow from the nomograph by 2x′ where x′ is the actual emitter discharge exponent. The procedure of using Figure 8.19 is as follows :

a. Choose 18 emitters at random in the submain unit and measure the lateral line pressure at the emitter.

b. P_{max} is the sum of the three highest pressures.

c. P_{min} is the sum of the three lowest pressures.

d. Find P_{max} on the vertical axis and draw a horizontal line to the right.

e. Find P_{min} on the horizontal axis and draw a vertical line up.

f. The estimated statistical uniformity due to hydraulics can be read at the point of intersection of the two lines.

The evaluation example Step 2 consists of a continuation of the example begun in the previous section and was written to illustrate the evaluation of the hydraulic pressure on emitter flow variation.

Evaluation Example : Step 2

Given : Eighteen randomly selected emitter pressures were collected from the evaluation example (see Fig. 8.18). The pressures (kPa) are : 40, 49, 35, 45, 59, 67, 75, 52, 70, 36, 39, 47, 52, 56, 51, 57, 60, 50. An emitter exponent of 0.4 was given by the manufacturer.

Required: Determine the statistical uniformity due to hydraulics.

Solution : 1. Calculating P_{max} and P_{min} using the three highest and three lowest pressures.
(Answer : P_{max} = 67+75+70 = 212, and P_{min} = 35+36+39 = 110)
2. Use Figure 8.19 to determine the statistical uniformity. (Answer : U_s = 90%)
3. Correcting for the emitter discharge exponent.
(Answer : 2(0.4)(0.1) = 0.08 or U_s = 92% ± 3%)

Comments : The estimated statistical uniformity due to hydraulic is acceptable. Also please note that it is the same as the design uniformity due to hydraulics determined in the design example.

3. Emitter Performance Variation

The emitter performance variation was defined because it includes a variety of factors affecting emitter performance such as emitter wear, manufacturer's variation, and partial emitter plugging. From a field evaluation perspective it is important to note that equation [8.35] also allows the differentiation of hydraulic and emitter performance related variables. Referring back to equation

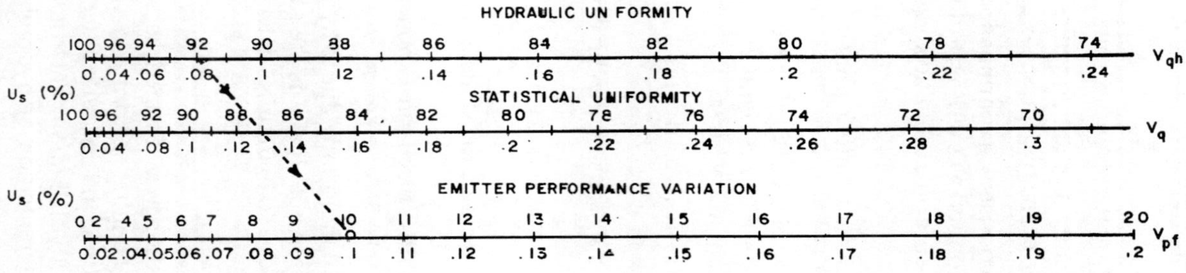

Fig 8.20. Nomograph for determining emitter performance variation based on the hydraulic and the emitter discharge variation.

[8.35], it can be shown that if any two of the three coefficients of variation are known the third is easily determined. Rewriting equation [8.35] we obtain

$$\hat{V}_q = [V_{pf}^2 - x^2 V_h^2]^{1/2} \qquad [8.42]$$

where all variables are as previously defined.

The utility of the information obtained in the previous two evaluation steps can best be seen when used in conjunction with equation [8.42]. Figure 8.20 is an example nomograph which can be used to determine the emitter performance variation given the emitter flow statistical uniformity and the statistical uniformity due to hydraulics.

The evaluation example Step 3 consists of a continuation of the submain unit evaluation begun in the previous two steps and was written to illustrate the evaluation of emitter performance variation. It is important to note that the result of Step 3 is an emitter performance variation of 10%. This value is the same as the value of the manufacturer's variation in the design example and confirms the relationship between the two examples.

Evaluation Example : Step 3

Given : The submain emitter flow uniformity (87%) and the statistical uniformity due to hydraulics (92%) as determined in Steps 1 and 2.

Required : Determine the emitter performance variation of the submain units.

Solution : Use equation [8.42] or Figure 8.20 to determine the emitter performance variation.
(Answer : V_{pf} = 0.1% or 10%)

Comments : The emitter performance variation is acceptable. Also please note that for a new system the emitter performance variation should be equal to the design manufacturer's variation and it is in this case.

4. Emitter Plugging

Field evaluation of emitter plugging is a very difficult task. If the time to fill a cup were used on plugged emitters it would take an infinite amount of time to fill the cup and determine emitter flow. If a decrease in submain unit flow is detected a 5% reduction relates to a totally unacceptable 25% level of emitter plugging (Bralts et al., 1982). For this reason the best assessment of the effects of emitter plugging on the statistical uniformity can be found by counting the percentage of emitters completely plugged and combining it with the coefficient of emitter flow variation found in the first step of this evaluation example. In this case Figure 8.13 of the Design Example can again be used with satisfactory result.

The evaluation example Step 4 consists of a continuation of the example begun in the previous section and was written to illustrate the inclusion of emitter plugging in the evaluation of the submain unit statistical uniformity.

Evaluation Example : Step 4

Given : The estimated statistical uniformity without emitter plugging is 87% as determined from step 1, and an emitter performance variation of 3%.

Required : Determine the statistical uniformity including emitter plugging.

Solution : Use Fig. 8.13 to determine the statistical uniformity including emitter plugging.
(Answer : $U_s = 78\%$)

5. Number of Emitters Per Plant

The number of emitters per plant in the evaluation phase has a similar positive effect on the final statistical uniformity as it does in the design phase. The evaluation example Step 5 uses the same nomograph (Fig. 8.14) to determine the improved statistical uniformity due to the number of emitters per plant.

It is important to note again at this point that the final uniformity estimate from the Evaluation Example was the same as that for the Design Example. Thus the utility of a statistically based design and evaluation procedure based on the statistical uniformity has been reasonably confirmed.

Evaluation Example : Step 5

Given : The statistical uniformity including emitter plugging is 78% as determined in the previous step.

Required : Determine the statistical uniformity when considering 2 emitters per plant.

Solution : Use Fig. 8.14 to determine the statistical uniformity including the number of emitters per plant.
(Answer : U_s = 84%)

6. Application Efficiency and Schedules

The assessment of soil water distribution and the resultant application efficiency are important components of any irrigation system evaluation. Since water in drip irrigation systems is usually applied at rates close to the consumptive water use of plants, soil water considerations are much more critical in drip than in sprinkler or flood irrigation. In the past, a 30 to 50% root zone wetting has been recommended for trickle irrigation. This value has been exceeded in tree crops particularly in the southeastern United States, where additional emitters have been added to improve uniformity. Under row crops, the soil wetted volume has sometimes gone as high as 70 to 80%.

The soil type is important because it affects the extent of available water and the distribution of water under an irrigation regime. The maximum application depth based upon soil type can be estimated from the relation

$$I_{max} = \frac{y \times AWC \times z \times p}{100} \qquad [8.43]$$

where I_{max} = maximum net depth of irrigation water, y = proportion of AWC to be depleted before irrigation, AWC = available water capacity, z = soil depth (root zone), and p = wetted volume as a proportion of the root zone, expressed as a decimal fraction of 0.01 to 1 (Karmeli and Keller, 1975). For field evaluation purposes, the maximum net irrigation depth is converted to the required irrigated volume (V_r) by multiplying irrigation depth by the irrigated area.

The application efficiency of an irrigation system is defined as the percentage of water applied that is actually stored in the root zone compared to the total water applied. When the root zone is not fully irrigated, or a specific irrigation deficit as a per cent of the

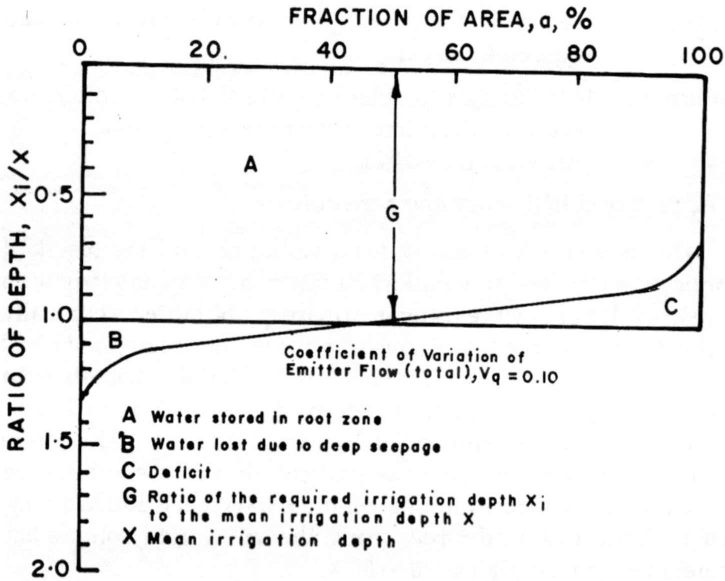

Fig 8.21. Soil water distribution for a coefficient of emitter discharge at V_{qs} = 0.10

total volume of water required is allowed, the irrigation application efficiency is defined by

$$E_a = 100\left[\frac{V_r(1 - P_D)}{V_a}\right] = 100\left[\frac{V_r(1 - P_D)}{3600\, Q_a\, T}\right] \qquad [8.44]$$

where V_a = total amount of water applied, P_D = irrigation deficit (expressed as a decimal), Q_a = actual discharge to the submain per second, and T = irrigation time in hours and all other variables are as previously defined (Wu and Gitlin, 1983). Equation [8.44] can be used to determine the application efficiency when the statistical uniformity is 100% and the irrigation deficit is known.

The relationships between the statistical uniformity, irrigation deficit and application efficiency are based upon probability and normal distribution function. For the special case where the irrigation deficit is equal to approximately 0.4 times the coefficient of variation for the submain unit. In this instance, the application efficiency can be determined by the equation

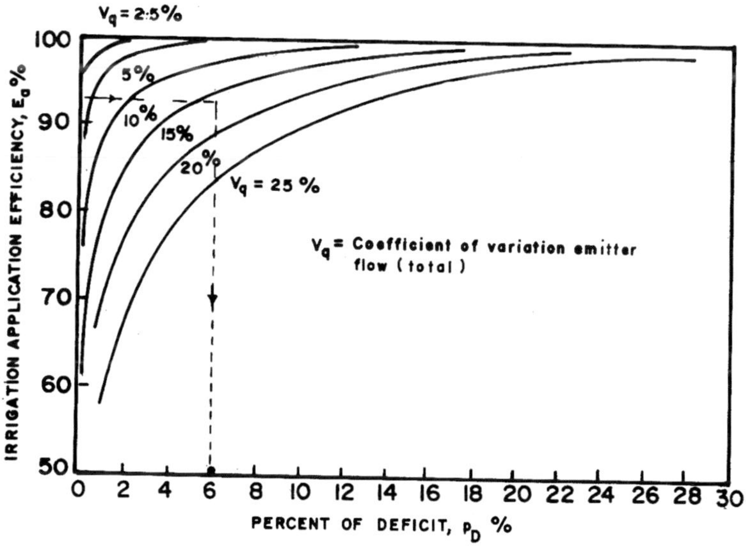

Fig 8.22. Relation between irrigation application efficiency, irrigation deficit, and the coefficient of variation of emitter flow (Wu and Gitlin, 1983).

$$E_a = 100\left[\frac{V_r(1 - P_D)}{V_a}\right] = 100\ [1 - 0.4\ V_q] \qquad [8.45]$$

A dimensionless plot of the cumulative frequency curve is given in Figure 8.21 which shows the required irrigation depth in the root zone for a coefficient of variation of 0.1 or a statistical uniformity of 90% and a deficit of approximately 4%.

The relationship between the application efficiency, irrigation deficit, and the coefficient of variation of emitter flow has been shown graphically in Fig. 8.22. It is evident from Fig. 8.23 that over irrigation to reduce the deficit will be at the expense of application efficiency. Wu and Gitlin (1983) also assume that an irrigation deficit of 0.5 per cent is small enough to be considered as full irrigation. An irrigation deficit of zero is basically meaningless when related to irrigation application efficiency. The relationship of excess irrigation application, irrigation deficit, and the coefficient of variation can be seen in Fig. 8.23. From a field evaluation perspective if the rate of the maximum irrigation required to the actual irrigation applied and the coefficient of variation of emitter flow are know, the application efficiency can be determined using Figures 8.22 and 8.23.

If the resultant efficiency is less than 90% (i.e., unsatisfactory), the irrigator can improve the statistical uniformity by the methods discussed in the previous section and change emitters or improving hydraulics design or if necessary increasing the acceptable deficit percentage. It should be noted that the irrigator could reduce the irrigation application to such a point as to obtain an application efficiency of 100%. However, this would put practical-

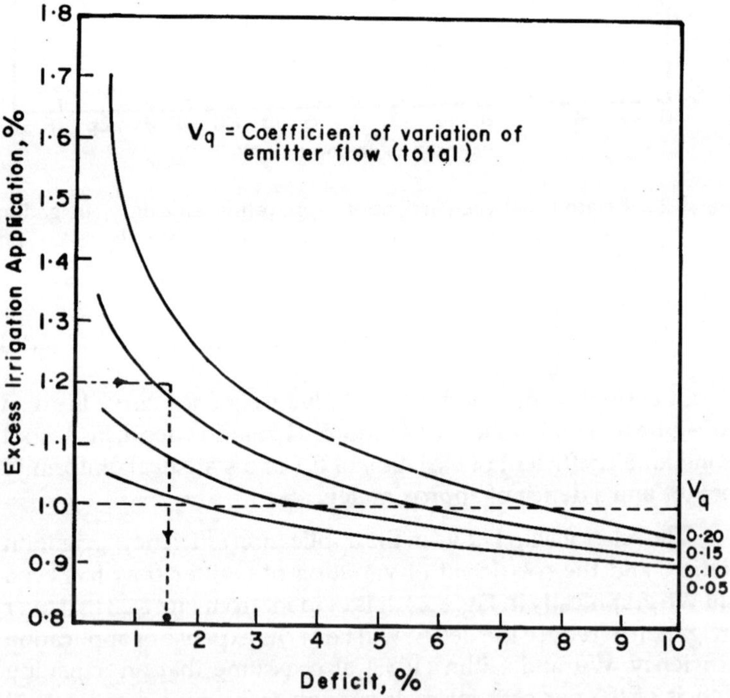

Fig 8.23. Relation between excess irrigation application and irrigation deficit for various coefficients of variation of emmitter flow (Wu and Gitlin, 1983).

ly the whole crop in jeopardy and is not a recommended practice. An allowable deficit in the range of 3 to 6% is recommended when the statistical uniformity is above 85% (Wu and Gitlin, 1983).

The evaluation example Steps 6 and 7 have been developed to illustrate the effects of the field evaluated emitter flow variation, and irrigation deficit on the irrigation application efficiency and schedules. From the examples it should be evident that over irrigation to compensate for nonuniformity will be at the expense of the application efficiency.

Evaluation Example : Step 6

Given : The total statistical uniformity for the drip irrigation submain unit as determined in the previous evaluation steps is 84% ($V_q = 0.16$).

Required : Determine the application efficiency and the per cent deficit when the irrigation application is equal to the required irrigation volume.

Solution : 1. Calculate the application efficiency using Equation [8.45] (Answer : $E_a = 94\%$)

2. From Fig. 8.22 the irrigation deficit for this situation can be found.

(Answer : $E_a = 94\%$, $V_q = 0.16$, and $P_D = 6\%$)

Comments : The submain unit application efficiency is acceptable.

Evaluation Example : Step 7

Given : The statistical uniformity for the drip irrigation submain unit is 84% ($V_q = 0.16$).

Required : Determine the per cent deficit and the application efficiency when the irrigation application is 1.2 times the required irrigation volume.

Solution : 1. Determine the per cent deficit for $V_q = 0.16$ and an irrigation application depth 1.2 times the required irrigation application using Fig. 8.23

(Answer : $P_D = 1.5\%$)

2. Determine the application efficiency of the irrigation using Fig. 8.22

(Answer : $E_a = 83\%$)

Comments : The application efficiency in this example is unacceptable because over-irrigation to make up for a large irrigation deficit or a low system uniformity decreased the application efficiency to an unacceptable level.

V. SUMMARY

In this chapter drip irrigation design and evaluation procedures based on the statistical uniformity concept have been developed. The coefficient of variation and the constant odds uncertainty formulation was used to relate hydraulic design and field evaluation procedures for drip irrigation submain units. In particular hydraulic variation, manufacturer's variation, emitter plugging variation, and the number of emitters per plant were related to submain unit performance. In addition the relationships of the statistical uniformity to application efficiency and irrigation schedules were discussed. Based on the above, a simplified graphical technique for hydraulic design and field evaluation of drip irrigation systems was presented.

The statistical uniformity concept like any other is only useful if actually applied by practicing engineers. The design and evaluation examples presented in this chapter are intended to stimulate that use. The authors have successfully applied statistical uniformity concept to both computerized and non- computerized hydraulic design and field evaluation procedures. The future of drip irrigation efficiency lied in our ability to apply the hydraulic and statistical relationships to practical hydraulic design and field evaluation procedures.

REFERENCES

Bralts, V.F., I.P. Wu, and H.M. Gitlin. 1981a. Trans. ASAE 24(1):113-119.

Bralts V.F., I.P. Wu, and H.M. Gitlin. 1981b. Trans. ASAE 24(5):1234-1240.

Bralts, V.F., I.P. Wu, and H.M. Gitlin. 1982. Trans. ASAE 25(5):1274-1281.

Bralts, V.F. and D.M. Edwards. 1983. Paper No. 83-2596. American Society of Agricutural Engineers, St Joseph, MI 49085

Howell, T.A. and E.A. Hiler. 1974a. Trans. ASAE 17(4):902-908.

Howell, T.A. and E.A. Hiler. 1974b. J. Irrig. Drain. Div. ASCE 100 (IR4):443-453.

Howell, T.A. D.A. Bucks, and J.L. Chesness. 1981. The Proceedings of the Second National Irrigation Symposium, October 20-23, University of Nebraska, Lincoln, Nebraska, ASAE Publication 6-81.

Hughes, T.C. and R.W. Jeppson. 1978. Water Res. Bull. 14:1159- 1166.

Jeppson, R.W. 1982. J. Hydraulics Division, ASCE. 108, No. HY1, p. 35-45.

Karmeli, D. 1977. J. Agric. Eng. 22:165-173.

Keller, J. and D. Karmeli. 1974. Trans. ASAE 17(4):678-684.

Keller, J. and D. Karmeli. 1975. Trickle irrigation design. Rain Bird Sprinkler Manufacturing Corporation, Glendora, CA, p. 133.

Kesner, C.D., B.R. Hahn, W.M. Klein, and V.F. Bralts. 1985. Proceedings of the Third International Drip. Trickle Irrigation Congress, Fresno, C.A.

Kline, S.J. and F.A. McClintock 1953. Describing uncertainties in single sample experiments. Mechanical Engineering, January.

Merriam, J.L. and J. Keller. 1978. Farm Irrigation System Evaluation : A guide for management. Agric. and Irrig. Engr. Dept., Utah State University, Logan, Utah, p. 271.

Moffat, R.J. 1982. J. of Fluids Engineering, Trans. ASAE 104, June, p. 250-260.

Soil Conservation Service. 1983. Trickle Irrigation. Chapter 5, Sec. 15, National Engineering Handbook, U.S. Department of Agric., Washington, DC, p. 129.

Sokal, R.R. and F.J. Rohlf. 1969. Biometry. W.H. Freeman and Company, San Francisco, CA. p. 776.

Solomon, K.H. 1977. Performance comparison of different emitter tubes. Paper presented at the 7th International Agricultural Plastics Congress, San Diego, CA, p. 97-102.

Solomon, K.H. 1979. Trans. ASAE 22(5):1034-1038.

Solomon, K.H. and J. Keller. 1974. Trickle irrigation uniformity and efficiency. Paper presented at ASCE Annual Meeting, Kansas City, MO.

Solomon, K.H. and J. Keller. 1978. J. Irrig. and Drain. Div. ASCE 104(IR3):293-306.

Watters, G.Z. and J. Keller. 1978. Trickle irrigation uniformity and efficiency. Paper presented at the ASCE Annual and Nation Environmental Engineering Convention, Kansas City, MO.

Wilcox, J.C. and G.E. Swailes. 1947. Scientific Agriculture 27(11):565-583.

Wu, I.P. and H.M. Gitlin. 1974. HAES Technical Bull. 96, University of Hawaii, Honolulu, p. 29.

Wu, I.P. and H.M. Gitlin. 1975. J. Irrig. and Drain. Div., ASCE 10(IR4):321326. Proc. Paper 11750.

Wu, I.P. and H.M. Gitlin. 1983. Trans. ASAE 26(1):92-99.

Wu, I.P., Howell, T.A. and E.A. Hiler. 1979. HAES Bull. No. 105, Hawaii, p. 80.

9

Nutrient Management In Response To Irrigation

Raymond W. Miller

I. INTRODUCTION

Planning your fertilization programme is like planning your chess-game--you can use some general plans and concepts to keep out of trouble and be effective, but you must make most decisions one by one, as there are changes in each specific situation (weather, available time, cost of fertilizers, and many more). Each nutrient management situation will have a crop, a soil, a climate, a general management scheme, and the particular interests of the owner or producer. In general, you will incorporate into the soil the needed phosphates and potassium before or during planting. Some nitrogen may be added at that time, also, but half or more of it is likely to be surface-applied later. If you are growing row crops, most of the fertilizer will be put into the soil in a band 8 to 20 cm deep and 5 to 10 cm to the side of the row. Fertilizer added to pastures and other non-row perennial crops will be surface-applied, usually. The amount of fertilizer needed may be determined from soil tests.

At this point in the crop-production design, modifications and compromises begin. Some of the considerations are the following ones.

1. What kinds of fertilizers should one apply? The possibilities include liquids, solids, and gasses.
2. Which nutrients are needed? Does one need micronutrients or sulfur?
3. Does the soil require lime?
4. Are nutrients the only growth-limiting factors? If the plant growth is limited by pH, cold, or drought, the addition of fertilizers will not solve all the growth limitations.
5. How will one apply the fertilizer? Will one apply the fertilizer in irrigation water?
6. How will one irrigate? Surface flow waters can leach out nitrates and urea in the upper part of the field. Sprinkler water can help move fertilizers into the soil.
7. Will surface-applied urea on alkaline soils be left long enough for appreciable ammonia to volatilize?
8. Will the wet clayey soil lose nitrate by denitrification?

The list of questions and the needed choices in a fertility-irrigation programme are numerous. Yet, each decision can be critical to the yield and economic success of each crop. It is expected that many "new" and unique fertilization-cropping- irrigation schemes will continue to appear. The environmental protection agency regulations are additional pressures. This agency will increasingly regulate fertilizer additions and cropping policies to effect more efficiency in fertilizer use. They will increasingly put restrictions on careless application methods of adding fertilizers. The increasing costs of energy will likely make" low input sustainable agriculture" (LISA) more acceptable to some producers.

II. THE GOALS OF NUTRIENT MANAGEMENT

Crop producers have at least three goals. The first goal, generally, is to increase their total crop yields. A second goal might be to raise each crop more cheaply per unit of yield. A third goal may be to improve crop quality, such as to increase the percentage of sugar or of oil, to increase protein content, to produce uniform size and shape, or to enhance the flavour. Defining programme to meet these three goals for the many crops, climatic areas, and diverse soils poses a formidable task for scientists. In addition, other goals, such as doing the needed operations more conveniently, developing more compatible systems, and double cropping for higher yearly production, may be of concern to some growers. The question should be asked, "When do nutrients become the limiting factor in plant growth? Two laws about growth and the factors affecting it have guided researchers. In 1862, Justus von Liebig, a German chemist, stated the law of the minimum. This law states that growth response to an added nutrient will be greater as the amount of that nutrient present (compared to its required amount) is lower. In this situation, growth would be increased by addition of this nutrient. The second law, given by E. A. Mitscherlich in 1909, was similar to that of Liebig but included all growth factors, not just nutrients. Mitscherlich claimed that yield could be increased some amount by increasing the available amount (or improving conditions) of every deficient growth factor, within the plant concentrations needed and favourable for growth.

Not everyone agrees completely with the growth laws of von Liebig and Mitscherlich. They do, however, agree generally that any growth factor (nutrient, temperature, water, pH, salinity, solar radiation, etc.) may, in deficient or undesirable amounts, control the yield of a crop. As we look at modifying the water and nutrient levels, we must recognize the possible importance of other growth factors that may be more growth limiting than nutrients or water.

In low rainfall areas, water and nutrients are two of the limiting growth factors whose supplies are most often critical to growth and yet are quite easily supplied. Good pest control and selection of high-yielding resistant varieties also are critical. It is assumed that the crop manager will use adequate pest control, make a wise selection of varieties, and follow other good management practices and timing. In this chapter the control of water and nutrients, as the most likely methods used to regulate yields, are discussed.

III. INTERACTION OF WATER AND NUTRIENT AVAILABILITY

Availability of nutrients to roots is directly proportional to soil water content. Nutrients become available to plant roots as dissolved ions and move in soil water. Although plants may seem to grow "normally" in drying soils, nutrient uptake and growth rates are usually, if not always, reduced by the lower water contents. When plants must depend on their deeper roots, in wetter soil, for adequate water, the plants also have temporarily become dependent on the nutrient supply in that deeper soil. Nutrient absorption from the fertilized and nutrient-rich surface 15 to 20 cm of soil requires that this enriched soil layer be wetted. A soil with a water content near field capacity has the maximum amounts of its soluble nutrients in solution; the nutrients will move to the root surfaces by mass flow and diffusion. Soil water content near field capacity is considered to be approximately the optimum condition for nutrient availability. Any less water reduces the mass flow and diffusion of nutrients to roots.

Because of other constraints, contents of water and nutrients are usually less than optimum in most soils. One practical constraint to keeping a high water content is that it is uneconomical to irrigate daily in order to maintain a high water content in the surface soil. Another limitation to keeping soil near field capacity is that a wet soil surface damages some produce (cucumbers, strawberries, tomatoes, others) because it favours molds and other diseases. One constraint in keeping available nutrient levels high is that fertilizers must be applied before or during planting in order to place phosphate and potassium into the soil. These placements may limit root access (tight bands, small soil volume), cause early nitrogen losses by ammonia volatization (chisel slits stay open), allow denitrification (in wet clayey soils), result in immobilization of nutrients into microorganisms, or cause reduced availability as the soluble phosphates precipitate and adsorb to soil minerals.

Thus, economics and practical management force us to use practices that are known to be less than ideal; we must compromise. Farming is, in fact, a continual succession of compromises, forced on us by many factors, including economics, time available, physical and chemical soil conditions, practicality, climate, and available equipment. Any programme finally selected will be the manager's conclusion of the best composite of the various compromises ranging from the perceived *optimum* conditions to what is practical to do.Irrigation can be done frequently; the costs and convenience of operating each system, however, will usually determine the frequency. Microbial and enzymatic inhibitors can be added with fertilizer nitrogen to minimize ammonia volatilization and nitrate leaching losses. Material costs are weighted against effectiveness. Banding phosphates often aids in making phosphate available early to seedlings and in a reduction in the amount of phosphate strongly adsorbed and precipitated. Split fertilizer applications, additions of nutrients in irrigation water, and foliar applications are all ways used to improve nutrient availability.

IV. TRENDS IN FERTILIZER USE

Costs, environmental pollution, and effectiveness are three concerns that initiate gradual changes in fertilizer use. Costs are responsible for an increased use of anhydrous ammonia and urea, while the use of the more costly ammonium sulfate and ammonium nitrate decreases. Pollution of the environment by nitrates and phosphates cause users to look for ways to obtain higher fertilizer efficiency and less leaching losses. Growers wish to produce optimum yields; yet, they need to reduce nutrient losses to surface and groundwaters by erosion and leaching.

Operational changes cause a need to look at new systems. An increased use of fertilizer in solution, rather than as solids, has increased the demand for water-soluble phosphoric acid and ammoniated phosphates but has decreased the use of superphosphates. Extensive use of sprinklers, particularly centre pivots, has increased the convenience of applying fertilizers in irrigation water (Johnson et al, 1987). Small increments of fertilizers can be conveniently added twice monthly or in other frequencies throughout most of the growing period, when added in sprinkler irrigation systems.

Changes in tillage, particularly reduced tillage, change fertilization needs. In these less tilled fields, less organic matter is incorporated. Fertilizer bands of previous crops no longer may be ploughed and mixed throughout the surface soil prior to the next crop, when no-till or some reduced tillage systems are used (Sander et al., 1990). Surface-applied materials are incorporated less extensively or not at all, when tillage is limited. Materials banded 15 to 20 cm deep are below even minimum tillage depth. Rolling point-injector (wheel spokes) "spot" applicators place small localized "spots" of fertilizer in undisturbed soil. In contrast to "spot" application, fertilizer also can be added to only the foliage (foliar fertilization). Foliar applications may be used as supplemental applications; the small amounts added act as growth stimulators.

The increased costs of fertilizers and the restrictive controls on nutrient pollution of waters both demand that more attention be given to fertilizer efficiency. Adding a little extra "insurance" fertilizer is becoming more expensive and its practice can increase the hazard of water pollution. Thus, many considerations make the improved efficiency of fertilizers imperative.

V. SOME GUIDELINES IN THE USE OF FERTILIZERS

1. Inorganic Nitrogen Fertilizers

The nitrogen cycle is a complex interrelationship, highly dependent on microorganism activities. Added nitrogen fertilizers do not reside in soil as slightly soluble materials. Urea, anhydrous ammonia, or ammonium forms dissolve readily and are oxidized to nitrate by soil bacteria. Either the ammonium ion (not very mobile in soil) or the nitrate ion (very mobile in soil) may be used by plants and microorganisms. The nitrate will move with water through soil. Volatile gaseous forms of ammonia, dinitrogen, and various nitrogen oxides may form and be lost from the soil.

The rapid and continual transformations and losses of soil nitrogen are important in retention or losses of nitrogen. One way to increase fertilizer efficiency is to add nitrogen in smaller increments by use of two or more split applications. Another way to increase fertilizer efficiency is to add it correctly, to reduce losses. Large losses of nitrogen can occur if nitrogen is added with poor techniques or in certain soil conditions (Howard and Tyler, 1989; Christianson, 1989; Buresh, 1987; and Fenn et al., 1987). For example, surface applications of urea or ammonium materials on alkaline soils may cause large losses of ammonia (Fenn and Hossner, 1985). The application of urea or nitrates followed by heavy rains or irrigations may leach urea or nitrate from the upper root zone. Wet and poorly drained (anaerobic) soils may lose large quantities of nitrogen by denitrification. Urea, the most used solid nitrogen fertilizer, can be used efficiently or inefficiently. Urea is immediately soluble and moves readily in percolating water. Thus, surface applications can be moved into the root zone by rain or irrigation. If urea is left on the surface of calcareous soils, its hydrolysis by urease forms alkaline ammonium carbonate. This reaction produces a high pH, allowing ammonia to be volatilized. If the ammonia is not volatilized, ammonium is oxidized to nitrate, becoming very mobile with soil water. Thus, urea is mobile in the first day or so, then is immobile as ammonium is formed, or is partly volatilized. Finally, after several days, the nitrogen becomes mobile again as nitrate.

The most used nitrogen fertilizer in the U.S. is anhydrous ammonia. It is the cheapest nitrogen source but also the most troublesome form. Efficient application of anhydrous ammonia depends on the absorption of ammonia into soil water. Ammonia is very soluble in water and the reaction is rapid. However, poor ammonia retention is likely in soils that are quite dry, particularly in sandy soils. Appreciable ammonia volatilization loss can occur, also, when wet clayey soils fail to cover back over the chisel slit, thus allowing escape of some of the gas. Because of the toxicity of ammonia to roots and seedlings, the application of ammonia usually is done several days to a week prior to planting. The injection chisel is drawn through the soil at a depth of 10 to 20 cm to minimize ammonia volatilization losses. The ammonia dissolves into the soil water and forms ammonium hydroxide of pH about 9 to 10 in the concentrated zones. This high pH solution will gradually be dissipated into surrounding soil in a few hours or days. Nitrification of the ammonium will produce acidity, thereby helping to neutralize the alkalinity produced in the solution.

2. Nitrogen Slow-release Fertilizers

Many materials added as sources of nitrogen do not have the nitrogen easily soluble in water. To become soluble, these materials require processes that are time dependent. Some of these fertilizers have low solubility or depend on diffusion of soluble nitrogen out small holes in pellet coatings. Other materials (urea-formaldehyde polymers) require microbial decomposition. Organic wastes, urea-formaldehyde polymers, sulfur-coated urea, and polymer-coated or petroleum-coated fertilizer materials are some slow-release substances commonly used (Hauck, 1985). Most of these slow-release materials are used only in specialty high-value crops (nurseries, golf greens, professional turf, landscaping, lawns, and indoor plants) (Boodley, 1981; Duble, 1978). Their costs range from about 25 to 300 per cent more than the simple inorganic materials.

The organic wastes include animal manures, processed organic wastes (dried blood, seaweed, animal by-products, food waste, sewage sludge, etc.), and crop residues These are used on crops mostly when they are free or cheap; the materials have varying values. Most organic materials are used, in part, to improve the soil's physical properties as much as they are for a nutrient source.

3. Phosphate Fertilizers

Most popular phosphate fertilizers used are moderately to very soluble phosphates but the phosphate does not stay soluble very long. Phosphates adsorb rapidly to soil particles; phosphates also form insoluble phosphates. The difficulty with added phosphorus materials is that phosphates have essentially no mobility and rapidly decrease in soluble phosphate concentrations. In soils of arid regions, calcium carbonate and iron oxides adsorb phosphates; soluble calcium forms low- solubility calcium phosphates (dicalcium and tricalcium phosphates). Superphosphates and ammonium phosphates have 88 to 96 per cent of their phosphate in soluble form (Young et al., 1985). These soluble phosphates soon form less soluble calcium phosphates. The solubility of common

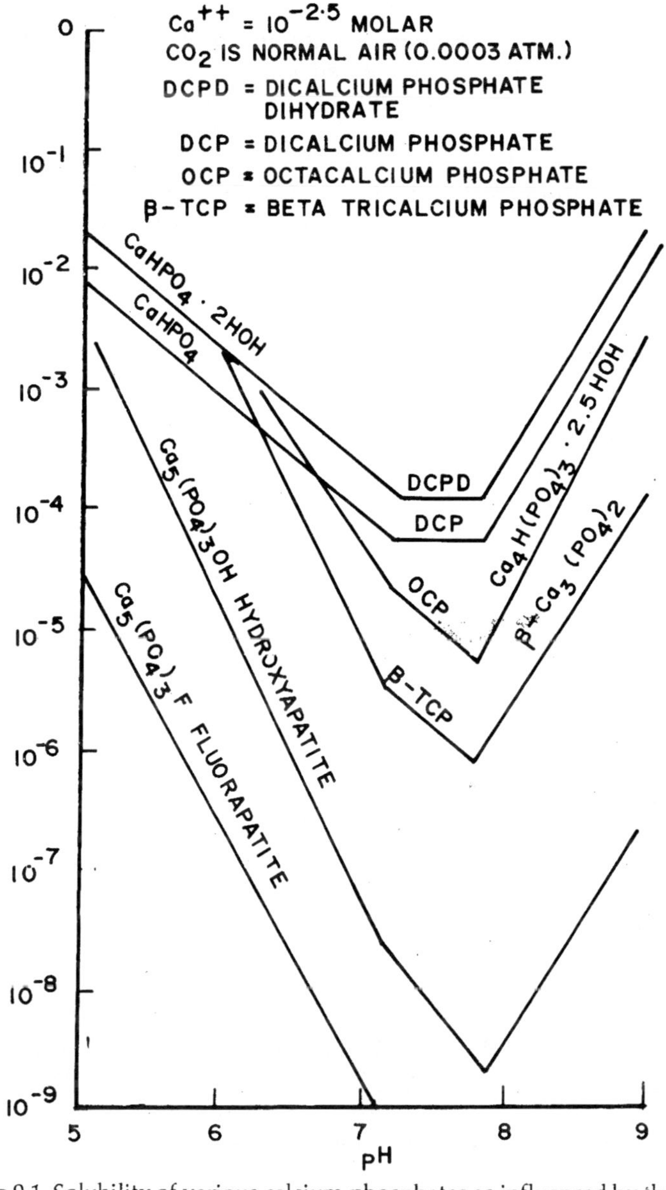

Fig 9.1. Solubility of various calcium phosphates as influenced by the pH of the media. Redrawn in part, from Lindsay (1979)

calcium phosphates is modified by soil pH (Fig. 9.1). Lindsay (1979) observed a generally lowered solubility of the various calcium phosphates as the soil pH increased up to pH 7.5 to 8. At pHs higher than about pH 8, the solubility again increased. In alkaline soils, the soluble orthophosphate added ($H_2PO_4^-$) forms the less soluble dicalcium phosphate dihydrate (DCPD) and dicalcium phosphate (DCP). Within three to five months much of these forms will reform into even less soluble octacalcium phosphate (OCP). Within the next 5 to 15 months much of the phosphate will be converted to the less soluble [Beta]-tricalcium phosphate ([beta]-TCP).

It is evident that maintaining available (soluble) phosphates in alkaline soils requires that we slow and minimize the formation of insoluble calcium phosphates. Generally, concentrating the fertilizer into bands lessens the contact of phosphate with soluble soil calcium. Periods of surface soil drying should speed conversion to the less soluble dicalcium phosphate and octacalcium phosphates.

The general approach to keeping maximum amounts of added phosphate in forms available to plants is to concentrate the fertilizer as bands, which are at least 7 to 15 cm deep into the soil. Phosphate rarely moves more than a few centimeters, with 10 to 12 cm often the maximum distance of downward movement in some sandy or porous soils (Young et al., 1988). Phosphates should be rapidly and optimally available to roots. Because of the need to place phosphate down into the soil and to keep it close to seedling roots, phosphate is usually placed in soil at the time of planting near to the seed. Some phosphate could be applied later during plant growth as a foliar spray or in irrigation waters. Neither of these latter methods has been widely acclaimed as methods to supply phosphate.

Heavy additions of phosphate are needed because of low phosphate solubilities, not plant phosphate contents. Plants contain much lower percentages of phosphorus (P) than of nitrogen or potassium. Nitrogen and potassium in plants may be more than 2 per cent of dry weight; phosphorus is closer to 0.1 to 0.3 per cent. For a high-yielding alfalfa crop of 10 Mg ha^{-1}, this crop would contain about 20 to 30 kg of total phosphorus (P) or about 60 to 90 kg of HPO_4^{2-} This is about 45 to 67 kg calculated as P_2O_5. Much of this phosphate requirement can be supplied by residual soil phosphates.

Single and triple superphosphates, mono-ammonium and diammonium phosphates, phosphoric acid, and polyphosphate are all highly soluble phosphates. The various materials produce some differences in soil pH around the fertilizer during dissolution.

4. Potassium Fertilizers

Potassium ions (K^+) have low mobility in soils. As they begin to move through soil, they replace other exchangeable cations and are retained as exchangeable cations in a small volume of soil near the fertilizer pellet. Potassium ions do not form insoluble precipitates but they do adsorb to cation exchange sites. They are only slightly more mobile in soils than is phosphorus. Over time, however, potassium can move in soil.

Potassium and phosphate are usually banded in soil at the time of planting. The major portion of this water soluble potassium becomes immobile, but largely available, as exchangeable potassium. A considerable portion of that potassium not used by the current crop may be available in subsequent years. Some of the added potassium is "fixed" in interlayer clay sites, particularly in soils containing illite and vermiculite clays. Excess potassium additions will cause increased fixation into clay lattices.

The limited leaching in arid soils leaves most potassium that is released from weathering minerals in the soil. Consequently, most arid region soils have adequate potassium supplies for many crops. Exceptional situations do need potassium (Weir et al., 1986). Responses to added potassium on arid irrigated soils are mostly on sandy soils, on sugar or starchy crops (sugarbeets [*Cheropo–diaceae*], potatoes [*Solnum tuberosum*], sugarcane [*Saccharum officinarum*), or on soils with small amounts of primary minerals which contain potassium. Heavily cropped soils may gradually have marginal amounts of potassium as the potassium released by centuries of weathering are "mined" from the soil by crops and new potassium supplies by weathering is slow. Particularly with potassium, subsoils may supply large portions of the amount needed by crops if the surface soil amounts are deficient.

5. Other Fertilizer Nutrients

In arid region soils, a few other nutrients might be deficient. Calcium and magnesium are usually in large excess. Sulfur is usually adequate except in soils from certain low- sulfur minerals, such as in sands. However, the alkaline soil pH may cause crops to be deficient in certain micronutrients, particularly in iron and zinc. Although deficiencies of sulfur, iron, and zinc are somewhat limited, these deficiencies are common enough to make it important to be aware of areas with inadequate amounts (Yoshida and Chaudhry, 1979; Sabbaich et al., 1986).

Sulfur is easily applied as gypsum, as a component in regular superphosphate, as potassium sulfate or ammonium sulfate, as ammonium phosphate-sulfate, as urea-sulfur pellets, as sulfur-coated urea, and as elemental sulfur in various forms (Beaton et al., 1985). Elemental and polysulfide sulfur will require from a few weeks to several months to oxidize to sulfate and the process produces some acidity (sulfuric acid). The SO_4^{2-} form used by plants is relatively mobile in the soil solution. In fact, irrigation waters may contain appreciable sulfate. Beaton et al. (1985) suggest that irrigation waters with 4 to 6 mg L^{-1} of S will supply enough S for most crops.

To supply needed iron to plants, iron chelate materials (amendments) are the most available forms. Iron oxides and hydrous-oxides are of very low solubility. The only known way to supply adequate iron is as *soluble organic chelates* or as soluble iron (such as $FeSO_4$) applied as a foliar spray. The chelates may be synthetic organic "sequestrene" materials or from various natural sources. Animal manures and decomposing humus produces chelating substances or complexing organic compounds.

Zinc inorganic materials ($ZnSO_4$ and ZnO) generally are adequately soluble to supply the small amounts of zinc needed. Zinc can also be supplied in chelates and in crushed (fritted) glass containing high zinc contents. Small amounts of zinc salts can be mixed with bulk blended fertilizers, if the Zn deficiency is sufficiently widespread to justify large bulk fertilizer production. In fluid fertilizers, the small amounts of zinc, such as 5 to 10 kg Zn per hectare, can be added into the liquid mixture.

VI. SOME NUTRIENT-MANAGEMENT SYSTEMS

Crop production is a continual sequence of choices and compromises. The agriculturalist must modify his plans as weather changes, as he changes crops, as he receives soil test data, as he modifies irrigation systems, as costs of fertilizers and water change, and as labour and market prices change. The remainder of this chapter emphasizes various irrigation systems, changes in the crops grown, and modifications or peculiarities of some sites (soil, geography, climate characteristics).

A. Surface Flow Irrigation

Regular surface flow irrigation by borders or furrows is simple and easy. Sometimes it is expensive in labour, although it is usually the cheapest method of irrigation. However, surface irrigation causes excess leaching in the upper part of the field and has high water runoff loss off the end of the field (low efficiency) unless runoff is collected and recycled. Flood or border irrigation alters fertility and plant survival differently than does furrow irrigation.

(1) **Flood and border irrigation:** Border and flood irrigation is adding water on a strip of graded soil. Planting is done on the flat and fertilizer is broadcast-incorporated or banded into the soil at planting time. Addition of irrigation water moves urea, nitrates, and sulfates deep into the soil, to deeper depths at the top of the field than occur in the lower (bottom) portion of the field. The upper portion of the field is usually overirrigated and nitrogen leaching loss can be large. Long application times makes the control of soil wetness a problem in many of these fields. White (1985), a farm advisor in California for 30 years, claims that irrigation drown-out is a major damage to alfalfa (*Medicago spp.*) in flood-irrigated fields. Soils with shallow hardpans, or those with high clay contents, are most likely to have problems.

In California, yields of durum wheat (*Triticum turgidum*) at Tulelake during four years, using flood irrigation, illustrates the

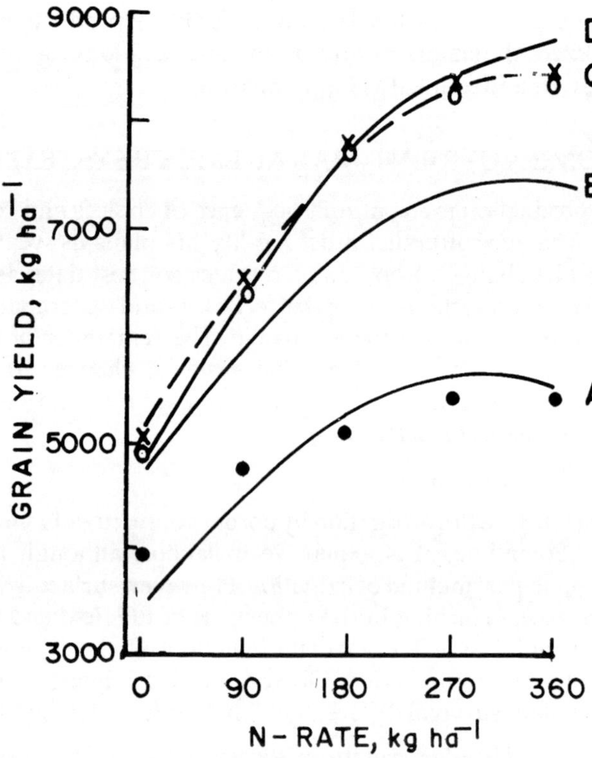

Fig 9.2. Influence of N rate and four irrigation regimes on yield of wheat. From Puri et al. (1988)

critical need for good water control, adequate nitrogen, and optimum seeding rate to maximize wheat yields (Puri *et al.*, 1988). Ammonium sulfate was incorporated into the soil by disking prior to seeding. Figure 9.2 shows the yields for four irrigation treatments averaged over four years and five seeding rates. Maximum yields on this peaty clay loam (12% organic matter, diatomacecous earth, pH 7.5) required three or four irrigations. Irrigation 'A' was one irrigation at pre-plant (12.5 cm deep); irrigation 'B' was 'A' plus a second irrigation at tillering (15.2 cm deep); irrigation 'C' was 'B' plus a third irrigation at booting stage (15.2 cm deep). With these irrigations, at least 180 kg ha^{-1} of nitrogen was very effective and 270 kg N ha^{-1} increased yields even more. A fourth irrigation

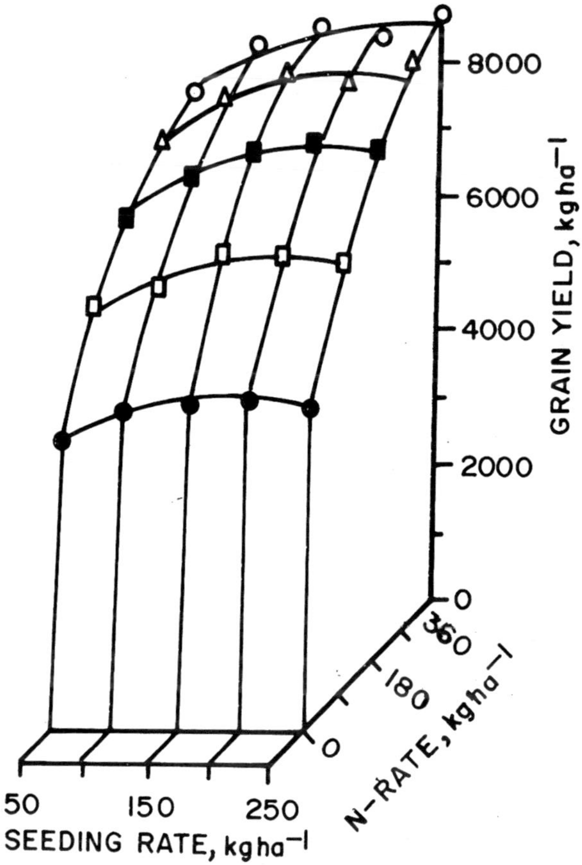

Fig 9.3. A 3–dimensional surface of wheat yield as influenced by seeding rate and N rate. From Puri et. al. (1988)

treatment, 'D', was the three irrigations of level 'C' plus wetting to 15.2 cm deep at early soft dough stage. This extra irrigation in 'D' increased yields in some treatments. Figure 9.3 shows a 3- dimensional graphed summary of average yields, as influenced by seeding rate and nitrogen fertilizer rate, when the mean of the three water irrigation levels (B, C, and D) are used. These are typical yield curves. Curves such as these will differ for each crop, location, and water treatment.

(2) **Furrow irrigation**: Furrow irrigation water flow has particular influence on nitrogen, sulfur, and broadcast or surface fertilization. Water flows downward in furrows, sideways into ridges, and upwards slowly into ridge tops. Mobile materials (urea, nitrate, sulfate, and boron) flow with the water. Mobile fertilizer put in upper layers on the surface of ridges will more likely move upwards into the ridge top during irrigation and drying (evaporation) cycles. Mobile fertilizer applied into furrows is washed deeper into zones with less roots; mobile materials applied to ridges may stay on the ridge tops or in the upper ridge layers (5-10 cm deep) where it was placed. In contrast, sprinkler water would tend to move materials downward from all surface or shallow application sites.

Deep wetting in the upper portion of the field and runoff water loss is a problem in furrow irrigation. In a study of water needs for wheat in Saudi Arabia, surface irrigation required about 15 to 25% more water than a centre pivot sprinkler (Mustafa et al., 1989). As the water was saltier (6 dS m^{-1}), the water needed for surface irrigation approached values of 40 to 50% more.

Careful attention to the width between furrows and the irrigating time period is important to assure that the ridges, where most plant roots exist, are completely wetted. If wetting is not adequate, even the banded fertilizer may be in soil that is somewhat dry much of the time. This problem is minimized by carefully grading fields, using shallow furrows, and employing wetting times long enough to allow most or all of the ridge to wet. Unless rainfall is common during the growing season, shallow band or surface fertilizer application on ridges of furrow-irrigated land is usually inefficient.

B. Sprinkler Irrigation

Most interest in specialized water-fertilizer innovations involves sprinkler and drip systems. The mobile centre-pivot systems are of particular interest because they irrigate a large area (50-60 ha or 130 to 150 acres) under one pivot. Fertilizers and other chemicals, such as pesticides, can be applied in the irrigation water. Even application of a polymer soil conditioner, Krilium (now modified and called polyacrylamide), at a rate of three pounds per acre in one-eighth to one inch of water (0.55 kg ha^{-1} in 0.3 to 2.5 cm of water) has been used (Wallace and Wallace, 1987). The polyacrylamide application increased lettuce (*Lactuca sativa*) and tomato (*Lycopersicon lycopersicum*) seedling emergence. The treatment increased soil aeration 75% and the weights of strawberry plants (*Fagaria X Ananassa*) 2.2 times. Costs of the soil conditioner could be as low as \$3 to \$10 per acre (\$7-\$25 ha^{-1}). Planting by applying seed in the water spread by centre pivots, as unlikely as that sounds, has been tried (White, 1985a). Turnips (*Brassica rapa*) were planted "perfectly" using a vegetable oil carrier. To see what extremes were possible, sweet corn (*Zea mays*, var. saccharata) was planted using "typical sludge nozzles". Although there are details to be learned, the possibilities of planting by putting seed through the sprinkler do exist for certain crops in certain conditions.

The addition of fertilizers in sprinkler-applied irrigation water is very effective. Peanuts (*Arachis hypogaea*) in Georgia, potatoes (*Solanum tuberosum*) in Idaho, cotton (*Gossyipium hirsutum*) in Arizona, wheat (*Triticum aestivum* L.) in Kansas and Utah, rice (*Oryza sativa* L.) in Arkansas, and corn (*Ze1 mays*) in Nebraska and Colorado are some examples of widely used sprinkler- applied fertilizers. Larsen (1985) listed these relative costs, assuming tractor sprayer and aircraft to be similar in cost:

Chemical applied	Sprayer or Aircraft	Centre pivot	Water applied cm
	Dollars ha^{-1}		
Fertilizer	6.20	5.60	1.3
Herbicide	14.00	4.50	1.0
Insecticide or fungicide	5.60	1.70	0.4
Nematocide	14.00	5.60	1.3

Some of the advantages of chemigation include these:

1. The uniformity of application is generally good,
2. It is possible to apply the chemical when it is needed,
3. The method is as cheap as ground or aerial application,
4. The amount needed can be applied,

5. Soil compaction is reduced,
6. Operator hazard is reduced and addomg pesticide in the mixture is easy, and
7. The efficiency of applied chemicals is good and well- documented.

Studies in Georgia's Coastal Plains prompted the statement that, "Our research with chemigation has involved a total of 26 crops, and never has chemigation failed to be at least equal to or better than conventional methods of chemical application." (Larsen, 1984). Adding 160 kg N per hectare in irrigation water through minisprinklers (one sprinkler per orange tree [*Citrus sinensis*]), that irrigated 70 per cent of the surface area, produced higher range yields than did a single, yearly, surface broadcast of fertilizer beneath the tree canopy (the conventional application) (Dasberg et al., 1988). When the nitrogen, phosphorous, and potassium was applied proportionally in each irrigation, the yields were good.

Even in wet areas commonly growing paddy rice (*Oryza sativa* L.), as in Arkansas, centre pivots have increased profits with or without increased yields (White, 1985). Pivots reduce costs by lessening the need for land surveying, levee construction, flood gates, levee ripping to harvest and re-form land, and damages and wear on that equipment. Under sprinklers, weed control must be given more attention and pivot wheels require "caterpillar treads." Nevertheless, overall costs are reduced. Two or three aerial fertilizer applications, often with high enough rates to burn somewhat, were applied to paddy rice. With the pivot, fertilizer can be applied in the irrigation water in more frequent but smaller applications, each application supplying about 25 to 30 kg of nitrogen per hectare.

The increasing pressure by the Environmental Protection Agency to reduce leaching of nitrate into groundwaters encourages use of centre pivots. As with any sprinkler system, it is possible to avoid wetting too deeply and applying excess water. Fertigation (adding fertilizer in irrigation water) also can be controlled to add nitrogen frequently but in small amounts as plants require it. This reduces possibilities of leaching nitrate into groundwaters. This is an outstanding asset in the irrigation and fertilization of sandy soils. For example, Sharma et al. (1990) studied winter wheat (*Triticum aestivum* L.) grown on a loamy sand of pH 8.0 The largest application of N (180 kg N ha^{-1}) correlated to the highest yield (4.1 and 4.9 Mg ha^{-1}) at the 1.2 irrigation-to-pan evaporation ratio during two years. Yet, careful control of applied water will minimize nitrate leaching at these high fertilizer rates.

Fertigation is not, however, without some limitations. The cheapest form of nitrogen--anhydrous ammonia-can have large losses by volatilization of NH_3 when applied in sprinkler systems. Anhydrous ammonia use is not recommended in sprinklers (Ratledge, 1988). Similarly, soluble phosphates, the most expensive phosphates, seldom produce the yield improvement equal to their higher costs. Phosphates move only a few centimeters deep into soils in irrigation water. Potassium is nearly as immobile over short time periods as is phosphorus; likewise, it is not often recommended in fertigation. However, on pastures and perennial crops and on sandy soils, chemigation with phosphates and potassium have been beneficial. In medium and fine-textured soils it may require as much as 3 cm of water to get some of the added chemicals to 1 or 1.5 cm into the soil.

Sometimes, non-uniform coverage of sprinkler-applied chemicals happens. This non-uniform coverage is caused by many factors, including wind, improper sprinkler design, plugged or incorrect nozzles, hills at larger angles than machine was designed to handle, and inadequate pump pressure (White, 1985b). Applications by most pivots should fit within a uniformity coefficient of about 85 per cent. This is a range of about 0.84 cm to 1.16 cm of water on different sites of the field (measured by collection cans) for an expected application of about 1.0 cm of water. Fertilizer uniformity applied in the water would be similar to the water distribution pattern. This does allow about 15% error above or below planned addition rates of fertilizer.

Sometimes growth plans may be designed to "lose less" rather than to "make more" profit. In Kansas, irrigated winter wheat (*Triticum aestivum* L.) is planned as a no-money-loss rotation crop to help control weeds in an area dominantly growing corn (*Zea mays*). Expectations are 100-bushel wheat (6.7 Mg ha^{-1}) using hybrids, sprinkle irrigation, and the correct fertilization (Gardner, 1984). Only 20 to 30 kg ha^{-1} of nitrogen is applied in the fall to limit the growth of plant size and minimize lodging later. Through the sprinklers, 50 to 60 kg ha^{-1} of nitrogen can be put on in early March. Another similar amount is applied at a later growth stage and a final application at flowering if information suggests that protein content can be boosted. A liquid starter fertilizer is based on soil tests and applied at the time of planting.

The use of sprinkler systems to add fertilizer has proven very effective in particular locations. Sandy soils have been a challenge to fertilize economically. The solution was often to add more fertilizer and expect to have considerable loss from leaching. With increasing concern about pollution, leaching losses must be reduced and kept low. The sandhills of Nebraska are one of the sandy soils irrigated by centre pivots. The goal on these sandy soils is to raise the crop on a basis of cheaper cost per unit of product (not just cheaper cost per acre, necessarily). To do this in the sandhills requires some planning. Crops in the sandhills face ovenish winds, sand blasting, and static-electricity damage to leaves from the dry moving sands. Compared to early years, when simple split applications of fertilizers were used, corn yields now involve more integration of fertilizer and water management (White, 1985), as described below :

1. Irrigate to keep the top 18 inches (45 cm) wetted. Usually a total of 24 to 30 inches (61-76 cm) of water are applied each season.
2. The sand is wetted, by rapid pivot speed (12 hours rather that 36 hours per revolution), whenever dryness becomes a problem.
3. About 50 to 60 pounds of phosphate (55 to 65 kg P_2O_5 ha^{-1}) and 12 to 15 pounds of potash (14 to 19 kg K_2O ha^{-1}) are injected about 8 inches (20 cm) deep prior to planting. A little sulfur and nitrogen may be added if soil tests or experience indicates they are needed.
4. The remaining 28 to 30 pounds of potash (31 to 33 kg K_2O ha^{-1}), most of the 185 lb of N (207 kg ha^{-1}) and 30 lb S (34 kg

S ha^{-1}) are added in irrigation water in as many as six separate applications.

5. Additional guidance is taken from soil tests, computerized ET estimates, a constant phone contact with flight service weather stations, a private weather station, colour and near-infrared aerial plots by a full-time pilot (who also sprays), and specialists in entomology and other needs.

What seems like a well-mechanized farm operation is in truth a highly integrated business with many components. The search, as done by any forward-looking management, is to find better and/or cheaper ways to do it all.

In Minnesota, corn (*Zea mays*) yields require similar concern. Careful management of one farm with low soil pH and porous soil produced the 1983 state-winning yields of 196.7 bu/a (12,348 kg ha^{-1}, first place) and 186.5 bu/a (11,708 kg ha^{-1}, second place). Their yearly programme was approximately as follows (Anonymous, 1984):

1. In 1983, six tons of lime (13.4 Mg ha^{-1}) and 300 lb potash (336 kg K_2O ha^{-1}) were applied. Also, 215 lb/a (241 kg ha^{-1}) of 15- 32-30-20S starter fertilizer were placed with corn.
2. A later side-dress of 200 lb/a (224 kg ha^{-1}) of "stabilized" anhydrous ammonia helped produce an average yield of 178 bu/a (11,174 kg ha^{-1}). The goal was 180 bu/a.
3. A third addition consisted of 30 lb N/a (34 kg ha^{-1}) in the irrigation water at the 12-leaf growth stage.
4. Plant populations were planned at 28,300 plants per acre (69,900 plants ha^{-1}).
5. Irrigation was by a low-pressure pivot making a 135-acre (54.6 ha) circle each 5 days and applying an inch (2.5 cm) of water.

The choices a producer makes vary with his interest, equipment, and experience. There is opportunity for many changes and modifications to any system. Soil texture and climate determine water needs but not how it is to be added. Fertilizer rates depend on total needs of the crop, leaching losses, rates of use, and kinds of materials. Usually phosphorus and potassium will be put on before or during planting. Most nitrogen will be split into several applications.

C. Drip (Trickle) Irrigation

Some of the most efficient and unusual fertilization schemes have been associated with drip irrigation. The water efficiency is high and little fertilizer should be lost by leaching. Although it is assumed that availability and uptake of surface- applied phosphate is proportional to shading of the soil by plant cover and the surface soil's water content, few studies have given direct proof of these factors. Simpson and Pinkerton (1989) conducted a greenhouse study using surface-applied phosphate on subterranean clover (*Trifolium subterraneum*). The phosphorus uptake increased sharply with increased frequency of surface watering. They concluded that adequate supply of phosphate can be maintained only when there is a high level of available water. This is in agreement with findings by Mouatt and Nes (1986) and with the general concepts of broadcast (surface) fertilization of pasture crops

Interesting information on the trickle irrigation-phosphate availability relationship was given by Bacon and Dovey (1989) in a field study on a vineyard in New South Wales, Australia. Placement of superphosphate at the trickle outlet increased phosphate availability to at least 22 cm deep. The soil volume containing contents of readily available phosphate greater than background levels increased as phosphate application rate increased. The soil was a moderately acidic to slightly acidic sandy loam (0-25 cm) over neutral clay soil below that depth. Broadcasting phosphate on the surface caused no increased available phosphate below the top 5 cm. By contrast, placement of 18 g of P at the trickle outlet increased available phosphate (Bray No. 1 test) a minimum of 15 cm beyond the placement site. When 73 g of P was added near the outlet, phosphate was available to 45 cm laterally and 40 cm below the soil surface. When placed around an outlet, the phosphate was spread in a square 25 cm x 25 cm beneath the trickle unit. Figure 9.4 shows the Bray No. 1 test values measured in soil samples taken one year after initially applying fertilizer. Water was applied every third day yearly during the 150-day irrigation season. The re-

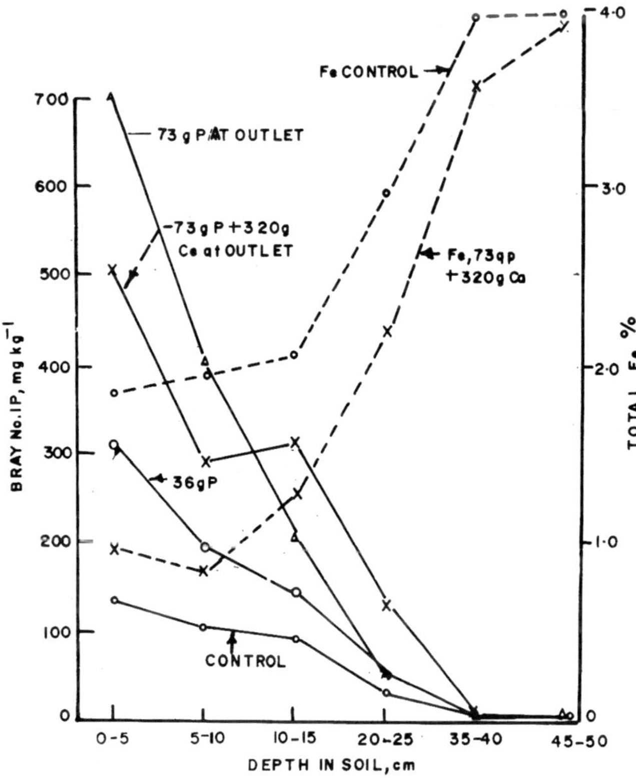

Fig 9.4. Concentrations of Bray No. 1 phosphorus and total iron at different depths under a trickle emitter with and without added phosphorus. From Bacon and Davey (1989)

searchers concluded that the reducing conditions in the soil during each irrigation cycle helped move iron the 35 to 40 cm deep and 15 to 45 cm laterally. This movement of iron and increased solubility of phosphate in water-logged soil would increase the chance for phosphate movement. Some phosphate appeared to have moved 10 to 25 cm (4 to 10 inches).

.The major reasons for using drip systems are to conserve water, to control surface wetting, to minimize damage from moderate salt levels in water or soil, and for irrigating steep lands. The modifications possible to integrate water and fertilizer applications are many. The few examples that follow only suggest some of the possibilities.

Sundance Farms, a large 4,500-acre (1,821 ha) producer between Tucson and Phoenix, Arizona, produces cotton (*Gossypium hirsutum*), pumpkin (*Cucurbita pepo*), small grains, squash (*Cucurbita maxima*), melons (*Cucumis melo*), and lettuce (*Lactuca sativa*) (Anonymous, 1988a). With such a variety of crops, the dry-climate, water-short producers emphasize water conservation-- they use drip irrigation. But, their drip line is buried 8 to 10 inches (20 to 25 cm) deep into the soil, for many years use. Lightweight tillage equipment has been designed to till only 3 to 4 inches deep (7.5-10 cm) to preserve the lines. To allow such long installation with minimal maintenance, water is carefully filtered through sand, and flushed regularly. To keep lines clean, small amounts of chlorine, sulfuric acid, and phosphoric acids are used to bring water pH to about 6.5 to 6.0 to minimize salt precipitation (especially calcium carbonate). Because the emitters are buried in the soil, all of the N, P, K, and some micronutrients can be applied by fertigation in the drip system. Foliar feeding is also used, particularly for zinc, iron, manganese, and some phosphate and nitrogen.

Although subsurface drip lines would seem to have problems of maintenance and installation costs, they do indicate some potential for yield increases and water and fertilizer reductions. Research in California (by C. J. Phene) demonstrated other facts about some subsoil drip systems (Anonymous, 1988b). Laterals were buried at a depth of 18 inches (45 cm) and 35 inches (88 cm) apart. High frequency irrigation (several times a day) was done, and irrigation done after 1 mm of water loss was measured. This was claimed to be more efficient in water and fertilizer than irrigating every few days (after loss of 25 mm of water, for example).

Usually the buried drip system caused the crop to have a smaller root zone than was produced when irrigating with regular sprinklers. In the buried drip system, large water and nutrient gradients could develop rapidly. Banding phosphate under the seed at planting (112 kg ha^{-1} of 11-48-0) followed by daily injection of N, P, and K through the drip system (totals of 301, 134, and 381 kg ha^{-1}, respectively) significantly increased tomato (*Lycopersicon lycopericum*) yields. Yields of tomatoes were:

lb $acre^{-1}$	(kg ha^{-1})	Drip System
206	231	SSD (high-frequency subsurface drip)
190	213	HFSD (high-frequency subsurface drip)
179	201	LFSD (low-frequency subsurface drip)

Percentage culls were SSD = 2.0%, HFSD = 2.8%, and LFSD = 4.5%. The somewhat unusual aspects of this study were the multi-irrigations daily and the daily application of fertilizer in that water. Such application should keep optimum water and fertilizer levels available to roots.

Papadopoulos (1988) claimed that fertigation in trickle- irrigated potatoes (*Solenum tuberosum*) on Vertisols (swelling clayey soils) produced 58,130 kg ha^{-1} of tubers when fertilized with 130 mg L^{-1}. of nitrogen every irrigation.

Drip irrigation is a watering scheme that needs to be better understood to use it optimally to increase yields. It usually wets a smaller soil volume, so the root volumes of crops are less. During irrigation, soil aeration of the wetting zone may be less than is good for optimum aeration. Herbicides break down more quickly under drip (Renn, 1985). Plugging of emitters by salts and microbial action (calcium sulfates, iron hydroxides, various phosphates, slimes and molds) makes water filtration critical.

Careful attention to costs of systems is important. Applying fertilizer in water requires purchasing higher-cost soluble phosphate and nitrogen materials. A study in the San Joaquin Valley of California (an area of high water table and salt problems) compared three surface flow and a subsurface drip system (Fulton et al., 1991; and Smith et al., 1991). One major cost was the 10-fold cost of fertilizer (high-cost materials, partly because of low amount purchased); other costs included the fumigant needed in the drip system and the system cost itself. The drip system did reduce potential drainage, a major concern, and increased production. However, the high particular costs of the subsurface drip caused an overall profit loss in the cotton production management system. Projected changes of wider line spacing, buying the cheaper individual fertilizers, and reducing fumigation could greatly reduce costs.

Drip (trickle) irrigation will not be suitable for all regions, soils, or crops. Its particular promise is for water conservation in arid regions, for use on sandy soil and moderate to steep slopes, and for widely-spaced and high-value fruit and vegetable crops. It will have value in many crops that respond well to high water levels and heavy fertilization.

VII. LOOKING AT THE FUTURE

The future in fertility management will include increasing efficiency, producing minimum pollution of the environment, and developing high-yielding production systems with less costly input. In effect, we have been trying to follow these guides for years. However, fertilizer costs, water shortages, environmental pollution concerns, and regulatory laws force us to make greater efforts and to reevaluate some of the priorities. Some of the needed research studies are given in the following paragraphs.

A. Improving Phosphate Efficiency

Phosphate moves little in most soils and soluble phosphate rapidly becomes adsorbed and precipitated to low solubility forms. Ways to keep P soluble longer and get more roots to its application sites are needed. It has been clearly shown that P in most soil situations moves only a few centimeters in several weeks (Lawton and Vomocil, 1954; Sleight et al., 1984). Yet, proliferation of roots in a small-volume phosphate band can supply a major portion of the plants' needs (Stryker et al., 1974; Anghinoni and Barber, 1980a, 1980b).

Eghball and Sander (1987, 1989) recently illustrated the restricted movement of phosphate in soil and the distance-plant age relationship. They placed the same total amount of ^{32}P- tagged ammonium polyphosphate solution in ten spot injections into the soil 4 inches (10 cm) deep and in a partial circle around five corn plants (*Zea mays*). The applied phosphate spots were placed different distances apart and at different distances from the corn plants. In the early stages of growth, corn within 6 inches (15 cm) of the phosphate sites used more of its phosphate from the fertilizer than from soil. By maturity at 90 days, the distance of fertilizer, even that placed 25 inches (64 cm) from the corn plants, had little influence on plant P taken from fertilizer. As spacing between application spots increased such that moved P did not overlap zones, the P taken up increased. This result was assumed to mean that when more soil was exposed to fertilizer P, it resulted in exposure of more P to more roots for uptake. This question between (1) the benefits of a concentrated P band to limit P adsorption-precipitation rates, and (2) the value of a large P-enriched soil zone for roots to exploit deserves additional study.

The low solubility of phosphates requires that the soil volume supplying phosphorus to roots be kept moist. In a greenhouse study with subterranean clover (*Trifolium subterraneum*), it was concluded that rapid phosphorus uptake occurred only during times when soil water contents were near field capacity (Simpson and Pinkerton, 1989). This would be caused by the need of high water contents to increase desorption and diffusion of P. The generally higher water contents in drip- irrigated soils may help increase phosphorus efficiency, even of added solid phosphates.

Placing phosphate into the soils of permanent cover crops (pastures, no-till grains) seems to be an improvement over surface applications. The question of placement raises the question of cost and damage to roots versus benefit. Coulters and converging disks make insertion less costly and damaging than occurred with chisel entries. Injecting liquid or solid phosphates into established pastures needs study. Peitscher (1991) claims that applications on established alfalfa (*Medicago sativa*) on low-P soils are profitable. Roots cut by the coulter were replaced by new roots about half an inch (1 cm) back from the cut, increasing roots near the bands. The caution is given, however, that damaged roots are more susceptible to disease.

B. The Nitrogen Problem

Improving efficiency of added nitrogen will be accomplished mostly by reducing losses. Losses of nitrogen are (1) leaching, (2) denitrification, and (3) ammonia volatilization. If losses by leaching are reduced, the problem of nitrate contamination is also reduced. The reduction in N leaching losses will be accomplished mostly by these techniques :

1. Use nitrification inhibitors with ammonium fertilizers. These keep nitrogen as ammonium longer, a form of low mobility in soil because it adsorbs to cation exchange sites.
2. Use frequent but small N applications. Losses of nitrogen added in just a few large additions are well-known. Fertigation in sprinklers and even daily applications in drip irrigation systems open up new possibilities to control both leaching water and nitrate concentrations in soil solution. An extensive literature review of nitrification inhibitors was published by Slangen and Derkhoff (1984) with 347 references; Meisisnger(1980) also published an earlier review.
3. Reduce ammonia volatilization loss, especially on alkaline soils,by good timing and management. Ammonia volatilizes as NH_3 from the soil surface and when pH increases above pH 8. Incorporated N (irrigation or soil cover) is not lost. Timing of application of soluble N and irrigation is critical. Application in subsurface drip systems should have no ammonia volatilization loss.
4. Reduce ammonia loss by using urease inhibitors with urea surface applications. The inhibitor allowed the urea to remain on the soil surface for several days before ammonia volatilization losses occur. Thus, more time was available for rain or irrigation to move the soluble N into the soil before losses occurred. Radel et al. (1989) studied eleven urease inhibitors banded in soil with urea. After three days, four of the eleven materials had permitted only 5 to 20% of the urea to be hydrolyzed. Control plots had only 86 to 95% of the urea hydrolyzed. O'Conner and Hendrickson (1987)

found that urease inhibitors reduced losses to nitrogen, even for several weeks time.

5. Reduce N_2 denitrification losses by better water control and concentrations of nitrogen. Denitrification losses occur in overly wet soil conditions, especially in clayey soils. Likely solutions appear to be (1) careful control in drip systems, (2) use many but small N applications, especially in rainy areas having clayey soils, and (3) manage clayey soils to loosen and aggregate them for better aeration.

6. Add nitrogen whenever possible into soil or get it there dissolved in water to avoid ammonia volatilization losses. New "point injectors" are being tried to insert gas or liquid nitrogen forms into soil before or during planting or as additions to growing crops (Wanzel, 1988). Insertion points are like small-diameter protruding spokes on a wheel and cause minimal damage to roots. However, seals in the equipment and plugging side-slot openings are still troublesome.

It is expected that new information will help in our control of soil nitrogen. Many persons have great hope that models will help; they will. But, models point to gaps in our knowledge; they don't provide knowledge. Reviewing predictions using fourteen nitrogen models, de Willigen (1991) found that only some of the models provided some good values, all using the same set of data. However, none of the models accounted for the striking phenomenon of the data set--a loss of mineral nitrogen immediately after a second and third application in late spring and early summer. de Willigen's conclusion was that a lack of ability to model soil microbial processes is the major deficiency in most or all of the studied models on nitrogen. These microbial processes are most of the processes in soil nitrogen.

VIII. SUMMARY

Irrigation-fertilization programme vary enormously. Differences in soils, climates, crops, and farmer's interests have altered what and how crops are managed. The grower must be sure he recognizes the limiting growth factors; they sometimes are not nutrients or water. Much of the time, however, the growth factors that can be corrected are deficiencies of nutrients or water. Inadequate water hinders plant growth and nutrient uptake; too much water may reduce nitrogen contents, moving nitrate into groundwaters. Nitrogen is most often the limiting nutrient and appropriate attention needs to be given to supplying nitrogen without permitting excess losses by leaching, ammonia volatilization, and denitrification.

Phosphates and potassium are deficient in many areas but less often than is nitrogen. Both phosphate and potassium have limited mobility in soil. Maintaining adequate phosphorus, because of low solubility of phosphates, is a major problem. Concentration (banding) and applications into the soil (5 to 20 cm deep) is usually most effective. Sulfur and other nutrients are less often deficient.

A management programme is a combination of choices that provides a good growing environment for the crop. Fertilizer and water is then added for high efficiency and minimum cost per unit of yield. The mechanisms to accomplish these many needs will change as methods of irrigation and crops are changed. Surface flow irrigation has the potential for high losses by leaching and inadequate wetting between furrows. Sprinkler irrigation wets foliage and fruits, often causing more fungus diseases. Costs are increased but fertilizers can be added in the waters.

Trickle irrigation offers unique possibilities to wet only parts of the soil, to avoid wetting the fruit or plant, to conserve water, and to fertilize and irrigate almost continuously, if desired. Unique conditions are formed: parts of the soil may be temporarily overwet (slightly anaerobic), root systems will be smaller and less extensive, and part of the soil can be left dry (unwetted). Thus, the surface soil can be dry by using subsurface tubes.

The next decade will require attention to minimize leaching of nitrate, allowing essentially no soil erosion, and efficient use of water. All of these needs will require less surface flow irrigation and more use of expensive sprinkling and trickle irrigation systems. Fertilizer nitrogen will need to be adequate but not in excess to avoid water pollution. Higher water efficiency will result in more salt accumulation problems. All of these essentials lead to the need for complete management schemes that involve (1) higher economical production by higher efficiency of fertilizers and water, (2) minimum pollution of the environment by controlling soil erosion and control of leaching, and (3) reclamation techniques to avoid salt accumulations.

The challenges in production agriculture have not lessened; they have only changed from many simple systems grouped together. The new challenges are to develop complex integrated land-use programmes which are designed to conserve the soil, protect the environment, and produce the foods needed by a demanding world.

REFERENCES

Anghinoni, I., and S. A. Barber. 1980a. Soil Sci. Soc. Amer. J. 44:1041-1044.

Anghinoni, I., and S. A. Barber.1980b. Soil Sci. Soc. Amer. J. 44:1016-1020.

Anonymous. 1984. Irrig. Age. 18(No.8):34-35

Anonymous. 1988a. Irrig. J. 38(No.4):12-16.

Anonymous. 1988b. Irrig. J. 38(No.1):12-14,16,18,20.

Bacon, P. E. and B. G. Davey. 1989. Fertil. Res. 19:159-167.

Bailey, L. H. and E. Z. Bailey. 1976. Hortus Third. Macmillan Publ. Co., NY. (Plant taxonomy) 1290 p.

Beaton, J. D., R. L. Fox, and M. B. Jones. 1985. Fertilizer Technology and Use, Third Edition. Soil Science Society of America, Inc., Publ., Madison, WI 531711. p. 293-322.

Boodley, J. W. 1981. Florist's Rev. 168(No. 4354, May 14):10- 11.

Buresh, R. J. 1987. Fertil. Res. 12:263-268.

Christianson, C. B. 1989. Fertil. Res. 19:183-189.

Dable, R. L. 1978. Turf-Grass Times 14 (No. 1): 14-23.

Dasberg, S., A. Bar-Akiva (deceased), S. Spazisky, and A. Cohen. 1988. Fertil. Res. 15:147-154.

Eghball, B. and D. H. Sander. 1987. Soil Sci. Soc. Amer. J. 51:1350-1354.

Eghball, B. and D. H. Sander. 1989. Soil Sci. Soc. Amer. J. 53:282-287.

Fenn, L. B. and L. R. Hossner. 1985. Adv. in Soil Sci. 1:123- 169.

Fenn, L. B. H. L. Malstrom, and E. Wu. 1987. Fertil. Res. 12:219-227.

Fulton, A. E., J. D. Oster, B. R. Hanson, C. D. Phene, and D. A. Goldhamer. 1991. California Agric. 45(No. 2).4-8.

Gardner, D. 1984. Irrig. Age 18(No. 6):64-65.

Hauck, R. D. 1985. Fertilizer Technology and Use, Third Edition. Soil Science Society of America, Inc., Publ., Madison, WI 531711. p. 293-322.

Howard, D. D. and D. D. Tyler. 1989. Soil Sci. Soc. Amer. J. 53:1573-1577.

Johnson, A. W., J. R. Young, E. D. Threadgill, C. C. Dowler, and D. R. Sumner. 1987. Ag. Age 31(No. 2): 8-9.

Larsen, R. 1984. Irrig. Age 19(No. 2):30-32.

Larsen, R. 1985. Irrig. Age 19(No. 6):6-7.

Lawton, K. and J. A. Vomocil. 1954. Soil Sci. Soc. Amer. Proc. 18:26-32.

Lindsay, W. L. 1979. Chemical Equilibria in Soils. Wiley- Interscience, NY. p. 162-205.

Meisinger, J. J. (ed.) 1980. Spec. Publ. No. 38, Amer Soc. Agronomy, Crop Sci. Soc., and Soil Sci. Soc. Amer, Madison, WI.

Monatt, M. C. H. and P. Nes. 1986. Aust. J. Soil Res. 24:435- 440.

Mustafa, M. A., K. A. Akabawi, and M. F. Zoghet. 1989. J. Arid Envir. 17:349-354.

O'Connor, M-J. and L. L. Hendrickson. 1987. Soil Sci. Soc. Amer J. 51:1062-1066.

Papadopoulos, I. 1988. Fertil. Res. 16:157- 167.

Peitscher, A. 1991. Solutions 35(No. 3): 29, 61.

Puri , Y. P., M. F. Miller, R. N. Sah, K. G. Baghott, E. Freres- Castel, and R. D. Meyer. 1988. Fert. Res. 17:197-218.

Radel, R. J., M. D. Crenshaw, B. R. Bock, and H. M. Williams. 1989. Fertil. Res. 20:123-127.

Ratledge, G. 1988. Irrig. J. 38(No. 3):32-34, 36.

Renn, L. 1985. Irrig. Age 19(No. 6): 48B.

Sander, D. H., E. J. Penas, and B. Eghball. 1990. Soil Sci. Soc. Amer J. 54: 1473-1478.

Sharma, B. D., S. Kar, and S. S. Cheema. 1990. Fert. Res. 22:119-127.

Renn, L. 1985. Irrig. Age 19(No. 6): 48B.

Sander, D. H., E. J. Penas, and B. Eghball. 1990. Soil Sci. Soc. Amer J. 54: 1473-1478.

Sharma, B. D., S. Kar, and S. S. Cheema. 1990. Fert. Res. 22.119-127.

Simpson, J. R. and A. Pinkerton. 1989. Fertil. Res. 20:101- 108.

Slangen, J. H. G. and P. Kerkhoff. 1984. Fertil. Res. 5:1-76.

Sleight, D. M., D. H. Sander, and G. A. Peterson. 1984. Soil Sci. Soc. Amer J. 48:336-340.

Smith, R. B., J. D. Oster, and C. Phene. 1991. Subsurface drip produced highest net return in Westlands area study. California Agric. 45(No. 2): 8-10.

Stryker, R. B., J. W. Gilliam, and W. A. Jackson. 1974. Soil Sci. Soc. Amer Proc. 38:334-340.

Subbiach, C. C., P. Manikandan, and Y. Hoshi. 1986. Plant and Soil 94: 35-42.

Wallace, A. and G. A. Wallace. 1987. Irrig. J. 37(No. 3):12- 15.

Walton, P.D. 1988. Principles and Practices of Plant Science. Prentice-Hall, Inc. Englewood Cliffs, NJ (Plant taxonomy) 439 p.

Wanzel, R. J., 1988. Solutions 32(No. 4):14 15.

Weir, B. L., T. A. Kirby, B. A. Roberts, D. S. Mikkelsen, and R. H. Garber. 1986. California Agric. 40(No. 5, Sept-Oct):13-14.

White, John G. 1985. Irrig. Age 20(No.2):6-7.

White, J. G. 1985a. Irrig. Age 19(No. 9):6-8.

White, J. G. 1985b. Irrig. Age 19(No. 6):18, 51, 52.

White, J. G. 1985c. Irrig. Age 19(No. 8):44-45.

White, J. G. 1985d. Irrig. Age 19(No. 6):68-69.

de Willigen, P. 1991. Fertil. Res. 27 :141-149.

Yoshida, S. and M. R. Chaudhry. 1979. Soil Sci. Plant Nutr. 25 (No. 1) : 121-134.

Young, R. D., D. G. Westfall, and G. W. Colliver. Production, Marketing, Fertilizer Technology and Use, Third Edition. Soil Science Society of America, Inc., Publ., Madison, WI 53711. p. 323-376.

10

Ecological Ill-Effects Of Arid Land Irrigation And Some Combating Measures

Vinod Shankar and Suresh Kumar

I. INTRODUCTION

The thirsty arid lands do need irrigation water to produce increasingly more of agro-products to sustain rising human and livestock populations. But consequent use of arid lands beyond their natural carrying capacity through heavy inputs of irrigation water and more of chemicals for plant production and protection has obvious fallouts. On this, much has been discussed in several national and international forums (Shankar, 1985a). The general consensus of opinion is that likely implications of irrigation on arid lands need to be well thought out at the project formulation stage so as to have a built-in environmental impact monitoring system in each project (Shankar, 1985b; Chatterji and Saxena, 1988).

Although several workers acknowledge the likely impact of arid land irrigation, much of the scientific knowledge gained is discipline-bound, disregarding the complex interactions within the system, and their impact on local populations. Retrieval of the diffused available literature on environmental implications of arid land irrigation is a formidable task. This chapter would, therefore, relate the arid land irrigation with such important aspects as climate and microclimate, soil-water-plant relationships, natural flora and fauna-both macro and micro, socio-economics and human health. Measures to overcome the detrimental effects, and the future research and management needs of irrigation projects are also discussed.

II. CLIMATE AND MICROCLIMATE

Irrigation brings about a chain of changes in different components of the command area. Such changes have been categorized (Dukhovny, 1977) as stable or unstable, and further as controllable or uncontrollable. Stable changes are sure to occur, but unstable may or may not occur, depending upon the combination of other natural factors. Controllable changes are those that have definite magnitude and trend, and so can be determined by man. Uncontrollable changes can not be so determined.

As a rule, stable uncontrolled changes, e.g., climatic have a beneficial impact on the environment (Babaev and Orlovsky, 1980). The interaction between irrigation and environmental conditions make the climate milder (Dukhovny, 1977). As the upper limit to evaporative demand, for successful crop production, is the potential evapotranspiration, irrigation provides a relatively small amount of vapour transport to the air masses. Thus, only insignificant changes are expected in the total amount and pattern of precipitation as a result of irrigation. This assumption was verified by the investigations carried out in an extensively irrigated part of Texas. Irrigation was found to alter the water-energy budget of the command area, resulting in the increased energy expenditure on evaporation that brought changes in air temperature and humidity. In another study on the irrigated areas of the deserts of the Middle Asia, Orlovsky and Utina (1977) observed that, in July, humidity of air at 2 m height increased 3 to 5%, while the air temperature decreased by 2.5° to 3°C. As a result, the temperature sum of the growing period decreased by 200 to 500 degree days (Orlovsky and Durdyiev, 1978).

III. SOIL-WATER-PLANT RELATIONSHIPS

Irrigation is supposed to alter the moisture content of the rooting zone, besides modification of the hydrological cycle quantitatively and qualitatively. These modifications can be classified as follows :

(a) Modification of the atmospheric aspects of the hydrological cycle.

1. Increase of actual evapotranspiration.

2. Rise in the atmospheric vapour content.

3. Moderation of atmospheric temperature.

(b) Modification of the groundwater level, migration of salt, and hydrological processes in the soil moisture.

1. Rise of water table in canal irrigated areas, and fall of groundwater table by over pumping in tubewell command areas.

2. Development of horizontal groundwater flow from irrigated areas towards the neighbouring non-irrigated lands, raising the water table of the latter and the development of 'dry drainage' areas; leaching of irrigated soils, transport of salts by groundwater flow, and the acceleration of salt accumulation under the dry drainage areas.

(c) Modification of water quality other than those occurring within the soil moisture zone.

1. Increase of salt concentration during storage, conveyance, and distribution of water due to evaporation.

2. Water pollution, salt intrusion, and deterioration of quality of water downstream by the effluent of the drainage systems having high salt content.

When we summarise modifications described under (a), (b), and (c) above, it is found that when water sees the signs of abuse, economic and ecological consequences are bound to occur. Mounting pressure on water use caused by rising population of man and livestock, adoption of green revolution technology, and growing demand of industries manifests in pollution, waterlogging, depletion of groundwater supplies, falling water tables culminating into damage to ecological system. In the semi arid region of New South Wales (Australia), irrigation has resulted in shallow water table, soil salinity, and deterioration of river flow quality.

Of all problems connected with the irrigation impact on arid environment, these are most vital : depletion of groundwater and salt intrusion, waterlogging, rise of water table, irrigation water quality, secondary salinization, return water quality, and relief variations (Postel, 1984; Zonn, 1982b).

A. Groundwater Depletion

In desert countries of the world, natural recharge to fossil acquifers is minimal. Water mining just supports only a fragile and short-term prosperity at best, because water becomes too salty to use, too expensive to pump to the surface, or runs out altogether. In the High Plains of United States, farmers are taking land out of irrigation because of rising pumping cost, diminishing water yields, and low commodity prices.

In the Indian State of Tamil Nadu, overpumping has caused groundwater levels to drop 30 m in just a decade. Considerable drop in groundwater has also taken place in Rampur-Mathania and Pal area in Jodhpur district of the Indian State of Rajasthan. Punjab and Haryana States and Delhi and Chandigarh among the Union Territories have exploited over 85% of their groundwater and further promotional work needs to be halted (Dhawan, 1989; Gangwar and Panghal, 1989). Fall in groundwater level has also taken place in four States of Texas, California, Kansas, and Nebraska of USA (Postel, 1989).

In Arizona, the Central Arizona Project aims at bringing water 540 km across the desert and preventing further depletion of groundwater (Anon, 1990). No more groundwater is withdrawn than is recharged has become an integral part of planning future water supplies in South Africa and Israel as well. Short-term strategy to allow overpumping to alleviate droughts should be discouraged.

Heavy pumping of groundwater for irrigation in coastal desert areas results in salt-water intrusion in the coastal acquifers. Israel, Syria, and the Arabian Gulf States are battling threats of salt-water intrusion. Arid and semi arid regions are underlain with saline water unfit for farming. In the Indian State of Haryana alone, 65% (2.8 million hectares) area is underlain by brackish to very saline groundwater (Gangwar and Panghal, 1989). Indian part of the Thar desert is a burning example of such problem. It has been found that excessive use of groundwaters causes the intrusion of saline waters into the acquifers, which the farmers come to learn quite late (Dhawan, 1989) when considerable damage has taken place.

B. Rise in Water Level and Waterlogging

Degradation of valuable cropland from poor irrigation practices is one of the least affordable consequences of irrational use of water. Where drainage is poor seepage of water from unlined canals gradually increases the water level, resulting in waterlogging the soil and entering of water level to crop root zone. In the Indian State of Madhya Pradesh, for example, a large irrigation project that originally was expected to increase crop production ten-fold led to extensive waterlogging and a reduction in maize [Zea *mays* L.] and wheat [*Triticum aestivum* (L.) emend Thell.] yields. Farmers there now call their once fertile fields as "wet deserts".

In Gang Canal and Indira Gandhi Canal command areas in Ganganagar district in the Indian State of Rajasthan, due to sandy terrain and seepage losses from the main canal, its branches and farm irrigation channels, the groundwater level is rising at an alarming rate of 43 to 83 cm per year (Shankarnarayan, 1985).

In part of Ghaggar flood plains, water table has risen by 6 m within the past 20 years. Caused by this rise of water level 6920 million cubic metre (mcm) water is being added to the acquifer annually leading to the development of salinity in many places. With the present rate of rise in the groundwater level 20, 40, and 60% of the canal command area is likely to develop drainage and salinity problems within 30, 30 to 50, and 50 to 100 years, respectively (Shankarnarayan, 1985).

Almost the entire command in the Mahi Right Bank Canal (MRBC) system in the Indian State of Gujarat is experiencing rise in water table over years. Nearly 40% of command is experiencing an average rate of water table rise of over 0.35 m a year (Shah, 1988).

In Pakistan an estimated 22% of the 8.4 million hectares (23 million acres) supplied by canal irrigation suffered from waterlogging and/or salinity damage as far back as in 1964 (Johnson, 1989). In India, about 7 million hectares of land - a large chunk of it being in the north-west (western Uttar Pradesh, Haryana, and Punjab) - is affected by salinity or alkalinity or both (Sawant, 1985). In Haryana alone the total annual loss caused by reducing water quality and rising salinity would approximate 5350 million Indian rupees (Rs. 535 crore) (Gangwar and Panghal, 1989).

In South Asian deserts of the former USSR, 3 to 5 years of irrigation in the Golodnaya Steppe caused water table to rise between 1 and 3 m below the surface. This figure in the Murgab oasis was 0.6 m, while in some locations it was 1 m.

C. Irrigation Water Quality

In central Asia, sustained growth of irrigated areas has brought considerable rise in water salinity of the Amudarya, Syrdarya, and Zeravshan rivers. In the western USA, potential adverse effects of irrigation on water quality have been identified as increased salinity, turbidity, colour, taste, temperature, nutrients, nematodes, bacteria, and viruses, and pesticide ingredients, all causing economic losses and deterioration of the environment (Hotes and Pearson, 1977). Some adverse effects of using poor quality water are described below.

(i) Effect on soil conditions : Use of poor quality water may lead to salinization, increase in alkalinity, swamping, destruction of soil aggregates, and soil cementation. These side-effects may be short-lived in the beginning, but may persist over a longer period.

Irrigation waters are strong and complex chemical agents. This has diverse chemical and physico-chemical reactions with soil minerals, their water solutions and colloidal compounds, soil organic substances, and with intrasoil biomass. The water of rivers, reservoirs, and artesian wells is always a solution of chemical compounds : diluted solution (0.1-0.4 g L^{-1}), weekly concentrated solution (1-3 g L^{-1}), and highly concentrated solution (5-8 g L^{-1}) due to dissolution of salt precipitates and intensive evaporation. Natural water always carries dissolved organic matter and relatively aggressive, polymerized solutions of the type of fulvoacids.

It is also saturated with gases - carbon dioxide, oxygen, sometimes hydrogen sulphide and methane at low temperatures. The polymerized solutions particularly with ready tetrahedrals and hydroxy and hydrogen groups are capable of destroying the crystal lattices of minerals and substituting metals in them. Weakly and moderately mineralized alkaline natural waters of rivers, lakes, and subsurface basins are more aggressive. The level of our knowledge of the interaction of fresh and mineralized water with irrigated soils is low for the time being.

(ii) Changes in the hydrophysical dynamics of soil : The air dry soil of arid and semi arid zones with an automorphic - evaporative type of water balance changes into automorphic-flushing-transpirational type following applications of irrigation water-10-100 mm per irrigation and over 50-1000 mm for the growing season. With groundwater rise, the soil water balance becomes hydromorphic-evaporative or hydromorphic-flushing (if drainage is good). In soil profile, capillary-suspended or capillary-backed moisture emerges and solution tends to flow upward to plant roots and soil surface. This, in turn, leads to considerable increase in the evaporation from the soil surface.

The temperature difference before and after irrigation amounts to 30 to 40° Celsius (the soil temperature - 60-70° Celsius before irrigation and water temperature - 20-25° Celsius). In this process 24 to 28 calories cm^{-2} of heat is released. The concentration of solutions and their osmotic pressure sharply drop first and then increase again; due to wetting of the dry soil that forces the air out. The silt in irrigation water settles on the soil surface, and during prolonged irrigation, the thickness of the silt layer may rise to several cm. This interferes with soil aeration, permeability, respiration of young roots, and emergence of seedlings.

(iii) Changes in the chemical and biochemical dynamics of soils : Irrigation water carries considerable amounts of dissolved substances to fields, and chemical interaction of irrigation water and soil occurs. Hydrolysis and destruction of primary minerals is accelerated. The weathering of primary and probably secondary minerals is intensified. In the course of evapotranspiration and drying, calcium carbonate crystals, silica gels and coagels of secondary aluminium silicates get precipitated in the soil due to reactions of absorption, substitution, and exchange of cations in the concentrated soil solutions. This leads to progressive clayezation of the soil in prolonged irrigation. Enrichment of irrigated soils by secondary minerals of smectite group is observed in such soils of the former USSR and Hungary and in the soils of Nile delta.

The composition of gases varies widely. The concentration of oxygen decreases from 20% to 18% in the upper 0-40 cm layer. On the contrary, the concentration of carbon dioxide increases from 0.5 to 2% and sometime by 4%. Oversaturation of soil air with carbon dioxide has also been observed especially when the groundwater level rises up to about 2 m below the land surface, or in over irrigation period when lenses of suspended water are formed in the soil. All these phenomena are new and unusual to the soil of deserts. For example, the stable oxidation regime typical of arid soils (+500 to +600 millivolts (mV)) for the upper soil layer decreases to +200 to +300 mV immediately after irrigation. It may even decline to +100 mV and sometimes -100 to -200 mV in the groundwater of irrigated soils. It is natural that shifts in Eh causes similar abrupt shifts in the migration capacity of iron, manganese, sulphur, and nitrogen. Alternating dissolution and precipitation of iron and manganese compounds in the form of oxides or silicates may also be a factor of irrigated soil cementation and self-compaction. In soils irrigated by groundwater, all potentially cementing agents - calcium bicarbonates, ferrous, manganese, mobile silica, and aluminium compounds - are also transported by the upward moving capillary fringe to upper soil horizons (0-100 cm). The effect of these cementing factors is added to the after-effects of soil structure destruction and the compaction effect.

If irrigation water contains week solutions of chlorides, sulphates, and sodium bicarbonates in particular, every water application leads to gradual increase in exchangeable sodium content in the absorbing chernozem complexes. The increase in the exchangeable sodium content turns structural chernozem into dense lumpy mass of poor permeability that is difficult to cultivate. In due course the substitution of exchangeable calcium with exchangeable sodium results in formation of gypsum or calcium chloride.

The most important after-effects of these processes of soil alkalization is intensification of hydrolysis, neosynthesis of mobile silicates and secondary minerals of smectite group with a mobile crystal lattice, very high hydrophilicity, and deterioration of soil physical properties.

The biochemistry of newly irrigated soils also undergoes appreciable change. It is known that desert soils have a small humus content and semi-sterility. Upon irrigation the humus content of sierozems and chernozems usually decreases further due to secondary salinization. The proportion of humic acids to fulvic acids undergoes appreciable change. The fermentative activity of steppe chernozems increases in the initial years of irrigation. In subsequent years, it decreases due to accumulation of toxic compounds resulting in dying off of the main root of the plants.

In general, optimum irrigation enhances soil biological activity and increases soil microbes and diversity of their groups. On the other hand, continued over irrigation stimulates the sparsely distributed microorganisms like soil anaerobes, including butyric, sulphate-reducing and nitrate-reducing bacteria. The enhanced activity of anaerobic micro-organisms tends to produce biological toxins including simple organic acids, which are gradually accumulated in soil and inhibit plant growth and activity of useful microflora.

Effect of irrigation on natural components has been tabularised by Zonn (1982 a); interaction between irrigation and environmental conditions by Dukhovny (1977); and water quality changes due to irrigation, hardness of river waters, and water quality constituent difference between applied and subsurface drainage water and pesticide concentrations in subsurface drain effluents by Hotes and Pearson (1977).

D. Secondary Salinization

In dry climates, waterlogging may be accompanied by salinization as water near the surface evaporates and leaves behind a damaging residue of salt. The problem is especially severe in India and Pakistan, the Valley of Mexico, the Helmand Valley in Afghanistan, the Tigris and Euphrates basins in Syria and Iraq, the

San Joaquin Valley in California, the North Plain of China, and Soviet Central Asia.

Farmers never pay the real cost of water they use. They would use less water per unit area if water is priced on the basis of volume of water used in agriculture. As a result, 2 to 5 tons of salt is believed to be added to cropland for ever 100 cm of good quality irrigation water per ha per year.

Gradually salinization begins to take place, causing damage to irrigated land (Table 10.1) and reducing yields.

Table 10.1. Irrigated land damaged by salinization, top five irrigators and world estimate, mid-1980s. From Postel (1989).

Country	Area damaged Million ha
India	20.0
China	7.0
United States	5.2
Pakistan	3.2
Soviet union (former)	2.5
Total	37.9
World	60.2

Areas under saline soils in Bukhara oasis and that under solonchaks in the eastern part of the Amudarya delta have expanded due to substandard irrigation systems. In the Kara Kum canal zone, 1400 ha area requires reclamation out of 1.9 million ha of lands available for. In the Karshi Steppe, 800,000 ha require reclamation out of 1 million ha and the soils of the Golodnaya Steppe could be completely developed if man-made drainage were provided (Anon., 1973).

A hard pan in upper soil surface creates drainage problem. As for example, in Indira Gandhi Nahar Project Stage II command area, there is a critical hard pan below 5 m depth over 0.07 million ha area, and below 5 to 10 m depth over 0.25 million ha. This area has been termed as critical area in so far as waterlogging is concerned, and suitable drainage arrangement will have to be done in case surface method of irrigation is introduced.

E. Return Water Flow

It is linked with the water cycle (Hotes and Pearson, 1977). The main constituents of the return flow during irrigation are : (i) seepage water losses from canals, (ii) surface runoff from irrigated lands, and (iii) water losses to deep percolation (Skogerboe and Law, 1971). Part of irrigation water gets back into river networks or groundwater aquifers with the return flow from where it may be again taken for irrigation (Dukhovny and Livtak, 1977). That is why the return flow and its constituents may be looked upon as a system having several subsystems as illustrated below :

The water supply subsystem includes water conveyance from the source to fields. Then comes the problem of controlling conveyance losses through seepage (Hocking, 1985), lest that may cause rise in groundwater level, waterlogging, or soil salinization, depending upon the depth to groundwater, type of drainage, and groundwater salinity.

The irrigated field subsystem comprises water application for irrigation until it is disposed off beyond the irrigated territory. The pattern of water use in this subsystem controls to a great extent the quality of the return flow. The use of modern system of irrigation (sprinkler, and drip irrigation), apart from water saving and the increased agricultural production, eliminates surface runoff and deep percolation that, in turn, reduces the content of salts, nutrients, suspended particles, and bacteria in the return flow.

The water disposal subsystem comprises the removal of surface and drainage water from the irrigated territory to a water receiving basin. The surface flow may be either recycled for irrigation or diverted into a water receiver. The subsurface flow may be either diverted by drains, or may percolate into deeper-lying horizons where it may add up in salt content at the expense of saline horizons and then get into the river network.

Pollution of receiving water bodies with the return flow is equally important as it poses a real danger to aquatic inhabitants, and in some cases, to animals and man. Severe water pollution in the return flow may cause decline in water use. It then induces not solely environmental implications but economic as well. This problem assumes more significance considering that irrigation is most often practised in water deficient regions, where small rivers suffer most from pollution with the return flow.

Now comes the relief variations. Irrigation transforms natural landscapes into agro-irrigational ones. Besides, it changes soil and creates a new microrelief (Babaev, 1982; Zonn, 1982 b).

The major mesorelief forms, where irrigated farming is practised and the above changes take place, are alluvial, socle and sheet plains, volcanic plateaus, valleys, terraces and river deltas, alluvial fans, individual mountain regions.

IV. PLANT AND ANIMAL LIFE

Irrigation changes the native ecological set up and creates new productive systems on arid lands such as crop fields, orchards, wooded areas, irrigated pastures, and house gardens, and in-water fisheries and aquaculture. These changes in the physical and biological environment lead to radical changes in the ecology specially affecting the terrestrial biota. One of the striking conclusions is that, whereas irrigation obviously increases the potential of biological productivity, it generally results in a reduction in the diversity of plant and animal life (Kassas, 1977). The more uniform habitat created by regular application of water allows some species to wax tremendously, while others disappear under this artificial form of 'natural' selection. Unfortunately the species that become dominant are not always those that are desirable from the social or economic viewpoints. Irrigation thus changes the character of soil and vegetation. Problems of salinization and other forms of soil fertility deterioration are in fact symptoms of manipulation exceeding the natural capacity of the ecosystem.

1. Changes in Terrestrial Biota

Modification of the terrestrial habitat by irrigation is reflected in the changes in soil fauna, soil microflora, birds, rodents, and large animal life and the weed flora (Ghabbour et al., 1977).

(i) Soil fauna : Soil fauna in desert are very sparse. This consists of specially adapted species able to withstand drought and take advantage of minimal gains in soil moisture from rainfall or whatever dew to feed and to breed. These species are mainly : harvester and carpenter ants that feed on seeds and on dead decaying wood; sand roaches that have the unique ability of absorbing water vapour from the soil atmosphere with relative humidities of 82%; a host of tenebrionid beetles that feed on decaying litter; and a number of predators such as carabid beetles, ant lions and centipedes. These animals usually have a daily rhythm of vertical migration from the soil surface to deeper layers. These rhythms are governed primarily by the changing temperature and moisture gradients that are inversed twice daily at sunset and at sunrise. They usually concentrate at places affording protection against excessive heat, evaporation, and predators; such as shrub covers. The population densities of these animals are closely correlated with seasonal variations in the moisture and temperature regimes, as has been found in deserts both in Egypt and in the United States.

Earthworms, on the other hand, are the main constituent of the soil fauna in irrigated soils, together with a host of collemboia, mites, nematodes, and other minor animal groups. The population densities of these animals are also dependent upon variations in soil moisture, a net outcome of soil texture, frequency of irrigation or the amount and distribution of rainfall. Studies on survival of earthworms in sandy and clayey soils of Egypt upon stoppage of irrigation revealed that earthworms could not remain active for long in sandy soils (of 28% water holding capacity) but remained active indefinitely in clayey soils (of 50% water holding capacity).

When sandy soils are irrigated, earthworms, nematodes, and other groups of soil animals invade the newly irrigated soils and their populations build up gradually at the expense of the original groups of soil fauna that cannot withstand the new high moisture levels. Populations of earthworms in newly irrigated sandy soils in Egypt jumped from 0.08 to 2.67 million ha^{-1}. If the plot is used for afforestation, they may reach 7.41 million ha^{-1}. These figures are attained in favourable conditions, which are permanent under orchards and artificial forests and transient under field crops, namely, an undisturbed soil, a regular and adequate water supply, a fine soil texture (to raise water availability), and a regular and adequate supply of organic matter.

The replacement of the original xeric soil fauna by ruderal species, does not necessarily mean that all native species are wiped out. On the contrary, some species of nematodes, bugs, etc., originally present, may become pests of the introduced crops - as has happened, for example, at Wadi Natrun in Egypt.

Because of differences in degree of water availability between sandy and clayey soils, sandy soils of low moisture content are optimal for species of earthworms that thrive in clayey soils of 25-45% moisture content, the determining factor is the pF and the relative humidity of the soil atmosphere, which is a function of pF.

Earthworms have been deliberately introduced in the southern parts of former USSR to accelerate reclamation of newly irrigated sandy desert soils by improving their properties and this has met with considerable success. The nitrogen excretion of earthworms is one of their contributions to soil fertility. Under experimental conditions, *Allolobophora caliginosa,* the common earthworm in Egypt, produces 6.5% of its fresh weight as urine each day. The average fresh weight of a normal population in sandy soils might be 74 kg ha^{-1}, which produces about 2 L of urine d^{-1}. This urine will contain 0.06 g ammonia and 0.8 g urea; the mucus produced will contain about 20 g protein; and further quantities of ammonia and urea are also contained in the casts. With increase in soil moisture levels, more ammonia and less urea are execrated by earthworms.

Remarkable changes have been observed also in the vertebrate fauna associated with the soil in newly irrigated areas of arid regions. For example, rodent outbreak is recorded from Ismailia, Faiyum, and Kom Ombo in Egypt. Under natural arid conditions, scarcity of food as well as predator pressure keep under control the rodent populations. Irrigation removes both constraints. This also happens in birds such as the desert sandgrouse (Pterocles) in Nubia and *Passer hispaniolensis* on the north-western coast of Egypt.

(ii) Soil microflora : Irrigation-induced physical and structural changes in soil, its moisture-air relationships also trigger changes in soil microbial population including fungi, actinomycetes, and bacteria. Tendency often is towards reduction of number of species and increase in population sizes.

Soil saturation or flooding is sometimes used in the control of soil-born plant pathogens because it restricts oxygen supply to the pathogen. This together with accumulation of carbon dioxide released in microbial respiration may destroy certain pathogens. Eradication, for example, of *Fusarium oxysporum f. culense* from banana fields by flooding is a supporting evidence. The greatest reduction took place in indigenous soil fungi, as 85% of the indigenous fusaria was destroyed in the first 40 days, and most of the indigenous fungus flora was eradicated 120 days after flooding.

(iii) Animal life : Changes from rainfed to irrigation farming profoundly influences population of wild animals. Wild animals in deserts, such as gazelles, ibexes, and hyenas, depend on drinking water from time to time. When such water is provided in the middle of the desert, populations of these large animals increases rapidly. This may be regarded as a favourable change that might be utilized in conservation measures for endangered animals.

The change from basin to perennial irrigation in Upper Egypt and the elimination of the annual Nile flood below the new regulation works has also brought favourable and unfavourable changes. The rat is no longer the serious menace over there, but it has been replaced by another rodent.

(iv) Weed flora : As water and nutrients become available upon introduction of irrigation to the reclaimed areas, weeds including native flora of the area, if not controlled in time, take a heavy toll on agricultural production by competing with crop plants for resources in the pool. For timely and effective control it is, therefore, necessary to have surveys of the native species of the area including flora and fauna.

2. Changes in Aquatic Biota

Water weeds and disease vectors in water bodies : Canals and distribution systems being rich in organic matter and nutrients are invariably invaded and sometimes choked by dense growth of algae and aquatic weeds. A large sum may be needed for clearing them out from canals.

Under adequate and reliable supplies of water in the network of canals, light and nutrients provide ideal conditions for growth of aquatic biota. They may float (Pistia and Salvinia) or be submerged (Ceratophyllum) in as profuse amount as to interfere with waterflow and cause obstruction.

In places other than canal, the presence of aquatic plants may serve as food for fish when alive, or provide protection for fish and invertebrates that are essential for the ecological balance of the medium. Submerged plants contribute to aeration of waters. Nevertheless, such advantages do not justify their presence in man made aquatic ecosystems where they are not watered for reasons stated above, besides high consumptive use of expensive imported water.

The association of aquatic weeds with the spread of certain vectors of disease in man and animals is another reason for the general concern over their presence in irrigation systems (Hill et al., 1977). Although weeds are not obligatory to the establishment of snails such as Bulinus and Biomphalaria that are intermediate hosts for schistosomiasis, for example, the snails tend to flourish well under their protection and support. In the Volta lake, offshore growths of submerged Ceratophyllum harbour more Bulinus snails than any of the other plant communities. Reachers of streams flowing into reservoirs, shallow shores, downstream areas of dams where the volume of water generally tends to be small, and irrigation distribution systems invariably favour the establishment of schistosome-supporting snails. Hence establishment or aggravation of schistosomiasis is almost an invariable features of irrigation schemes in tropical areas.

Modification of aquatic ecosystems may also increase disease transmitting mosquitoes. The development of vegetation like *Pistia* and *Polygonum* favours the genus *Mansonia* whose larvae and pupae attach to roots, stems, and leaves. Stable waters, especially when infested with weeds and polluted with organic matter, provide a suitable breeding place for *Culex*. The *Anopheles gambiae* complex prefers transient, shallow, sunlit weedless water bodies like irrigation pools, whilst *A.funestus* is associated with ditches and swampy areas As will be noted in Human Health Impacts, conditions created by flooding of poorly graded fields are suitable for the breeding of mosquitoes, and irrigation pools created by the use of brackish water support mosquitoes that tolerate saline water. Cultivation of densely growing crops encourages mosquitoes by providing shady resting places with suitable humidity that may increase the longevity of adults. There are several instances where rice fields under irrigation have increased malaria vectors and especially *A. gambiae*. In the Kano Plains of Kenya, 65% of the mosquitoes that attack human beings in rice fields are A. gasmbiae. In a similar but non-irrigated area, the species complex was represented by only 1% (Hill et al., 1977).

Mosquitoes populations breeding in irrigated areas also transmit arboviruses that cause human disease. Forms of encephalitis causing viruses have been isolated from *A. funestus* and *Culex tarsalis* both being favoured in irrigation schemes. *C. tritaeniorhynchus* is a major rice field mosquito in the Orient and *C. annulirostris* in eastern Australia.

Many aquatic plants are troublesome weeds in irrigation canals and reservoirs and may cover extensive areas. In natural conditions, the extent or even the presence of suitable aquatic habitats is often seasonal and many of these plants have evolved to grow rapidly when good growth conditions exist.

Unlined irrigation canals and shallow reservoirs generally provide suitable conditions for the growth of aquatic plants, which may therefore become a serious problem and interfere with the planned use of water.

SOCIO - ECONOMIC IMPACTS

A brief account of socio-economic impacts is given as a part of ecosystem analysis. The impacts of irrigation and associated drainage development upon water, soil, aquatic, and terrestrial ecosystems, and human health extend throughout the social and economic fabric of local and national societies. The way in which those societies in turn undertake improvements in irrigation and drainage has a powerful influence upon the character of the environmental effects (Worthington, 1977) A nation's aims and its mode of analysis of possible means of translating those aims into irrigation schemes set the framework within which choices are made among social, economic, and environmental consequences (Ehlers, 1977).

National aims in irrigation and drainage may include national economic efficiency, gaining of foreign exchange through cultivation of export cash crops, settlement of nomadic people, drought evasion, stabilization of agricultural systems, and modernization of the rural economy. Often these are mixed. Sometimes they follow each other in distressingly rapid sequence : the government support of irrigation projects in the Sabi Valley of Rhodesia moved from drought relief to crop production to stabilize agriculture over 20 years. Depending upon national resources and aspirations, irrigation may be viewed not so much as a means increasing not production per ha or per person or per cu m of water but as a means of enabling an agricultural livelihood to maintain itself in harmony with the environment (Shankarnarayan and Shankar, 1985).

Increased Density of Population

A new project stimulates the growth of towns or cities as well as increases the density of the rural population. Moreover, the immigrants are likely to be relatively young adults who will establish households and raise families. With enhanced food supplies, increased employment opportunities, improved amenities, reduced flood hazards, and the likelihood of improved utilities such as potable water, sanitation facilities and electric power, population grows rapidly.

Employment, Income, Distribution of Wealth, and Inflation

Whether improved productivity will be shared by the bulk of the population or concentrated among a few will depend in part on other economic circumstances that surround the financing, operation and disposition of the project's net output. Careful integration of engineering and agronomic activities with economic policies may be required if the project benefits are to accrue to those judged most urgently in need.

In state-owned projects, the distribution of income can be determined by administrative action : whether in cash or kind, and if the latter, over what range of good and services, and according to what rules of equity ? Where projects are developed for privately owned land, these matters are likely to be mediated by market processes, and governmental action is usually limited to fiscal and monetary controls and establishment of rules or constraints over markets for labour, land, goods, and agricultural inputs.

V. HUMAN HEALTH IMPACTS

1. Beneficial Effects

Irrigation can improve health in four major ways. Increased availability of potable water makes it possible to improve domestic supplies for the inhabitants. Increase in agricultural production provides opportunities for better nutrition. The development of new settlements and communities allows better planning of housing and health care facilities. The general rise in level of living resulting from an economically successful irrigation scheme can reduce major determinant of disease and poverty. But in all four cases the benefits are not automatic. They are opportunities to be seized, dependent on attention being given at the planning stage of new schemes, and not as an after thought once the irrigation works are completed.

2. Deleterious Effects

Establishment of an irrigation scheme may render a region more vulnerable to diseases. Water can carry toxic chemicals and many communicable diseases, serving both as a transfer medium and as a habitat for vectors and intermediate hosts. The major diseases are transmitted by mosquitoes and simulium, and born by snails and freshwater crustacean. Mention of those diseases and their carriers does not fall within the realm of this book on irrigation.

It may be seen that diseases not existing in an area before, may appear after irrigation. Their low prevalence and less intensity can grow to continuous and massive infections. Severe and disabilitating infection then reduces the workers capacity, with socio-economic consequences. Malaria, schistosomiasis and onchocerciasis have caused the greatest damage in countries where irrigation has been developed with no consideration for public health effects.

In arid lands where schistosomiasis existed at a relatively modest prevalence of, say, 10-15% of the population, and with a correspondingly low intensity (and therefore gravity) of infection, it is common to see prevalence grows quickly to 80-90%. As with onchocerciasis, clinical gravity, and therefore socio- economic impact of schistosomiasis, becomes important when worm loads become heavy due to intense transmission (Amin, 1977).

In the South Chad irrigation project area of Borno State in Northern Nigeria pre-construction surveys showed schistosoma haeimatobium (Bilharz) to be endemic in the project area with prevalence rates in the region of 40% in the larger communities sampled. S.mansoni Sanbon appeared to be an uncommon infection with only two cases among 610 persons examined. Due to irrigation, population of snails-intermediate hosts increased and so also the cases of schistosomiasis (Batterton, 1984).

VI. OTHER ENVIRONMENTAL IMPLICATIONS - AN ANALYSIS

For convenience of discussion, the impacts of irrigation on ecological balance can be separated into three groups in accordance to the site affected. Firstly, ecological imbalance occurring in the catchment area the project is located. Secondly, ecological disorder taking place on account of seepage all along the way of unlined or improperly lined water distribution system. Thirdly, on-farm damage occurring as a result of mis-management of water that brings about waterlogging and salinization.

As other aspects have been described in the preceding section, this section will have its centre of focus on ecological problems arising out of submergence of agricultural and forest lands and resettlement and rehabilitation of people displaced from the area submerged. In this context a part of the National Song, as given below, sang by Sir Bankim Chand Chatterji in 1882 is symbolic to the type of environment and agriculture India, for example, used to enjoy a little over 100 years ago.

"Sujlam Sufalam Malayaj Sheetlam
Sasya Shyamalam Mataram Bandemataram"

(I bow my head to my motherland India that has been blessed with rivers and ponds full of clean soft water; trees decorated gracefully by bearing bunches of sweet fruits; environment endowed with a gentle, cool, sweet-scented breeze; and agricultural fields laden with lush green crops with their earheads bearing plumpy shining grains).

Nearly after a century since then India is reported to be in for a major water famine that could drastically affect its food production and drinking water supply (Postel, 1989). The country's water problems stems largely from gross mismanagement and degradation of its land. Deforestation and denuding of critical watersheds have upset the natural water cycle, causing more rain to runoff in damaging floods and less to percolate into the ground to recharge aquifers. The Vice President, Ms Sandra Postel, of Washington-based Worldwatch Institute quoted Jayanta Bandhopadhyaya of the Research Foundation for Science and Ecology in Dehradun as saying that this largely explains why high rainfall areas now petition for drought relief.

River and pond water supplies, found to be abundant some 100 years ago, are now shrinking. Tens of thousands of villages across the subcontinent are living with water shortages, and their numbers are increasing. In this context Ms Postel has rightly said "Water is only a renewable resource if we respect the ecological processes that maintain and give stability to the water cycle. This, India has signally failed to do".

To revive our water supplies near to what Sir Bankim Chand Chatterji has inculcated in his National Song, the Government of India has given a call for a new integrated policy for the development and the use of water resources in the country, both for drinking and irrigation purposes (Hindustan Times (HT) 19 January 1991). The new strategy would mean a mix of both small schemes that could yield results in short periods, and large dams for storage of monsoon rainwater to be diverted and used for the period of water shortages.

In preparing such plans, care to be taken of environmental factors has been proposed. In so doing three distinct areas have been identified. One, people's participation in both preparing the schemes and the implementation of such schemes for better results. Two, the aspect related to inter-basin of river waters. It would imply large storage of surplus water in areas fed with rains and flood waters, which could be diverted to water deficit areas like Rajasthan, Tamil Nadu, and Saurashtra. The scheme to divert flood water from the Ganga river to the Yamuna river for utilization in dry region of Haryana and the Thar desert in Rajasthan is one such example. The planned diversion of surplus water would overcome the problems created by uneven distribution of water resources in the country. Three, the aspect related to "Water for Future", including environmental issues and policy matters relating to the availability of water based on the perception of needs of the people, and regulation of water resources for better management environmentally safe.

Thus, India perhaps best illustrates the need for a rethinking of ends and means. As much of the progress made by agriculture is attributed to irrigation, the nation seems to be in an all out race to harness its entire irrigation potential estimated at 113 million hectares. Huge amounts are thus being spent on large surface-irrigation projects - such as 260.5 m high earth and rockfill dam on the Bhagirathi river and 135 m high Sardar Sarovar Project (SSP) on the Narmada river in the Indian States of Uttar Pradesh and Gujarat, respectively though there are controversies with regard to fulfilment of goals set for many such projects, apart from project-associated some vexed issues as listed below.

One, the delay in clearance - budgetary and environmental and implementation of the big surface-irrigation projects similar to SSP. Experts have taken almost three decades before reaching consensus on various aspects of this project including the dam height (HT 24 January 1991). According to the World Bank analysis the irrigation facilities to be created are "expected" to start flowing in 1998-99, and the project completion is expected not before 2008 (HT 15 February 1991).

Second, the resettlement and rehabilitation of oustees of the Narmada valley project described as the biggest of its kind in the world : 30 big dams with 100 medium and 1000 small dykes are to cover the project (HT 5 June 1991). Environmentalists led by Baba Amte launched a campaign with a focus on the threat to the life of poverty stricken tribals whose land and dwellings would come under submergence when the Narmada Sagar projects are fully implemented. About 134,000 hectares of land are projected to be submerged in the Indian States of Madhya Pradesh, Gujarat, and Maharashtra. This included 54,000 hectares in Madhya Pradesh (HT 24 January 1991). Nearly 0.3 million people, mostly tribals, in and around dam site are likely to be displaced (HT 30 December 1990). The resettlement and rehabilitation of oustees from the Narmada valley are estimated to cost Indian Rs. 3500 million.

To minimise the menace, environmentalists are demanding lowering of dam height with the plea that by reducing it from 455 to 410 feet (135 m to 121 m) some 16,188 hectares of land from submergence that includes 10,522 hectares cultivated land, 5,666 hectares of forest area, and displacement of 240 villages may be avoided (HT 17 January 1991).

Three, the ecological damage the project entails that may require identification of alternative plots of land for compensatory afforestation, to make good the loss of forest caused by submergence (HT 16 February 1991). However, a recent analysis has shown that environmental loss is not as alarming as some environmentalists have claimed. When fully implemented the project will submerge 4,500 hectares, less than 2% of the total forest area in Gujarat (Ghosh, 1991). According to World Bank report, most of the area submerged has degraded forest. There are no endangered or rare species likely to suffer caused by construction of dams.

Four, whether large investments on this project is justified in terms of the flow of tangible benefits, and the economic rate of return ? An economic analysis by World Bank reveals that the rate of return is expected to be lower than the original projection yet the investments on the project if properly implemented are worthwhile. It may be remembered that in water scarce regions life is more important than the cost.

Thus it seems that submergence of cultivated land, damage to ecosystem due to submergence of forest area, resettlement and rehabilitation of dam oustees, ex-gratia payment for transportation of belongings from the site of submergence to the site of settlement, construction of houses, etc. have been the major issues for the current debate on big dams versus small ones.

According to Sandra Postel (1989), what India needs today are not large irrigation projects but small, efficiently-managed and cost effective ones. The World Bank experts maintain a view quite different from what Postel (1989) has recommended. According to them small dams in the Indian context can not be taken as "alternative", but can be regarded only as a useful compliment, and "they cannot deal with the scale of India's needs".

Like Sandra Postel (1989), Bhumbla (1991) has also criticized the implementation of large irrigation projects on the basis of his observations in canal irrigated, high rainfall areas. Under Sharda Sahayak canal system in the Indian State of Uttar Pradesh and the Tawa Irrigation project in Madhya Pradesh, a large area of land reported by him to have gone out of irrigation as a result of rise in the level of groundwater.

In areas receiving high rainfall, groundwater level usually remains high. On top of this, irrational distribution of canal water and the farmers' tendency of over-irrigating triggered by existing practice of realising the water cost on the basis of area irrigated further aggravate the rise in groundwater level. Thus, a combination of a planned water distribution system, irrigation only on irrigable land, and water cost realisation on volume than on area basis can avert what all problems Bhumbla has mentioned.

In addition, the experience gained from high rainfall areas cannot be replicated in low rainfall areas like the Indian States of Gujarat and Rajasthan where the groundwater by and large is deep, undrinkable, and unfit for agricultural uses on account of high salinity level, and where drought as severe as of 1987 is a recurring feature. Such recurrences adversely affect the man- land-animal-water-vegetation balance the nature has established since ages in conformity with whatever meagre rainwater supply available in the region. Under these circumstances import of water from the neighbouring high rainfall areas, or more efficient planned use of river water (as the Narmada river in Gujarat) through construction of large dam or a mix of small and large dams becomes an urgent necessity.

There is another dimension to the problem. Statistics shows that the target before the country for the production of foodgrains by the year 2000 will have to be 300 million metric tons a year to feed the projected population of 1000 million by the year 2001. As populations would grow, so would the need for water. Now the question comes how to fulfil this production goal set for the nineties ?

An answer in sight may be to accord the highest priority in decade of the '90s to the planned development and the use of our water resources that can bring about prosperity similar to what the large Bhakra dam has brought to the people in the Indian States of Haryana and the Punjab (HT 17 January 1991). The Aswan Dam in Egypt; the Grand Coulee Dam in Washington State; the Hoover Dam in Nevada and Arizona; the Central Arizona Project, aiming at pumping life into the desert; the Nile water known to have flourished the Great Cotton Farm in Sudan; and the Gang Canal that has turned Ganga Nagar district in the Indian State of Rajasthan into a granary of the region are other similar examples of prosperity the large irrigation projects have brought about or will continue to bring.

Instead, the Sardar Sarovar-Narmada Sagar projects are under attack by a group of environmentalists led by Veteran social worker Baba Amte. No one denies the fact that such issue as environmental protection needs be looked into in depth before work on big project is undertaken, but Baba Amte and his green brigade in the meantime owe an answer to the nation on how accelerated development has to be ensured maintaining ecological balance. In its farm policy draft the government of India has given priority to the effective utilization of the irrigation potential through implementation of only those projects that have master plans for development of both catchment and command areas with ecological safety (Banyal, 1991).

Thus, the country's water problems will have to be seen in a wider perspective, planning, development, and resource use with ecological safety. Bearing this in mind India is now thinking to change the philosophy of its water resource development policy from a single-purpose use to an Integrated River Basin Planning on lines similar to Tennessee Valley Authority (Howard, 1991).

The new policy also envisages to include the best possible use of water whether the need is for irrigation, drinking, electricity, or industries; with least damage to environment. This is significant because through more efficient planned use of water it is possible to increase the overall benefits to everyone out there in the river basin, or even in a region where water is a scarce resource (Howard, 1991).

Lastly the desirability of undertaking a high-cost large irrigation scheme is weighed on the scale of the basic principles what the nature has been employing since ages in maintaining in an ecosystem the community of different plant species commensurate with the amount of soil water available at a given period of time. As the water stress develops in the root zone, the nature thins out the load of plant population by allowing a neighbouring plant to die for another one to survive. Similarly in implementation of large irrigation project some area in the catchment will have to be sacrificed upon submergence for another area in the command to flourish. Distribution of social benefits among people can be ensured through generation of more employment in the command area. Care for the improvement of environment in the countryside can be taken through compensatory plantation of forests in wastelands as also of trees all along the main canal and its distributaries.

There is another dimension to the problem in totality. In the decade beginning 1981 the India's population has risen from 685 million to 843 million (HT 11 July 1991), and to 865 million as of now (TV telecast 30 May 1992). Sixteen per cent of world's 5.3 billion population is packed in 2.4% of India's land area (World Population Day, 11 July 1991). Can India sustain it?

The growth rate of 2.11% would make it 1000 million by the end of the century and 1700 million by 2024 AD (Ved, 1991).

Such population explosion is bound to subject our limited resources under stress. The country may not be able to meet even the basic needs of food, water, housing, education, employment, and health care. Upsurge in urbanisation is likely to encroach upon finite agricultural lands and sewage-sludge and city garbage disposal problems will be on a steady increase especially in and around metropolitan cities. Even today Delhi, the capital of India, is described to be one of the highest polluted cities around the world with its pollution rating 871.92 metric tons (HT 13 August 1991). The Central Pollution Board, monitoring the ambience levels of pollution due to exhaust pollution, has found it ranging as high as 1500 mg at times against a permissible level of 1000 mg.

As populations have increased, so have the needs for water, food, cloth, shelter, employment, environment control measures etc. Urbanisation and industrial growth are bound to encroach upon the agricultural lands besides the growing need for environment control measures suitable to handle the problem of increasing urban and industrial wastes as also the smog problem especially in metropolitan cities.

To feed the rapidly increasing number of people, India has achieved an increase in food production from 50.8 million metric tons to 172 million metric tons during the last four decades, but major stride took place from integrated use of better seeds, fertilizer, water, plant protection, and better crop husbandry practices that enabled the country to celebrate Green Revolution in the early seventies.

Standard breeding programme led to simultaneous release of high- yielding varieties of wheat and upland rice. Continuous use of monocots-flowering plants with one seed leaf-rice and wheat in rotation undoubtedly has enhanced the farm production, but at the same time has increased our dependence on environmentally damaging agrochemicals, besides impoverishment of soil fertility as a result of removal of plant nutrients by both crops from the same upper surface layer of the soil. Soil degradation brought about in this way has now caused a great concern about maintenance of subterranean environment.

The present food production level at 172 million metric tons that India has achieved at the end of '80s is considerably lower than its estimated requirement of 300 million metric tons per annum to support the burgeoning population in the '90s. Thus the major challenge faces the agriculture today is how to achieve self- sufficiency at the projected food front with production practices carefully so controlled as they do not result in environmental problems.

The above production level appears to have occurred through vertical increase in yields by application of intensive agricultural practices on irrigated areas that account for merely 30% of the total cultivated land in India. Relying any more to improve agricultural production by an attempt to achieve the near potential yield from this meagre 30% of irrigated land by deliberate release of genetically engineered crop varieties is likely to set an ecological precedent that may have risks. Liberal use of water, fertilizers, and plant protection in cultivation of such varieties is bound to aggravate the degradation of environment that could occur through enhancement in non-ET water uses, dependence on environmentally damaging high doses of fertilizers, and pesticidal residues.

High-yielding varieties of field crops that are responsive to applications of water and fertilizers also attract pests and diseases. It was for this reason that plant protection assumed significance in the pool of practices the scientists designed for bringing revolution in Indian agriculture in the early seventies. With our growing concern about endowment of ecological safety into farming system the genetic engineers, now working with many of the major food plants, are exploring genetic changes that improve the food value and crop yields, reduce our dependence on environmentally damaging chemicals by giving crops an ability to protect themselves against pests, confer drought resistance or help plants tolerate saline water, and introduce weedlike resistance into agriculturally important food plants (Cowen, 1990).

Moss (1976) stated that "yield can be no greater than the net photosynthesis from which it is derived. Thus, it seems logical that one way to increase yield is to build into crop varieties the capacity for greater photosynthesis". Efforts therefore are underway to develop crop cultivar with physiological functions genetically engineered to provide their plants with an ability to photosynthesise at the rate equal to upper limit of the range of net photosynthesis.

Grain crops such as rice (*Oriza sativa* L.), wheat (*Triticum aestivum* (L.) emend Thell.), and barley (*Hordeum vulgare* L.) have C_3 photosynthesis with maximum net photosynthetic rate of 15 to 35 mg CO_2 ha^{-1} dm^{-2} of leaf in the optimum day temperature range of 15 to 25°C. Other field crops like maize (*Zea mays* L.), pearl millet (*Pennisetum americanum* (L.) Leeke), and grain sorghum (*Sorghum bicolor* L.) are C_4 plants with maximum rate of net photosynthesis at 40 to 80 mg CO_2 ha^{-1} dm^{-2} of leaf in optimum day temperature range of 30 to 45°C (Black et al., 1971).

High yielding varieties of these C_3 and C_4 crop species released for cultivation are reported to photosynthesize at their potential rates. Therefore the only alternative to increase agricultural production through an increase in photosynthesis is to insert genes of C_4 plants into C_3 plants that would enable the latter to improve their net photosynthetic rate from 15-35 mg CO_2 ha^{-1} dm^{-2} of leaf to 40-80 mg CO_2 ha^{-1} dm^{-2} of leaf. If sink capacity increases in the same proportion yield shall in all likelihood be doubled. The new genotypes thus developed must have disease and insect resistance, acceptable quality, lodging resistance, and other desirable agronomic traits. Stated precisely the new breeding programme will have to be based on physiological and biochemical traits and should lead to a more metabolic system that should bring about efficient use of nutrients, water, CO_2, and solar energy (Schrader, 1980). Efficient use of these by plants should increase yields of agronomic crops, and help us advance further up the yield curve.

Yield is the product of genotype and environment. The latter composes soil, air, and management. Moment a crop cultivar adapted to the given region is selected, it integrates the effects of soil and climate. About 49% of increases in wheat yields is attributed to genetic improvement and 51% to improved technology (Jensen, 1978). In maize, breeding accounted for 57% of observed gains in yield and 43% to improved technology (Duvick, 1977).

In applications of genetic engineering to farming, scientists must design those applications with all the possible ecological consequences - both desirable and undesirable - strongly in mind so that ecological safety in agriculture is ensured and suitable measures against undesirable consequences are taken. However, the technology is likely to start having a significant effect on farming by the turn of this century. Until such time will the gap between the present level of foodgrain production and the estimated food need to feed the increasing population of India in the decade of nineties shall remain impending ?

It is probably this gap between what India needs and what it can produce from its limited 30% irrigated areas that has prompted the attention of farm scientists to improve the production by application of response farming in environmentally fragile resource-poor rainfed areas that account for 70% of cultivable land in the country.

There is another vital field for moving up the yield curve. As we know, leguminous crops fulfil their nitrogen requirements for growth through fixation of atmospheric N in symbiosis in which the host plant (i.e., ability to provide photosynthate to the nodules) and the symbiont (i.e., potential ability of rhizobia to provide nitrogen via nitrogenase) benefit each other. In the absence of such mechanism in nonleguminous crops to make their own fertilizer, world's farmers apply nearly 80 million tons of nitrogen fertilizer to their crops annually (Bylinsky, 1991).

Using imaginative combinations of molecular biology and electrophysiology, Ms S.R. Long (cited by Bylinky, 1991) has been able to decode the rhizobium-alfalfa conversation. In the symbiotic process, alfalfa (*Medicago sativa* L.) plant says "rhizobium come to work". They tell the plant : "We got your message and we are on our way. Start preparing nodules". Eventually, "genes that enable plants to hold such dialogue could be inserted into crops that don't make nodules for rhizobium now. That, in turn, could change the face of agriculture" (Bylinsky, 1991), with ecological safety that is often endangered due to heavy applications of nitrogen fertilizers.

A. Excess Pesticides Harm Ecology

Irrigation has made and would continue to make green revolution possible in many Asian countries. The uncritical use of pesticides in the wake of the green revolution has upset the ecological balance. Writing the UN Journal Development Forum, In Steele

(HT 4 November 1991) contends that though the green revolution brought spectacular gains in yield, the role of pesticides in the equation was greatly "oversold".

According to him, the most dramatic evidence of this was recorded in Indonesia where the rice brown Planthopper - a minor pest before the green revolution - infested rice crops in Java and North Sumatra in 1976-77. In a bid to control the outbreak, the aircraft blanketed more than one million rice fields, but to no avail.

The lesson out of this Indonesian experience is that pesticides combined uncritically with fertilizers was in fact a recipe for brown Planthropper. Following a renewed bout of this pest outbreak in Indonesia, the country tried a new approach by quickly cutting the subsidies on pesticides and banning many pesticide products from use in the rice fields. This measure coupled with Integrated Pest Management (IPM) had resulted in saving the government close to 150 million dollars a year and raised crop production from six to seven metric tons ha^{-1}, besides gaining considerably in public and environmental health.

IPM is defined as regular observation of individual crop fields to take stock of the pest problem so that unnecessary insurance type applications of pesticide in the absence of actual problem is avoided.

VII. SOLUTION TO IRRIGATION PROBLEMS

1. Combating Waterlogging

Several unconventional suggestions have been made. Some of them are : (a) Reduce canal water release at the head as water táble rises to critical levels (Chambers, 1988), (b) Divide the entire command in two or more segments and stagger canal irrigation in each segment as is done through the "turn method" in the Periyar Vaigal System in the Indian State of Tamil Nadu (Shah, 1985; Rajagopalan, 1988), or replicate the Chinese experience in conjunctive use in which farmers in the upper canal reaches have increased their reliance on groundwater, while those in the lower reaches rely on a system of ponds and reservoirs (Kramer, 1989), (c) Promote intensive minor irrigation in the canal command well in advance of the implementation of the irrigation project as a means of "substitutional irrigation" (GWRDC, 1986; Desai, 1988), (d) Ban canal irrigation along a length of the main canal (Pant, 1988), (c) Use "augmentation wells" to pump groundwater, especially where saline, in to canals as is done in the Indian State of Haryana (Mahodaya, 1983; Michael, 1983) and in the Khairpur Irrigation Project in Pakistan (O'Mara, 1984), and (f) Intensive groundwater development throughout the command by offering farmers incentives to invest in wells and pump sets and to irrigate with groundwater (Michael, 1983; Dayal, 1988). Many of these suggestions have been made on the ground that they encourage "conjunctive water use", a term often used with joint or supplemental use (Desai, 1988). Conjunctive water use views the entire command as a watershed and aims at optimum use of ground as well as surface waters at a macro-level.

Empirical investigations made by Shah (1988) led him to conclude that the instability in the water resource in a canal command arises primarily from the equilibrium properties of the water resource system model. This means that rise in the water table must decline as one moves farther and farther from upper reach (core command) of the canal. Thus, fundamental to optimal water use in a canal command are policies that discourage canal irrigation and stimulate groundwater irrigation in the core command as part of the strategy of securing extensive spread of canal water resources. This kind of strategy would result in serving the tail-enders, in the peripheral command, as best as possible under design constraints. It also minimizes the need for drainage and lining, ensures maximum possible equity in the spatial distribution of access to water; and finally, mitigates the diseconomies of waterlogging and soil salinity.

Setback to above optimal policy is the "default" pattern that tends to emerge, which is opposite of "equilibrium" pattern. Left to themselves. farmers located in the core command would depend heavily on canal irrigation and be least interested to invest in and use of groundwater irrigation. Away from the core, on the other hand, groundwater based irrigation becomes the mainstay of farming. There highly complex social institutions operate for private exploitation and sharing of groundwater. Further, if State Electricity Board switches from metered power tariff to flat rates, then state-wide reduction in groundwater sale prices by private well-owners may take place. That may result in substantial redistribution of 'irrigation surplus' in favour of the generally resource poor water buyers.

In the event of active groundwater markets, much more water gets pumped. Active water markets greatly enhance the value of private tubewells in supply vertical drainage in core command suffering from high water table.

2. Replacement of Inefficient Irrigation Method

By surface method of irrigation much more water is applied to crops than is necessary for them to grow. Since it is the least- cost system to install, this method has widely been practised. Continued use of water by this method has caused poor drainage problems at many places in many countries. In problematic areas of high groundwater table, drainage is carried out in a variety of ways. Gupta (1984) has suggested the possibilities for water table control by subsurface horizontal drainage. Imperial Valley of California has the best underground drainage system in the world. Water is drained into inland Saltan Sea. Dukhovny (1977) has reported that more than 3000 km of closed drainage system is being constructed on irrigated lands annually.

3. Lining of Canals

Elaborate drainage system involving large resource commitment can diminish the amount of unwanted water entering the aquifer but by itself cannot make this water available for irrigation in far flung areas of the peripheral command. The lining of canals, e.g., as done in the case of Indira Gandhi Nahar Project, may be somewhat more useful in this respect. It will reduce seepage and also make more canal water available at the tail-ends. In sandy terrain, lining of canals is a must. In medium to heavy soils, huge investments in lining would not be needed at all if surface water seepage is adequately matched over space and time by energizing tubewells in the canal command for groundwater withdrawals. For the same reason, lining of farm irrigation canals may not be necessary at all if groundwater withdrawal matches the water entry through seepage.

4. Improving Irrigation Efficiency

Practices of deficit irrigation (i.e., irrigation much below the level of potential evapotranspiration), and increase at least by 10% the efficiency of the surface method of irrigation are the other methods to alleviate a lot of waterlogging and salinization problems.

5. Introduction of Modern Irrigation System

Another solution to the problem of waterlogging is to introduce sprinkler irrigation system. This system has been helpful in curtailing desertification as, for example, in Mahendragarh district of Indian State of Haryana. The sand hummocks has been gradually but steadily levelled after the introduction of sprinkler irrigation. It is particularly suitable for sandy areas where percolation losses from surface irrigation are high and where frequent light irrigation is preferred because of poor water holding capacity of the soil.

Sprinkler irrigation system uses more energy than surface method and requires a larger capital investment to install. But it has brought irrigation to rolling and steep lands otherwise suited only for dryland farming. In Israel that is one of the driest countries in the world and that has been able to make the best of a limited water supply, "partial root zone wetting" method of irrigation by the sprinkler system is practised. On an average, the rooting zone of the most field crops is 1 m. In partial root zone wetting method, it is advisable to irrigate to wet only the one-fourth of the crops rooting zone by the sprinkler system. This practice of irrigation is not feasible in the tubewell command especially in the arid and semiarid regions having water salinity. Because, soon a salt layer will be built up in the potential rooting zone, and to leach it one to two-year of fallowing is necessary depending upon the amount of rainfall. In canal command areas with freshwater, this type of irrigation by sprinkler system is feasible. High frequency sprinkler irrigation is economical in countries with cheap labour.

The less you apply water to crop fields to optimize yields, lesser are the problems of waterlogging and salinity build up. Drip irrigation is the most modern system of applying water directly to the crops root zone in an amount equal to the daily crop water requirement.

Studies have shown that, in drip irrigation system, a high water content extends to a depth of 30 cm. Mutual action between the evapotranspiration demand and a small amount of water applied by daily application from the drippers prevents the downward flow of water into the deeper soil layers. Quantitatively, until 5 hours after irrigation the water content in an upper 15 cm layer of soil has been found to vary between field capacity and saturation point under the emitter, close to field capacity at the point 15 cm from the lateral, and well above 50% of the available water at the point 25 cm from the lateral (Singh et al., 1989).

At the full grown stage, 74% of the total root weight was confined to the top 15 cm soil layer, and root growth in this layer was independent of soil moisture content in the lower layers (Singh et al., 1989). Under these conditions, a significantly highest yield of 90 tons ha^{-1} of tomatoes by daily drip irrigation equal to evapotranspiration and application of fertilizers to supply 224, 88, and 168 kg ha^{-1} of NPK, respectively, disproves the profile wetting concept, i.e., scheduling irrigation to charge the rooting zone profile to field capacity in each irrigation. Hence under drip irrigation system there could never be a problem of waterlogging. And when the supplies of all growth factors are adequate in the resource pool in the just 15 to 30 cm upper layer, there would never be a sagging effect on crop yields. Israel has attained impressive gain in water-use efficiency by adoption of sprinkler, drip systems, and optimum management of water.

Irrigation efficiency depends as much on the way the irrigation system is managed as on the type used. The average efficiency is 40 to 50% for surface irrigation method, 80 to 85% for sprinkler system, and close to 100% for drip system of irrigation. Farmers using surface method of irrigation can cut their water demands by 34% by on-farm better management of water, and 50 to 60% by switching over to sprinkler and drip irrigation systems that would otherwise percolate down and hasten waterlogging.

6. Deficit Irrigation

Farmers can also reduce their water demands and delay the process of waterlogging by resorting to deficit irrigation scheduling. Deficit irrigation means applying water less than the amount needed to reach potential yield by meeting potential evapotranspiration of each crop over each hectare. It has been found (Singh and Mann, 1979) that, in the case of wheat Kalyansona, the water level to attain the most profitable yield is below the level needed to obtain the maximum yield, whilst maximum production per unit application of water was obtained at about 50% of potential evapotranspiration. The irrigation efficiency, i.e., evapotranspiration as per cent of applied water was 100%. Hence farmers can avoid or delay the process of waterlogging.

Earlier the irrigation department had banned digging of tubewells in canal command areas. With passage of time, rise in groundwater table has necessitated energization of tubewells and conjunctive use of groundwater and canal water. This has been discussed earlier.

7. Cultivation of Salt Tolerant Crops

Another way for better management of water is to grow salinity tolerant crops with brackish water in order to save freshwater for less salinity tolerant field crops or for domestic uses. In Saudi Arabia, brackish water is used to irrigate salt tolerant crops such as sugar beat [*Beta vulgaris* L.], barley, cotton [*Gossypium hirsutum* L.], spinach [*Spinacia oleracea* L.], and date palm [*Phoenix daetylifera* L.], thereby saving the best quality water for drinking and other household uses. Chatterji and Gupta (1969) and Tanwar (1984) have recommended planting of salt tolerant species of trees belonging to the phreatophyte group on the sides of farm canals, field boundaries, and roads that will help to act as wind break, besides helping in lowering of groundwater level.

8. Positive Aspects of Arid Land Irrigation in Combating Ecological Imbalance

Development of water and soil management programme matching with specific conditions of an area, for example, Stage II of the Indira Gandhi Nahar Project, is essential. Such programme should aim at (a) greater reliance on crops with deep root system like cluster bean [*Cyamopsis tetragonoloba* (L.) Taub.], raya [*Brassica juncea* (L.) Czern. & Coss.], and safflower [*Carthamus tinctorius* L.], (b) soil compaction to reduce infiltration rate and increase retention capacity, and (c) use of slow release granulated fertilizers to minimize leaching losses of applied nutrients. To take advantage of the inevitable deep percolation losses, introduction of deep rooted trees of economic value will be beneficial. Agroforestry, agri-horticulture, and farm forestry approaches are the useful land use systems in this regard.

Every suggestion has its own merits and demerits, so are the technical package given under section (a), (b), and (c) above. The first basic principle in planning strategy for managing a limited water supply in arid region is the choice of crop most responsive to and efficient in use of water. Studies by Singh (1984) have shown that, although reduction in yield per unit reduction in evapotranspiration (ET) was least in cluster bean, this crop was the poorest among dryland crops in efficient use of water. The most efficient in use of water was hybrid pearl millet cultivar BJ 104.

Second consideration for managing water in an arid region is crop earliness. A crop maturing in 60 days will have water management requirement 30 to 60 days lesser than other maturing in 90 or 120 days. Safflower is a much too long duration crop, its maturation coincides with high ET demand period from April to May, and possibly for this reason it was replaced by mustard and wheat in the optimal cropping plan under a given water supply on the farm in northwestern India (Singh and Mann, 1979).

Soil compaction to reduce infiltration rate and increase retention capacity is difficult to apply on a farm scale. A more feasible alternative to this approach is the optimal programming of the date and depth of all seasonal irrigation in such a way that partial rooting zone is wetted in an irrigation. This is what we call an under irrigation approach that always seeks to apply a small amount of water at higher frequency of irrigation so that only a part of the potential rooting zone restricted to the upper soil layer is wetted in each irrigation. In this way no soil moisture shall ever be available to loss caused by gravity, beyond the rooting depth in soil profile.

When we think of crop earliness and light frequent irrigation, preferably by sprinkler system (best suited to light soil controlled light irrigation), choice of fertilizer should rest on one that is available to plants as soon after application as possible, so that sufficient nutrients are stored in plants in the vegetative stage, and later move to developing seeds (high sink capacity) according to distinct physiological priority. In wake of this requirement suggestion for the use of slow release granulated fertilizers to minimize leaching losses of applied nutrients hardly has any bearing.

Wind erosion is a continuing menace in the sandy arid plains of Rajasthan. Experience has shown that irrigation in areas suitable for arable farming can help reduce the vulnerability of soil to wind action in many ways. Data gathered from studies made at CAZRI, Jodhpur show proportionate increase in biological yields, hence a similar increase in root biomass, with increasing supply of water to hybrid pearl millet (Singh, 1989). On the contrary, harvest index (grain yield as per cent of aboveground biological yield) tended to decrease with increase in seasonal ET. Harvest index at 40% is considered to be optimum in most field crops. In the subsistence agriculture in which animal husbandry is basic to overall agricultural system, 25 to 27% harvest index that he found is within the acceptable range.

Extensive root system resulting from the optimal irrigation would help bind the soil particles through exudates of roots and addition of organic matter. Aided by frequent wetting and drying cycles soil structure would improve in the long-run. A common practice in northwestern India is one of summer ploughing of irrigated fields after the harvest of winter season crops. This leaves large clods on the surface resistant to wind action. In rainfed lands, on the other hand, summer ploughing is a sin since it exercises a reverse effect by forming a pulverized surface highly vulnerable to wind action. The advantage of irrigation in soil management is thus apparent.

Another favourable feature of irrigated farming is the optimal use of fertilizers. This too helps indirectly in the addition of root biomass as shown in Fig. 35 from Singh (1974). Addition of soil organic matter takes place further upon introduction of better crop husbandry like crop rotation. It has been reported that after a 10-season cropping of irrigated wheat in rotation with rainfed pearl millet, the soil organic matter in the surface 0-15 cm layer was 72% higher in plots that received 15 t ha^{-1} of manure than in those that did not (Singh, 1985).

Cluster bean is the poorest in efficient use of water, but in rotation with pearl millet under rainfed conditions it was found (Singh et al., 1985) to maintain soil productivity and save 60 kg ha^{-1} N that energy intensive monocropped pearl millet otherwise required in the rotation cycle. This is one of the goals what sustainable agriculture seeks to achieve.

In many areas of the desert, it is now increasingly evident that even some of the hardier species do not get established and succumb to moisture stress under aberrant rainfall conditions. As such, assured water supply in the form of protective irrigation is required for tree plantation in the arid zone.

The foregoing thus brings out that irrigation can do more good than harm in the management of arid lands. The need is for an efficient management system that can sustain production, without any degradation of the environment, on a sustained basis.

The negative effects are primarily due to water delivery and distribution through unlined canals and mismanagement of on-farm irrigation water.

9. Irrigation System Using Pipelines

In recent years, irrigation system using pipelines has become more popular especially in developed countries. This system reduces infiltration losses and the extent of unused lands to 10%, facilitates the delivery of water to irrigated areas, and does not hamper machine operations.

10. Return Water Flow

Detrimental effects of the return flow can be prevented by a variety of ways. Important among them are diversion of return flow to the specially reserved place(s), the return water treatment before removal, recycling for irrigation, and dilution with sweet water that may prove beneficial in regions with surface good quality water. The combative measures will be more effective if these are carried out in all three subsystems that, along with the positive agricultural effect, will considerably diminish the damage caused by return flow.

The return flow does two important functions of linking. First, it links the water receiving basin (rivers, lakes, reservoirs, etc.) with new sources of irrigation water. The water quality in the latter determines to a great extent the quality of the return flow and of the water in the receiving body. Second, the return flow links (geochemically) the receiving basin with new territories found under man's active impact (irrigated lands). The resultant effect of this is the removal, with the return flow, of salts, fertilizers, chemicals, organic remnants, bacteria, pathogens, and microelements into the basin of a water receiver.

The most widespread method to control the return flow salinity in the water supply subsystem is canal lining. Assessing its economic validity, account should be made not only of the benefits provided by water saving, but also of the slackening and hazard ous effects of seepage from canals.

The effective method in the irrigated field subsystem is the improvement of irrigation technique on a farm level. No less vital are effective agricultural practices, including regulation and speed of water application, overcoming the unevenness in water distribution, and controlling deep percolation.

Methods in the water disposal subsystem comprise the construction of facilities for tapping surface and subsurface irrigation flows and facilities for the tapped flow treatment and disposal. Deep percolation can be controlled by means of water impermeable membranes.

The increased water salinity of receiving bodies caused by the return flow from irrigated fields is not a sole problem connected with the general effect on surface and groundwaters. In some cases no rise in salinity may be recorded or its role in the changed water quality of the receiving body may be insignificant. However, one must not think that the return flow is harmless in these cases.

11. Measures to Overcome Soil Degradation

The measures to overcome soil degradation are : (1) Regular application of high doses of organic manures, (2) Introduction of grasses and legumes in crop rotation, (3) Use of crop varieties that can take the advantages of irrigation and fertilization, (4) Use of fertilizers that develop acidity upon transformation with application of nitrogenous, phosphorus, and potassium fertilizers generally balanced for new ecological conditions, (5) Use of gypsum to reclaim alkalinity, (6) Optimum tillage at optimum time, (7) Inter-cultivation in irrigated row crops, (8) Weed control measures, (9) Regular deep subsurface loosening, (10) Night irrigation in the arid regions, (11) The experience of the USA, former USSR, and Hungary has shown that deep breaking (50 to 70 cm) of structureless soils, having exchangeable sodium in particular, in combination with application of manure and large doses of gypsum or sand contributes to improvement of soil physical properties. This practice is least feasible in developing countries, and (12) Where possible, use of underground pipelines for conveyance of irrigation water to fields.

12. Remedies to Environmental Consequences of High-Yield Based Farming System

High-tech agriculture integrating better seeds, fertilizers, water management, and liberal use of pesticides brought green revolution in many Asian countries. Although there has been considerable increase in yield and food supply in many countries of the world where green revolution has been a success, liberal use of fertilizers and excess pesticides have harmed ecological balance. The U.S. Environmental Protection Agency (US EPA) has identified agriculture as the largest nonpoint source of water pollution. Pesticides and nitrates from fertilizers have been found in the groundwater. The issue of pesticide residues remains unresolved.

A package of practices environmentally safe includes : genetic improvement of crops to resist pests and diseases and to use nutrients more effectively; crop rotation; diversification of crops grown; integration of livestock and crop operations (e.g., mixed farming); reduced use of agrochemicals; cultural practices including tillage and optimum time of planting; summer ploughing, where feasible, to kill soil born pests and diseases; IPM; the use of leguminous green manure crops and rotation of erosion control and as a nitrogen source; etc

Crop rotation provides many economic and environmental benefits to farmers (Shrader and Voss, 1980; Baker and Cook, 1982; Kilkenny, 1984; Voss and Shrader, 1984; Heichel, 1987; Power, 1987). To name a few, it improves yields of grain crops beyond yields obtained from energy intensive monoculture (Singh et al., 1985; Heichel, 1987; Power, 1987).

Increased soil moisture, pest control, and the availability of nutrients contribute to the rotational effect on crop yields. The most important part of this effect is the insect and disease control benefits of rotations (Cook, 1984, 1986). The healthy root system therefore is better able to absorb nutrients in the soil, which can reduce the rates of fertilizers needed (Cook, 1984).

Increase in soil organic matter (Singh, 1985; Singh et al., 1985) may be the basis for the improved physical characteristics of soil. Deep rooted leguminous crops may use nutrients from deep in the soil profile. In the process, these plants may bring the nutrients to the surface making a part of them available to a shallow rooted crop upon decomposition of root residues. It has also been reported that without any additional nitrogen fertilizers, leguminous nitrogen can support grain yields (Voss and Shrader, 1984; Heichel, 1987). All rotations involve diversification, and therefore, ensure the farmers against vagaries of crop prices, production inputs, pest infestations, and weather (National Research Council, 1988).

Researchers have discovered biological and genetic alternatives to the use of pesticides. They have devised a variety of cultural and biological control methods.

The IPM programme gives priority to non chemical controls such as host plant resistance and the use of insect predators to safeguard food and other crops against damage by insects. According to Ian Steele (HT 4 November 1991), the Indonesian model has attracted wide interest among experts from India, China, Bangladesh, Bhutan, as also from Africa and Central America.

Quoting an entomologist with IPM programme in Indonesia, Steele says there are normally several hundred species of insects in a rice field, and that for every pest there are many more insects that help the farmer. They are natural predators that do not damage the rice crop but feed on the pests.

Steele further claims that experts from FAO now say that 15 years of data gleaned by the International Rice Research Institute and its collaborating networks in Asia demonstrate no positive yield response to insecticides. Worse, the immediate use of pesticides had upset an ecological balance between pests and their predators that had evolved over three centuries of rice cultivation in the region.

New virus that kills crop insects has been isolated by an Indian biologist, Dr. V.P. Chaudhary, presently working as director of the Antibody Engineering Laboratory at the University of California, Davis. Working with his American colleague, he inserted the scorpion toxin gene in the genetic structure of the butterfly virus in order to create a genetically engineered version of the virus that quickly kills crop-destroying insects (HT 8 September 1991).

Following refinement and successful field trials, the biopesticides are now ready to tackle pests damaging major crops including cotton, cabbage (*Brassica oleracea* L.), tobacco (*Nicotiana tabacum* L.), and tomato (*Lycopersicon lycopersicum* (L.) Karst.) in India and elsewhere.

Spiders can rid earth of pests. During his participation in the Indian Science Congress held at Indore in the Indian State of Madhya Pradesh, Professor Om Prakash Mittal from Punjab University reported that of the 30,000 known species of spiders only two are poisonous-Brown Recluse and Black Widow, not found in India (HT 4 January 1991). Remainder of the known species that live only on insects could be put to extensive use for pest control. Earth would have witnessed major disasters caused by pest explosion had there been no spiders that averted them without man ever knowing it.

Some member of species is found in warehouses in New York, where it preys upon pests of stored grains. The banana spider (*Heteropoda venatoria*), a tropical species, keeps a check on cockroaches and other destructive insects. The 'Jumping Spiders' hiding under the foliage of plants attack insects from a distance. These spiders have been doing an excellent job in the Fiji Islands, where coconut plants are frequently attacked by a moth. During each pest outbreak, the jumping spider (*Ascyltus crygodes*) increases rapidly in numbers and attacks the caterpillars and pupae. The American scientists had identified spiders as an important factor in checking cotton worms, gypsy moths, and other destructive insects. Thus, if these spiders are put into practical use for checking pest explosion, it may be possible to avert man-made disasters to a certain extent.

Not all spiders are good insect hunters. However, through gene therapy a lot of spiders could be improved to induce the same genetic make-up as the crab spiders. If this is done, spiders can do wonders in helping man in his economy.

VIII. FUTURE RESEARCH AND MANAGEMENT FOR IRRIGATION PROJECTS

Changes in the physical, biological, and social components of the irrigated arid agro-ecosystem, need to be monitored in an integrated manner. The multidisciplinary team should be orchestrated under a single authority so that the whole programme is designed to feed data to each other discipline, to combine discipline bound inference and finally to arrive at a synthetic picture of the arid ecosystem - how it is functioning and what corrective measures are required ? It is to emphasize here that nothing short of a long-term programme on a regional/national scale can bring answers to the whole array of questions in basic as well as applied research. Detailed proposals for action are available in a variety of reports from UNESCO-MAB, the International Commission on Large Dams (ICOLD), International Soil Science Society (ISSS), and the International Commission on Irrigation and Drainage (ICID). A series of reviews has been made on agricultural water research needs (Consultative Group on International Agricultural Research, 1973), which recommend research on specified goals in hydrology, engineering, soils, agronomy, biology, health sciences, geography, economics, systems analysis, organizational management, and related fields. Any research that deepens knowledge in these fields will increase the capacity of the people to cope more effectively with possible environmental effects from irrigation and drainage.

The significance of filling many of the gaps in information about arid land irrigation can be gauged from the detailed description of various environmental implications in the preceding pages. Two points of great relevance are, however, emphasized here. One, there has to be specific budgetary provision in any irrigation project to take care of the problems of human health as a regular feature. Two, both at planning as well as implementation stage a multi-disciplinary team of experts including entomologists, medical men, socio-economists, agronomists, soil scientists, hydrologists, and specialists in the field of environmental analysis, should be involved.

IX. EPILOGUE

The main objective of any human activity since the beginning of the long history of tool making was to develop and control the environment. The relatively slow increase of man's influence was accelerated by rapid population expansion and development of the technical civilization, which started in the last century and reached a stage when its quantitative increase turned into qualitative change. Two basic causes initiated this problem: the demand to have a truly comfortable way of living (the consumptive society) and the power and the technological knowledge to execute large-scale modifications of natural conditions. Activities aiming to achieve better living conditions cannot be stopped but better planned. It is in this context that the phrase protection of environment is not absolutely correct. The reasonable development of the environment is often more important than its protection. It is, therefore, necessary to survey all possible side effects, their socio-economic influences, and to determine supplementary measures required for minimizing undesirable changes. Arid lands provide mankind with the largest unexploited reserve for agricultural production. This potential resource cannot be utilized, however, without water. What are the side effects that arid land irrigation brings in its wake ? These have been classified and discussed here as these pertaining to climate and microclimate, soil-water-plant relationships, natural flora and fauna-both macro and micro, socio-economics, and human health. Measures to overcome the detrimental effects, and future research and management needs of irrigation projects are also discussed. Organic farming including replacement of high-tech agricultural practices by practices for sustainable agriculture has been highlighted to arrest other sources of ecological imbalance.

REFERENCES

Adree, K. 1920. Geologie des meeresbodens. 2nd ed. c.f. Kovda, V.A. 1980. Problems of combating salinization of irrigated soils. Centre for International Projects of GKNT, Moscow, 274 p.

Amin, M.A. 1977. *In* E.B. Worthington (ed.) Arid land irrigation in developing countries : Environmental problem and effects, p. 407-412, Pergamon Press, Oxford.

Anonymous. 1973. Karakum Canal and Environmental changes within its zone, Nauka, Moscow.

Anonymous. 1990. Span 31(3) : 24.

Babaev, A.G. (ed.) 1982. Combating desertification in the USSR : Problems and experience. Centre for International Projects of GKNT, Moscow, 120 p.

Babaev, A.G. and N.S. Orlovsky. 1980 *In* J.A. Mabbutt and S.M. Berkoviez (eds.). The threatened drylands - Regional and systematic studies of desertification, p. 57-72, School of Geography, P.O. Box 1, Kensington, N.S.W. 2033, Australia.

Baker, K.F. and R.J. Cook. 1982. Amer. Phytopathological Soc., St. Paul, Minn.

Banyal, S.S. 1991. The Hindustan Times, 19 January 1991.

Batterton, E. 1984. J. Arid Environ. 7 : 43-57.

Bhumbla, D.R. 1991. The 78 session of Indian Sci. Congr., 3-8 January 1991, Indore, India.

Black, C.C., G.E. Edwards, R. Kanai, and H.H. Mollenhauer. 1971. *In* 2nd International Congr. on Photosynthesis, p. 1745- 1757.

Bylinsky, G. 1991. Span 32(6) : 2-9.

Chambers, R. 1988. Managing canal irrigation. Oxford and IBH, New Delhi.

Chatterji, P.C. and S.K. Saxena. 1988. *In* Iswar Prakash (ed.) Desert ecology, p. 223-258, Scientific Publishers, Jodhpur, India.

Chatterji, P.C. and R.K. Gupta. 1969. Ann. Arid Zone 8 : 247-250.

Consultative Group on International Agric. Res. : Technical Advisory Committee 1973. Report of research review mission to the Near East and North Africa. Consultative Group on International Agricultural Research, Rome.

Cook, R.J. 1984 *In* D.F. Bezdicek and J.F. Power (eds.) Organic farming : Current technology and its role in a sustainable agriculture, special publication no. 46, p. 111-127.

Cook, R.J. 1986. Plant Disease 70(9) : 894-898.

Cowen, R.C. 1990. Span 31(6) : 8-9.

Dayal, P. 1988. Workshop on the development and management of groundwater resources in eastern Uttar Pradesh, Narendra Deva University of Agriculture and Technology, Faizabad. 5-6 April 1988.

Desai, C.G. 1988. IFPRI-TNAU Planning workshop on policy related issues in Indian Irrigation. Ooty. 26-28 April 1988.

Dhawan, B.D. 1989. *In* V. Ballabh and T. Shah (eds.) Efficiency and equity in groundwater use and management. 53 p., Institute of Rural Management, Anand 388001, India.

Dukhovny, V.A. 1977. United Nations Conference on Desertification, 29 August - 9 Sept. 1977, Nairobi, Kenya.

Dukhovny, V.A. and L. Livtak. 1977. *In* E.B. Worthington (ed.) Arid land irrigation in developing countries : Environmental problems and effects, p. 265-276, Pergamon Press, Oxford.

Duvick, D.N. 1977. Maydica 12 : 187-196.

Ehlers, E. 1977. *In* E.B. Worthington (ed.) Arid land irrigation in developing countries : Environmental problems and effects, p. 85-98, Pergamon Press, Oxford.

El Gabaly, M. 1977. *In* E.B. Worthington (ed.) Arid land irrigation in developing countries : Environmental problems and effects, p. 239-250, Pergamon Press, Oxford.

Gangwar, A.C. and B.S. Panghal. 1989. *In* V. Ballabh and T. Shah (eds.) Efficiency and equity in groundwater use and management. 53 p., Institute of Rural Management, Anand-388001, India.

Ghabbour, S.I., M. Imam, M. Mahi-Ali and M.S. El-Abyad. 1977. *In* E.B. Worthington (ed.) Arid land irrigation in developing countries : Environmental problems and effects, p. 329-334, Pergamon Press, Oxford.

Ghosh, A. 1991. The Hindustan Times, 17 January 1991.

Gupta, O.P. 1984. Appraisal document on studies for the use of saline water in the command areas of irrigation project, Haryana, India. p. 11-19.

GWRDC. 1986. Performance of GWRDC Tubewells : 1985. Gujarat Groundwater Resources Development Corporation (GWRDC), Ahmedabad.

Hagan, R.M., H.R. Haise, and T.W. Edminster. 1967. Irrigation of agricultural lands. ASA, Wisconsin, USA, 1180 p.

Heichel, G.H. 1987. *In* J.F. Power (ed.) The role of legumes in conservation tillage systems, p. 29-35, Soil Conser. Soc. Amer., Ankeny, Iowa.

Hill, M.N., J.A. Chandler, and R.B. Highton. 1977. *In* E.B. Worthington (ed.) Arid land irrigation in developing countries : Environmental problems and effects, p. 307-316, Pergamon Press, Oxford.

Hocking, D.L. 1985. Geog. J. 15 : 54-57.

Hotes, F.L. and E.A. Pearson. 1977. *In* E.B. Worthington (ed.) Arid land irrigation in developing countries : Environmental problems and effects, p. 127-158, Pergamon Press, Oxford.

Howard, C. 1991. Span 32(3) : 28-31.

Jensen, N.F. 1978. Science 201 : 317-320.

Johnson, R. 1989. *In* V. Ballabh and T. Shah (eds.). Efficiency and equity in groundwater use and management. 53 p., Institute of Rural Management, Anand 388001, India.

Kassas, M. 1977. *In* E.B. Worthington (ed.) Arid land irrigation in developing countries : Environmental problems and effects, p. 335-340. Pergamon Press, Oxford.

Kilkenny, M.R. 1984. M.S. thesis, Univ. of Minn., St. Paul.

Kramer, R.A. 1989. Workshop on efficiency and equity in groundwater use and management held at Institute of Rural Management Anand, Gujarate, 30 Jan.-1 Feb. 1989.

Mahodaya, M.M. 1983. Bhagirathi 30(4) : October 1989.

Michael, A.M. 1983. Irrigation theory and practice. Vikas Publishers Pvt. Ltd., New Delhi.

Moss, D.N. 1976. *In* R.H. Burris and C.C. Black, Jr. (eds.). CO_2 metabolism and plant productivity. Univ. Press, Baltimore.

National Research Council. 1988. National Academy Press, Washington, D.C.

O'Mara, G.T. 1984. The World Bank Staff working paper No. 707. World Bank, Washington, D.C.

Orlovsky, N.S. and A.M. Durdyiev. 1978. Problemy Osvoenia Pustyn. 4 : 70-76.

Orlovsky, N.S. and Z.M. Utina. 1977. Problemy Osvoenia Pustyn. 5 : 3-12.

Postel, S.L. 1984. Worldwatch Institute, Paper 62, p.17.

Postel, S.L. 1989. Worldwatch Institute. Paper 93, p. 17.

Power, J.F. 1987. Amer. J. Alternative Agric. 2(2) : 69-73.

Rajagopalan, V. 1988. IFPRI-TNAU planning workshop on policy related issues in Indian irrigation held at Ooty. 26-28 April 1988.

Sawant, S.D. (ed.). 1985. *In* Indian agricultural development since independence. Oxford & IBH Publishing Co. Pvt. Ltd., New Delhi.

Schrader, L.E. 1980. *In* Moving up the yield curve : advances and obstacles. ASA special publication no. 39 : 25-43.

Shah, T. 1985. ODI Irrigation Management Network Paper No. 110, Overseas Development Institute, London.

Shah, T. 1988. Managing conjuctive water use in canal commands : Lessons from the analysis of the Mahi right bank canal, Gujarat. 24 p.

Shankar, Vinod. 1985 a. *In* S.D. Singh (ed.) Irrigated agriculture in arid areas. p. 142-172. WAPCOS, New Delhi.

Shankar, Vinod. 1985 b. Symp. Env. Raj., University of Jodhpur, Jodhpur, India.

Shankarnarayan, K.A. 1985. *In* S.D. Singh (ed.) Irrigated agriculture in arid areas, p. 180-189, WAPCOS, New Delhi.

Shankarnarayan, K.A. and Vinod Shankar. 1985. *In* S.D. Singh (ed.) Irrigated agriculture in arid areas, p. 202-209, WAPCOS, New Delhi.

Singh, S.D. 1974. CAZRI (Jodhpur) Annual Rep. 1974. p. 150.

Singh, S.D. 1984. *In* Proceedings Indo-US workshop on arid zone research, p. 485-492, 9-14 January 1984, Jodhpur. Published by DST, Govt. of India, New Delhi.

Singh, S.D. 1985. J. agric. Sci. 55 : 104-107.

Singh, S.D. and H.S. Mann. 1979. CAZRI (Jodhpur) Res. Bull. 1, 88 p.

Singh, S.D., R.C. Bhandari, and R.K. Aggarwal. 1985. J. agric. Sci. 55 : 274-278.

Singh, S.D., Y.V. Singh, and R.C. Bhandari. 1989. Canadian J. Plant Sci. 69 : 991-999.

Singh, Y.V. 1989. Ph.D. thesis, University of Jodhpur, Jodhpur India.

Skogerboe, G.V. and J.P.J. Law. 1971. Office of Res. and Monitoring, U.S. Environmental Projection Agency.

Shrader, W.D. and R.D. Voss. 1980. Crops and Soils Magazine 7 : 15-18.

Tanwar, B.S. 1984. Appraisal document on studies for the use of saline water in the command areas of irrigation projects, p. 20- 27, Haryana, India.

Ved, M. 1991. The Hindustan Times, 18 August 1991.

Voss, R.D. and W.D. Shrader. 1984. *In* D.F. Bezdicek and J.F. Power (eds.) Organic farming : Current technology and its role in a sustainable agriculture. Special publication no. 46, p. 61-68. ASA, CSSA, SSSA.

Worthington, E.B. 1977. Arid land irrigation in developing countries : Environmental problems and effects, Pergamon Press, Oxford.

Yaron, B., E. Danfors, and Y. Vaddia. 1973. Arid zone irrigation. Vol. 5, 434 p; Springer-Verlag, Berlin, New York.

Zonn, I.S. 1982 a. Natural resources potential of arid and semi arid regions and principal indicators of desertification. *In* I.P. Gerasimov (ed.) Integrated regional schemes to combat desertification - Guidelines, p. 1-29, Centre for International Projects of GKNT, Moscow, 88 p.

Zonn, I.S. 1982 b. *In* A.G. Babaev (ed.) Combating desertification in USSR : Problems and experience, p. 75-90, Centre for International Projects of GKNT, Moscow, 120 p.

11

Arid Land Irrigation And Desertification

Vladimir S. Zaletaev

I. INTRODUCTION

Irrigation is one of the main ways to develop the arid lands. The almost all ancient civilizations of the globe : Harappa and Mohenjo-Daro in India, the civilizations of Mesopotamia and the ancient Egypt of the 5th millennium B.C., Peru and Mexico, the valley of Yangtze in China, and the Geocsur oasis of the 4th millennium B.C. in the rivers delta of Tedgen in the Central Asia, irrigated their crop fields through irrigation canals. Since then irrigation technology has made a great progress, but the essence of the problem of interaction of irrigated agriculture with the nature of arid regions remains unresolved. While it is true that the life cannot develop without water and the soil cannot realize its potential productivity, it is also well known that the arrival of water in the arid territories causes complete transformations in the whole natural complex. These reforms culminate in cultural oasis and the unusual high biological productivity of artificial agricultural complex ultimately desertify these lands.

Nevertheless the mankind cannot do without irrigation, and also without the knowledge of the laws of the interactions between water and soil, water and salt, water and vegetation, and between all other components of natural complex. Nowadays the area of irrigated lands in the world is about 222 million hectares (World Resources), and in the 1975 it was up to 223 million hectares, according to FAO statistics.

In Asia, there is 156.5 million hectares (m. ha) of irrigated lands including 20.5 m. ha in the former USSR, 11 m. ha in Africa, 16 m. ha in Europe, 27.5 m. ha in the North and Central America, 8.5 m. ha in the South America, and 2 m.ha in the Australia and Oceania. From 1700 km^3 up to 2500 km^3 of water is being borrowed to the irrigation and the irretrievable consumption is about 1500 km^3. The arid regions of the world are provided with less resources of surface water that is why the 20% of river water is being borrowed for the irrigation, and in the Central Asia (former USSR) it is up to 50%. With the increase in the spatial redistribution of the moist area, the ecological contrast between new irrigated lands and neighbouring arid lands with natural desert ecosystems arises. Specific ecosystems in transitional zones - ecotones are forming. The rolė of the latter in the development of biotic complex, in the preservation of biological diversity as well as their role in the possibility of the development of negative processes such as secondary soil salinization and the distribution of ruderal species of plants and harmful species of animals is undoubtedly significant.

The complex of ecological changes caused by irrigation has its natural development. In the beginning these are the changes of ecosystems of hydromorphic series, but after the accumulation of a large number of negative changes, which level the economic-valuable effect of irrigation, the degradation processes in the land ecosystems develop, which are accompanied by sharp decline in biological productivity. Further irrigation becomes economically inexpedient. Then the hydromorphic ecosystems begin to transform to the automorphic state and evolve according to the type of zonal or ecotone desert ecosystems. This phenomenon is known as "the irrigation desertification of lands".

Thus problem is not so simple. We can reveal the series of irrigative desertification has its straight, secondary, indirect, quick, and remote ecological consequences, resulting from the interaction between the nature of arid zones, the water and the complex of anthropogenic factors.

It will be pertinent to examine here those changes of natural systems that desertify a territory, anthropogenically, through some examples of degradation of irrigated lands and its biotic complexes in different regions of the world. One of the most recent such example is of a large ecological crisis caused by erroneous strategy of irrigation in the basin of the Aral Sea.

Finally, the problems of ecological destabilisation will be examined in the light of some common laws of dynamics and peculiarities of the ecosystems development upon spatial desertification. The knowledge of these laws can be useful for the prognosis of ecological changes in the regions of irrigation.

II. DESERTIFICATION AND ITS ECOLOGICAL SENSE

Ever since the term "desertification" was suggested by Aubreville in 1949, it has been defined in various ways. But in all the cases desertification or results of this process were connected with decrease or loss of bioproductivity of some territory in the arid, semiarid, and subhumid regions. Desertification is, in fact, degradation of lands from normal condition to complete destruction due to anthropogenic activity (Glantz and Orolovsky, 1983).

According to Dregne (1978), desertification is the impoverishment of terrestrial ecosystems under the impact of man. It is the process of deterioration in those ecosystems that can be measured by reduced productivity of desirable plants, undesirable alterations in the biomass, and the diversity of the micro and macro fauna and flora, accelerated soil deterioration, and increased hazards for human occupancy (Dregne, 1978). He pointed out that under desertification following processes take place: the impoverishment or the overwhelming destruction of plant cover, water erosion and deflation, crust-formation, condensation of ground, and decrease in fertility of soils, their salinization and waterlogging.

From the view of modern ecology relating to ecosystem, most of existing definitions may be clubbed with those processes that are developing in the primitive ecosystems and influencing the structure and transformation of biogeocenotic and landscape cover. It is the way that helps to understand mechanisms and biogeocenotic sense of desertification, especially of irrigation desertification.

Desertification includes many processes that cause varying levels of transformation of natural systems, the rate of development, duration and summaries of ecological effects. All these processes have degradational trend, i.e., decreasing biological diversity and productivity of biological communities, accelerated short cycles of successive changes in an ecosystem that is predominantly primitive in structure and composition. It causes ecotonization of biocenotic cover, aridization of conditions and hydromorphization. Mechanisms of degradation are especially effective at the biocenotic level of organization and the slightest imbalance of system may activate processes leading to self-destruction. New species that invade in such a system serve as agents of future functional destabilization of the natural system.

The loss of structure-functional completeness of biogeocenotic system is the main ecological aspect of desertification phenomenon. The loss of species diversity potential and bio-productivity capacities is the first indicator of beginning of desertification. The disfunction of self-rebuilding mechanism leads to the destruction of stability of the whole biogeocenotic cover consisting of a little number of nearby genesis, and jointly functioning biogeocenoses that organize the system of mutually functional supplements. Destruction of the system of supplements in the group of interlinked biogeocenoses makes the process of desertification irreversible.

Thus desertification needs to be understood at three levels: 1) the study of disruption of the structure-functional organization of ecosystem, 2) the prognose of changing of inner and interlinks (with determination of destabilized agent-factor) in the system, and 3) identifying ways of meliorations and regeneration of ecosystem.

Overall character of disturbance of arid environment causes transformations that appear as new ecological phenomena. Some of important ecological phenomena are: 1) self-broadening of centres of destability of the biogeocenotic cover, 2) vast spreading of "short-time" processes, that are specific group of quickly going and stopping successions, 3) development of peculiar quick forms of desertification such as "gradual desertification" taking place at the dried bottom of the Aral Sea, 4) ecotonisation of biogeocenotic and landscape cover, 5) increasing of ecological differentiation, mosaic and contrasts of environment, 6) strengthening of processes of animal and plant settling and permanent mixing of species in communities, and 7) development of "geographo-ecological transmissions", the fact of sending impulse to change environment from the region of heavy anthropogenic transformations to another region where direct anthropogenic effects do not occur (Zaletaev, 1989a).

All known transformations in the nature of arid region ultimately destabilize it. It is easy to suppose that the development of natural systems in these conditions may go by new ways, and new fundamental laws will be opened in the evolution of organic world (Zaletaev, 1976, 1989b, 1991).

Since desertification is a phasic process, the knowledge of ecological laws of stages of replacement and the duration of each stage is extremely important for practice for the prognosis of the changes in environment.

And the phasic character of desertification makes it difficult to determine it scientifically, because each stage of development has its own characteristic features and when the stage change each other the replacement of active factor occurs and the mechanism of transformation of intersystems relations changes.

These processes are operative desertification due to irrigation, i.e., additional amount of water. This alters hydrologic and hydrogeologic regimes of territory and brings fresh certain changes in moistening either more high or more low.

Irrigational desertification is thus degradation of natural systems (ecosystems and landscapes), with the reduction of biologic production and deterioration of its quality as a result of deep rebuilding of intrasystem and intersystems relations, and in the end as the alteration of composition and the structure of ecosystems that leads to new, in essence, way of further development of natural systems of region that has undergone as a result of the direct or indirect influence of irrigation.

Irrigational desertification alters the regional inputs to natural systems as well as geochemical balance of biosphere and reduces the production potential of soils, though initially high biological production characterises newly irrigated desert lands.

Agrolandscapes occupy 18 million km^2 land area on earth, of which 2.2 million km^2 (2%) is irrigated. More than 12% of all irrigational lands in the world fall out of the sphere of agricultural usage and are now completely desertified. Various estimates reveal that 50 to 60% of all irrigational lands experience desertification of varying degrees (Babaev and Orlovsky, 1982; Kovda, 1984; Rozanov, 1984; Kharin et al., 1986). Thus irrigational desertification is vast in occurrence and spread and has become an important problem of world's society especially International Geosphere-Biosphere Programme "Global Change".

Irrigational desertification is manifested in development of variety of secondary processes : 1) "direct irrigational desertification", especially on the marginal patches of irrigational territories; 2) "secondary attendant irrigational desertification", developing on the adjoining territories of irrigated traces by altering hydrologic regime or microclimatic conditions, and 3) "desertification transmissional-irrigational" that develops on the territories distant from the region of irrigation, but linked with the spatial redistribution of water resources.

The last form of irrigational desertification became widely spread in the second half of former USSR, as a result of sharp increase of the scale of modern irrigation. While the technology of irrigation remained at the low level, the ecological strategy of irrigation has also been considerably erroneous. This is exemplified desertification in the Aral Sea region of Soviet Central Asia: on the drained bottom of the Aral Sea, on the desert pastures of north-west regions of the kyzyl-kum desert, and in the eastern part of plato Usturt. Here, salts get precipitated and are blown by wind from the drained and deserted bottom of the Aral. There has been slow steady degradation of natural lands and expansion of the area of different negative changes by the pattern of desertification. These negative changes include loss of species variety and biodiversity, decline in volume and quality of bioproduction in areas that earlier had high production. The waters of great rivers the Amu Darya and Syr Darya that earlier used to come to these lands are now used almost completely for the irrigation mostly in the middle currents of these rivers.

III. STAGES AND MECHANISM OF IRRIGATIONAL DESERTIFICATION

It is important to understand that different components of ecosystems - soil, plant cover, animals, and others-change at different rates, have different levels of conservation and their roles vary from stage to stage in irrigational desertification. At the initial stage of irrigation, an increase in humidity of territory as well as direct influence of irrigated agriculture cause rapid changes in the composition of plant cover and animal communities. These changes must not be called desertification. Nevertheless, the process of desertification begins as the lands in arid regions are supplied with water. Increase in plant communities of marginal habitats of irrigated fields have been observed in the role of native mesophytes and freatophytes: *Alhagi persarum, A. cannescens, A. pseudoalchagy, A. camelorum, Karelinia casjoica, Atriplex tatarica, A. rubrum, Chenopodium album, Erianthus purpurescens, Artemisia scoparia*. They often form true thicket. Abundance rises, and habit becomes higher in such arid species as *Salsola paulseni, Astragalus vilosissima*, and youth of *Haloxylon aphyllum*. New territories are invaded and settled by hydrophytes and mesohydrophytes such as *Phragmites australis, Aster tripolium, Paramierorinchus* sp., *Polygonum hydropiper, Cynodon dactilon*. Then ruderal species appear: *Erygeron canadensis* and even *Solanum nigrum*. At the first stage of irrigation the *Glicophyte* species are of primary importance, due to large supplies of water and leaching of soils. Changes in soil begin from its moisture content that leads to geochemical changes, but essential modifications of structure do not occur. This first stage of natural change in the system under irrigation may be named as "stage of primary hydrogenic phytocenotic regroup" or "phytocenotic stage".

With rising groundwater table, close to irrigational channels some arid species of rodents, such as *Rhombomis opimus* which build deep holes, begin to leave these places. The depth of holes reaches 2.5 m and even deeper, and humidity in these holes hardly rises. On the other hand in these conditions expansion of *Nesokia indica* and *Mus musculus* occurs. Number of these rodents quickly increases. These changes in plant cover and animals at the initial stage of irrigation are reversible and ecosystem can return if water supply is stopped.

The next stage of ecosystem-changes concern with increasing salinity of groundwaters and soils. The salt from the deep horizons rises to the surface, changes the properties of the soil, its fertility, edaphotops of wild and cultured plants and habitats of animals. Soil transformation begins qualitative changes that later on constitute the desertification process at the second stage.

Irrigation with solonetz like phenomenon appeared leading to the formation of the surface crusts and structural changes. These phenomena hamper cultivation of cotton (*Gossypium hirsutum* L.) and lucerne (*Medicago sativa* L.), and in marginal habitats cause specific halohydrogenic successions of natural plants in desert or steppe.

Essential change of soil properties takes place when the adsorbed sodium is 20% of bases, and the content of colloids in clays reaches 46%. In such cases it is impossible to see saline fields with the crop, stable to residual salts and solonetz-like phenomenon and early ripening, that it makes possible in south region of the former USSR (Azerbaijan and Middle Asia) to undertake deep ploughing in July. The latter promotes coagulation of colloids.

Winter crops, especially barley (*Hordeum vulgare* L.) does not permit to leach salt-lands properly because of early dates of sowing (not later than October in these regions), when large supplies of water cannot be provided. Thus winter crops themselves promote indirectly resalinization of soils, leading to desertification.

Changes in the soil properties and structure mark the beginning of the next stage of transformation, when changes become irreversible. It is determined by total change of the biotic system-compound-replacement of the type of ecosystem organization and structure, destruction of old functional links making new ones between system components and a change of geochemical process character, that means quality modification of the ecosystem.

Irrigation can alter rate of soil formation that may result in the formation of another soil profile, different from the original natural soil profile, which was the result of long period of interaction with lithogenic base, climate, precipitation, surface and subsurface runoff, evaporation, transpiration, complex of biogenic and geochemical processes, etc. Formation of original soils under irrigation is called methapedogenesis. It is characterized by higher rate of soil formation. In this process alluvial soils of the river plain and arid deltas can be transformed into solontchaks, an indicator of desertification. Under periodically flooding alluvial soils can transfer into gley soils. Cultivation of rice (*Oryza sativa* L.) in humid tropic regions may cause formation of plictites with latosalts in condense horizon. In tropical monsoon climate using irrigation waters, rich in silica may lead to formation of grumisoils. In arid climate such soils after stopping of irrigation become the base of development of the whole complex process of desertification.

The way in which the irrigational methapedogenesis influences plant cover in arid lands needs to be known. Before the salinization of soil suppresses the plant growth, it leads to appearance of dwarf forms and creepers, such as *Phragmites australis*. The height of this species remains 10-20 cm instead of the normal 300-500 cm. Sometimes teratogenic changes occur in some individuals. Bioproductivity, quantity and quality of seeds of plant communities are reduced. The dates of periods of ontogenetic cycle are disturbed, reproductivity processes are delayed and sometimes total cessation occur in some local populations. Numerous examples of this phenomenon can be observed in the Aral region, in the Syr Daria and the Amu Daria deltas and on the dried bottom of the Aral Sea.

Suppression of the organism and plant growth may occur either as a result of decrease of osmotic potential, or due to effluence of redundant concentration of specific ions. Moreover for some plant species subject to an equal concentration of ions, sodium chloride salinization may be more dangerous than calcium sulfate one.

When soil salinization increases, plant transpiration and the water uptake by the root system is decreased. This phenomenon creates effect of "physiological drought" that causes suppression of plant.

However, some plant species behave differently on salinization of habitats and hence result in differentiation of plant communities. Consequently a series of halohydrogenic successions begin by the way of environmental changes from stage of innersoil processes, change of soil structure, halogeochemical and microbiological modifications and become irreversible.

Thus, desertification penetrates into the most secret links and functions of plant species, influences their growth, development, vitality and productivity, on the one hand, and tends to decrease composition and structure of plant communities, their bio-productivity, fullness, projective coverage, on the other. Reduction in albedo changes the temperature regime. Deflation is intensified. As a result, the whole system is converted into new dynamic condition. Its structure, inner and inter-systems links must undergo destruction and then reconstruction at the new background in the new conditions, changed by irrigation and its numerous consequences.

Since severe desertification destabilises the homeostasis, it is important to prevent desertification at the first phytocenotic stage, so that it is arrested at localized level and search methods of adaptation of human activity to changed environment.

Amelioration of ecological situation on desertified lands is difficult such as that in the Sachel where after catastrophic drought in the early 70s, the humid climate cycles had been formed. But, in spite of special efforts, desertification process has not stopped. However, regions of irrigated agriculture show two contrary processes : 1) desalinisation of lands and freshening of water in channel influencing the zones, and 2) progressive salinization of soil and increase of groundwater mineralization in the marginal parts of oasis. That causes transfer to solontchak desert with highly mineralized groundwaters. These salt affected lands often occur along the irrigated fields. The territory differs sharply in degree of salinization, for example, Kura-Arax lowland in Azerbaijan, at the periphery of the Buchara oasis and near Hauskhan reservoir in Turkmania, in the basins of Chinob and Jhelum rivers in Pakistan, in Iraq (Kovda, 1984), and in many other regions of modern and ancient irrigation.

Thus, different centres of irrigational desertification in one region are in different stages of development, which need to be tackled individually.

Irrigational desertification occur in consecutive stages, some of them may overlap each other in the time scale of their origin.

1. The Beginning Stage

This relates to primary disturbances in the functional regime of ecosystems at the beginning of irrigation, e.g., input of additional amounts of water, the rising of the groundwater level, disturbance of the integrity of phytocoenotic cover and technogenic disturbances of soil surface during the irrigation related constructions.

2. Hydrogenous-Phytocenotic Stage

Under this comes the beginning of regrouping of species in the phytocenotic composition, caused not only by technogenic disturbances of the environment, but also by the direct influence of the changes in the hydrological and hydrogeological regimes of the territory.

3. Stage of Metapodegenesis

This stage relates to changes of water and salt regime of the soil, transformation of initial natural soils into the original semi-natural soils of irrigated lands, the changes in the character of the soil-formative process that leads to the rebuilding of the soil structure and the halogeochemical processes. These result in the alteration of soil section, the progressive increase of salt in soil and groundwater, the worsening of the structure of soil thickness, especially in the root, floating, gleying and compacting of soils. All these processes are accompanied by the loss of soil fertility.

4. Stage of Halogenous Physiological Drought of Plants

The origin of this effect is determined by the soil salinity that disturbs the water balance of the plant species as a result of which the symptoms emerge similar to those of drought. The growth and development of plants retard, the biological and seminal productivity reduces in some cases, the plants do not reach fertility stage; they wither or dry up. In essence, this desertification of vegetation becomes apparent on the level of species. The stage is marked by the active changes in the microbiological processes.

5. Stage of Halogenous Successions of Vegetation

The basis of such successions is the differential tolerance of species to salinity of the environment. At this stage the species and biological diversity of vegetation tend to reduce.

6. Stage of Zoo-complexes Successions

It is followed (and superimposed in the time scale) by the stage of halogenous successions of vegetation and based on the direct influence of soil salinity on the animals as well as the indirect influence of changes in plants composition, which are the feeding objects. The species diversity can increase in some conditions, but the main process leads to total reduction in species and biological diversity.

7. Stage of Active Biogenocenotic Rebuildings

Increased dynamic stage of biotic complexes promotes new species of the animal and plant communities. Reciprocal links of biota and soil become stronger on the basis of the biogenic environmental formative processes. At this stage the main process is the total rebuilding of the structural organization of ecosystem, created by the previous stages of soil-geochemical process development and by the changes in biocomplex. The active ecotonization of natural system, i.e., the forming of ecotonal communities and young ecotonal systems takes place. This stage is rather longer and is the step of development of dissipative processes, of the search of new forms of organization of natural systems, which has been thoroughly changed under the irrigation influence and has become seminatural. The laws of ecologically destabilized environment become apparent.

8. Stage of Comparative Stabilization of Structural and Functional Rebuildings of Desertifying Ecosystems

This new ecosystem of different species composition slowly moves to attain stability. As the water-heat-soil regime stabilizes, the halo-geochemical processes slow down. Succession activity in the biotic cover descends. Hydrogenous and halogeochemical processes give way to the processes of edaphic and climatic aridisation that adds force to the ecological effect of desertification. Transition of ecosystems from the sphere of hydromorphic processes into the sphere of automorphic processes is possible.

9. Stage of Sparing Reversion of Ecosystems

The first stage of the natural systems transformations due to irrigation is reconstruction of biogeocenoses composition and structure. Complexities result from diversity of the lithoedaphic conditions under irrigation as well as the broad dispassion of fluctuations of the hydrological and hydrogeological regimes of the territory and the influence of technogenic and agro-technological process of environment. On irrigated territories a network of large and small canals is built and the fields are periodically flooded. Natural vegetation is destroyed and protective afforestation belts are being set up. Moistening of the territory and percolation from the irrigation system result in waterlogging, meadowing, and secondary salinization. In the majority of newly-irrigated districts the soil-vegetative cover is extremely diversified.

On newly-reclaimed desert ranges in the Ili, Zaravshan, Tedzhen, Murgab, the Amu Darya, the Syr Darya river basins and in the Fergana valley, as well as under the conditions prevailing in the semidesert and steppe zones the diversity of vegetation changes become apparent even over small areas.

In the irrigation zone, on ranges adjoining to canals, collectors, discharge lakes, and water reservoirs there also occur ecotone-type biocomplexes in different stages of their development. Total bioproduct of such young ecosystems is occasionally quite large but its nutrient value for cattle is low. The cases in point are altered biocomplexes in the Akkum sands (Hamangan region, Uzbek SSR), on sand ridges lying in the irrigation zone of the Tedzhen delta. Under these conditions moisture-and salt-loving species become wide-spread. They find their way into communities of desert shrubs: *Calligonum caput-medusae, C.setorum, C.turkestanicum, Haloxzlen ammodendron, Salsola richteri, S.paletckiana,* with the soil covering compose of *Heliotropium, Horaninivia, Bromus tectorum, Schismus arabicus* and other psammophytes. In these communities there occur groupings of *Alchagi,Karelinia, Atriplex, Zigophillum, Euphorbia, Salsola, Suaeda, Aeluroous letoralis, Phragmites communis,* etc. In their turn, desert plants, above all those from sub-crown communities of *Haloxylon ammodendron* and desert shrubs, appear in the fields where they acquire the role of weeds. Such are *Senecio subdentatus, Malcomia, Coelpinia, Lappula, Salsola paulseni, Hordem leprinum,* etc. Quite typical is the spreading of introduced weed species *Erigeron canadensis,* particularly along Kara- Kum canal route.

Significant changes are taking place in animal population within the canal influence zones. The relationship of vertebrate species is changing in the communities and replacement of dominant species is taking place. The causes lie, both in direct effect of the rising level of groundwaters and the consequent flooding of the holes of rodents, for instance *Rhombomis opimus,* and the indirect effect due to the appearance of new biotypes, new covers, new accessible foods and watering sites, thus promoting the concentration of certain fowl and mammal species.

All these processes are finding their classical expression in the zone of ecosystem transformations on the Kara-Kum canal. Within 0.5-1 km wide strip from canal bed, where groundwaters have come close to the surface, and in sandy massifs wherein inter-ridge depressions occur at 50-100-120 cm depth, the colonies of *Rhombomis opimus* become less since this large rodent, active during daytime, makes holes up to 2 to 2.4 m deep. Its place in biocenoses near the canal is being taken by *Meriones libycus.* The population of this smaller night-time rodent is rapidly growing, because it prefers more humid conditions and its holes penetrate to 60-100 cm, rarely 120 cm deep, not exposed to flooding. Furthermore, the main soil horizon inhabited by the animals is 20-30 cm deep. This is the way the dominant rodent species are being replaced near the canal. The third rodent species, *Meriones meridianus,* preserves its more stable position in the biocenocis because its holes are relatively not deep 80 100 150 cm and the soil horizon predominantly inhabited by the rodent reaches to 30-50 cm depth. The colonies of *Meriones meridianus* are observed to be moving upslope on sand ridges.

Within the zone of abundant development of hydrophilic vegetation in depressions, near the banks of filtration lakes and on irrigated cultured lands, the numbers of *Nesokia indica* are observed to be growing. This rodent, harmful to gardens and melon fields is inhabiting most humid areas where rising groundwaters occur at the 20-100 cm depth.

Wesokia indica, Lobivanellus indicus aigneri, *Insecta* and *Arachnoidea* are observed to have been settling along the Kara-Kum canal.

The effect exerted by large man-made water reservoirs in the arid zone, by water receptacles and lakes of drainage and discharge waters on adjoining territory is similar to that of canals. It moistens only the coastal strip and the rise of groundwater level within a relatively narrow strip of land. Their softening effect on the microclimate is insignificant on account of the high general dryness of the air. On the banks of water reservoirs, and along canals, specific vegetation groups of mixed composition are being formed with the predominance of initial xerophytic communities to mesophytic and finally to hygrophytic ones.

Water reservoirs and filtration lakes have become breeding sites for ducks (*Anatinae*). chiefly *Anas strepera, Netta rufina,* and for fowls of the ralline family- *Ralliformes,* mostly *Fulica atra, Callinula chloropus, Perzana parva, Porzana pusilla.*

On the lakes in the Kelief uzboi, on the Khauzan and other water reservoirs in winter time there appear thousands of migrating water fowl staying here for winter.

Of quite a different character is the dynamic of land ecosystems in those areas where the changes in the hydrological regime are accompanied by the deepening of groundwater level and the prevalence of drying processes, thus leading to aridization and desertification. These phenomena are particularly manifest in the coastal zones of the north-eastern part of the Caspian, on the Aral Sea coasts, and in the Syr Darya and Amu Darya deltas.

Succession processes associated with the receding Aral Sea level are particularly manifest on the alluvial delta plain of Syr Darya and Amu Darya rivers and on the eastern and southern sea shores, which are involved in the process of desertification nowadays.

IV. SHORT REVIEW OF LAND SALINIZATION AND DESERTIFICATION IN REGIONS OF IRRIGATION

Details about secondary salinization in relation to irrigation are given in Chapter 10. Here in this Chapter, a brief account of salinization is given; in relation to irrigational desertification.

Irrigational desertification can be well revealed on the lands of ancient irrigation, especially in Mesopotamia. Here in the inter-river region of the Tigris and the Euphrates, irrigation had been practised since the 5-4 millennium B.C. and in the 3rd millennium B.C. During the period of Shumerian culture big main canals were built. To the 3rd millennium B.C., belonged the facts of secondary land salinisation, that could be the sign of desertification. So the whole period of postirrigational transformation of environment that has the nature of desertification on the territory of Lower Mesopotamia counts up to 4.5 thousand of years. Nowadays the irrigational desertification become catastrophically large in some irrigational desertification become catastrophically large in some places. In Iraq more than 70% of irrigated lands are subjected to desertification. The fallow system of agriculture called in Iraq "niru-nir", adopted for centuries is one of the main causes of the irrigational desertification. Predisposition of natural systems to the process of desertification, e.g., the low natural soil permeability to water; the slow flow of groundwater as the result of plain conditions; the plain relief; and low natural drainage of territory, etc. have promoted desertification. As a result, the accumulation of salts in the soil and deflation has developed.

As a result of accumulation of vast amounts of silt deposits on the fields, which come with the irrigation waters, the soil physical properties are deteriorated. When these fields are abandoned and the soil is not irrigated it becomes takyr with foliated and platy structure, especially in the upper layers. The porous surface crust is forming often. Intrasoil processes include the destruction of humus, microaggregates, dispersing of clay enrichment by magnesiale silicates, and increased migration of clay mass in the condition (Alekseeva and Kurakova, 1989). As a result of these processes, the irrigated fields are transformed into the takyr deserts.

Subsequent development of these territories can follow two ways passing through several stages.

In one case when the irrigational influence on takyrs and takyrlike soils are absent for a long time, the desalinization of the upper layers of soil begins and on the patches along the abandoned canals, where there are more loose technogenic soils, the processes of deflation intensify. Then begins the stage of deflation of takyrs and the gradual increase in sand cover. As a result of this, the secondary sandy deserts are formed, which also pass through several typical stages beginning with the primary barchan sands to the semistabilized and stabilized by vegetation and ridge sands. The long range of soils transformation ranges from the hydromorphic irrigational to the automorphic desert soils. Destruction and deflation of soil leads to complete transformation of the soil profile. And the forming of aeolian sand massifs marks the beginning of sandy desert. The latest stage is hardly connected with irrigation because the signs of irrigation are buried under the sands.

The other way of territory transformation takes place on the marginal patches of the irrigated massifs and on those lands that have been irrigated and where the saline drainage waters flow. In these conditions the irrigated landscapes have passed through a series of transformation stages have been turned into the salt desert with specific extremely poor primitive biocomplexes.

The process of desertification in the Lower Mesopotamia nowadays is characterized by stirring up. This process is promoted by the flow down of drainage waters with the average salinity of $12\ g\ L^{-1}$ into the rivers and the rise of the groundwater level on the irrigated patches as well as on the neighbourng places. About 25

Photo 11.1. Cosmic photo of Lower Mesopotamia taken in 1978. From Glushko and Maslennikova (1987).

thousand hectares of irrigated lands fall out of agricultural usage annually because of irrigational desertification Nowadays we can consider the Lower Mesopotamia according tot he opinion of some geographers and ecologists (Alekseeva and Kurakova, 1989) as "salt plain which is sprinkled with sand and takyr deserts, swamps. Small oasis in 10-15% of the area is used for agriculture".

Photo 11.1 is a cosmic photo of lower Mesopotamia taken from orbital station "Salut" in 1978 (scale 1:2400000). Various forms of irrigational desertification different by age are reflected on this photo. The interpretation of this photo revealed the vast areas of ancient desertification (dark grey colour in the central part of interstream area of the Tigris and the Euphrates) as well as lands of new irrigational salinization and desertification of lands (white colour) (Glushko and Maslennicova, 1987).

On the Syrian territory, according to the same workers, in the valley of the Eupharates over 50% of irrigated lands are salt affected and bogged up and in the valley of the Habur they are over 65%. Sulfate and chloride-sulfate salinization prevail.

On the Arabian peninsula the secondary salinization of ancient irrigated lands is mainly visible in the oasis of E1-Harge, E1 Haric, in the lowland E1-Hasa from E1 Catifa to Hufufa and Gabrine, in Higace, on the Red Sea coast, along vadies in the Rubel-Hali, where chloride salinization is of high degree.

There are highly salt affected soils in the irrigated tracts in intermountain depressions of Iran's highland on the Husistan alluvial plain and on the seaside depressions that have soils, ground and unconfined groundwaters with high natural salinity. Chloride and in some places sulfate-chloride salinization prevails. The 25 millions of hectares are involved in the secondary salinization and irrigational desertification. In Afghanistan, the secondary salinization is known on the irrigated lands in river basin of the Gilmend and the Herirud (Ajekseeva dn Kurakova, 1989).

Shown in Fig. 11.1 is the scheme of interpretation of processes of anthropogenic desertification of Lower Mesopotamia (Scale 1:1000000) (Glushko and Maslennikova, 1987).

Groups of processes are :

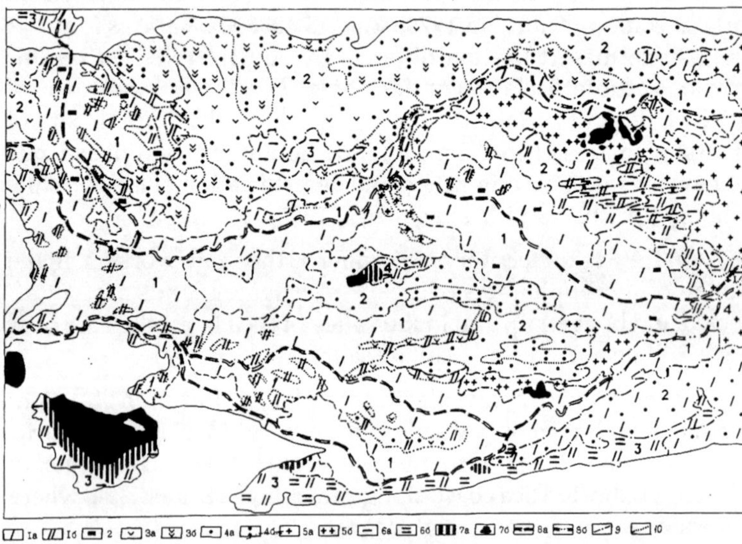

Fig. 11.1. Scheme of interpretation of processes of anthropogenic desertification of Lower Mesopotamia (scale 1:1 000 000). From Glushko and Maslennikova (1987).

1 - irrigational desertification;

2 - desertification on the lands of former irrigation;

3 - flooding by drainage waters; and

4 - drying of natural lake-bog complexes.

Processes :

1a - secondary salinization;

1b - formation of solonchaks;

2 - irrigational erosion;

3a - spreading of takyrs;

3b - development of takyrs;

4a - spreading of sand massifs;

4b - development of wandering sands;

5a - drying of lakes and bogs;

5b - formation of solonchaks depressions;

6a - bogging up;

6b - development of bogs;

7a - spreading of lakes;

7b - development of lakes;

8a - reduction of rivers flow;

8b - lack of flow in summer time for irrigation;

9 - boundaries of groups of processes; and

10 - boundaries of process.

Another area of irrigation desertification includes south of Asia, many regions of India and Pakistan. There secondary salinization of soil occurs in ancient irrigated lands and in regions of new irrigation practices (Table 11.1). Now in India, up to 90% of total water use is being borrowed to the irrigation, and the area of the secondary salinization in the new land of irrigation practices is near 940,000 ha.

Table 11.1. Extent of secondary salinization on the lands of new irrigational development in India. From Yadav (1986).

Irrigation project	State	Area (1000 ha)	Share of the whole irrigated land (%)
Sriramsagar	Andhra Pradesh	1.0	0.8
Tungabhadra	Andhra Pradesh and Karnataka	24.5	6.8
Gandak	Bihar and Uttar Pradesh	400.0	40.0
Sara Sahiak	Uttar Pradesh	30.0	4.7
Ramganga	Uttar Pradesh	352.4	59.6
Mahi-Cadana	Gujarat and Rajasthan	35.8	7.3
Tava	Madhya Pradesh	6.6	3.8
Chambal	Madhya Pradesh and Rajasthan	40.0	8.2
Rajasthan canal	Rajasthan	29.1	5.4
Total		939.4	

Secondary salinization in South Asia tends to increase from coastal regions to inner parts. Especially salinization is spread where evaporation is 5-10 times more than precipitation, and where plane surface of territory does not promote drainage. In these arid conditions, area of secondary land salinization may be more than the area of irrigated lands. This phenomenon has aggravated after irrigation systems construction in the valleys of Hang, Pendgable, Indostan (Nira river). on alluvial terraces of Decana. Construction of Sukkur dam on the river Ind at the beginning of the 20th century promoted salinization on vast areas in Sind. According to existing estimates about 7 million ha of India's lands in 9 states have undergone to salinization (Mathur, 1986) and desertification to some degree. In the Indian state of Uttar Pradesh, 1.3 million ha, in Gujarat - 1.2 m.ha (6.4%); in West Bengal - 0.85 m.ha (9.7%); in Rajasthan – 0.73 m.ha (2.15); in Punjab - 0.69 m.ha

(13.5%); in Maharashtra - 0.54 m.ha (1.8%); in Haryana - 0.53 m.ha (12.1%); in Orissa - 0.41 m.ha (2.6%); and in Karnataka - 0.41 m.ha (2.1%) salinity. Soil salinization is very high in between the Yamuna and the Gomati region (Mathur, 1986).

Figure 11.2 shows the sketchmap of interpretation of natural-anthropogenic processes and destructive forms in the landscapes of Central Rajasthan. The processes linked with the development of irrigation and other forms of anthropogenic influence are reflected. Deforestation, salinization, and desertification are the most typical (Glushko and Shestakov, 1989).

Data on reduction of irrigation area due to soil salinization in five countries widely using irrigation are given in Table 11.2 (Postel, 1990).

Table 11.2. Area of land salinized due to irrigation in five countries of the world (Postel, 1990).

Country	Area of wasted lands due to soil salinization, m. ha	Area of wasted lands %
India	20.0	36
China	7.0	15
U.S.A.	5.2	27
Pakistan	3.2	20
USSR (former)	2.5	12
Total	37.9	24

Supposing that situation with the soil salinization in irrigated region in other countries is similar to five countries, the total wasted area is up to 60.2 m.ha; 24% of irrigated lands (Postel, 1990).

This value can look very high, but in some countries soil salinization in irrigated regions is near the critical level. In Egypt about 50% of all irrigated lands have undergone to salinization which has led to annual decrease in crop productivity. In Mexico decrease in crop productivity is equal to 1 million tons of grain per year.

Scheme of interpretation of natural anthropogenic processes and destructive forms in landscapes of the Central Rajasthan shown in Fig. 11.2 is given below :

Leading natural anthropogenic processes : A - deforestation; Б – deforestation and savanisation; B - savanisation; Г – deforestation, savanisation and desertification; D – savanisation and desertification; E - desertification.

Particular natural-anthropogenic processes :

1 - speeded up erosion; 2 - speeded up deflation; 3 - disturbance of structure and decline of soil fertility; 4 - reduction of concentrated flow; 5- stopping of concentrated flow; 6 - sedimentation; 7 - lowering of the groundwater level; 8 - rising of the groundwater level; 9 - groundwater mineralisation; 10 - salinization of soils; 11 - reduction of biomass and lowering of biological fertility of vegetation; 12 - pasture digression.

Destructive forms :

13 - ravine-balka network, badlands; 14 - semistabilised sands; 15 - wandering sands; 16 - cattle destruction of vegetation around the wells and on cattle routes; 17 - dried river valleys; 18 - boundaries of groups of particular processes (a) and leading processes (b). Direction of expansion; 19 - deforestation; 20 - savanisation; 21 - desertification.

V. IRRIGATIONAL DESERTIFICATION IN CENTRAL ASIA AND KAZAKHATAN

(example of ecological crisis in Aral Sea region)

The modern and most interesting example of different forms of the irrigational desertification are observed in the Aral Sea region, which includes the alluvial delta plains of the Syr Darya and the Amu Darya rivers, sea shores, the postaquatic lands, and the neighbouring territories of Kyzyl-Kum, Usturt deserts, and other arid sand tracts, located in the Aral Sea basin.

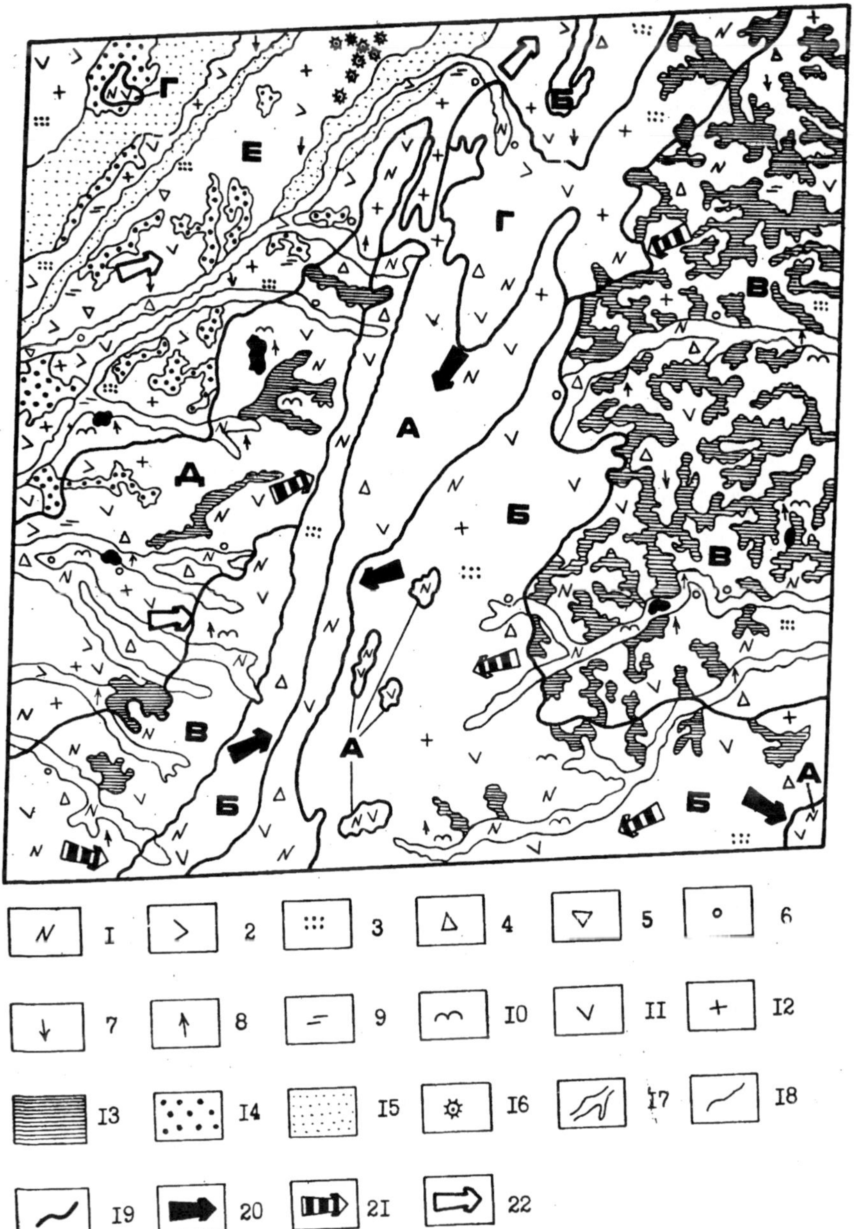

Fig. 11.2. Scheme of interpretation of natural-anthropogenic processes and distructive forms in landscape of the Central Rajasthan. From Glushko and Shestakov (1989).

1. The Syr Darya Delta

Changes in the ecological situation in the Syr Darya delta set in somewhat earlier than on the Amu Darya. As early as the 1960s the flow abstraction coefficient in the Syr Darya basin approached a unity (Rubinova, 1983) and in the 1970s it exceeded a unity because of the multiple use of return water.

A further increase of the river flow abstraction went together with a decrease of the inflow of water to the delta head. In the latter half of the 1970s the Syr Darya delta received on the average only 4% of resources from the catchment area annually. Spring floods on the Syr Darya stopped already in 1971. Simultaneously, there occurred a rapid fall in the level of the Aral Sea leading to the progressive desertification in the delta plain. By 1978, 114000 ha of alluvialmeadow soils had desertified and turned into solonchaks, 532000 ha of bog and meadow-bog soil had dried up, 31000 ha had desertified, and 55000 ha had turned into solonchak. The 732000 ha land virtually been withdrawn from agricultural use. Productivity of grass-forb and forb meadows decreased by 3 times.

Lands rich in hydrophilic vegetation that had served as hay and forage lands were devastated over the greater part of the delta area. The total forage reserve reduced from 1200000 to 500000 t. The area of tugai vegetation reduced by twice and the preserved-tugais strongly degraded and desertified. The desertic animal and plant species began to expand into the delta and the flood-plain and to form impoverished ecosystems of the desertic type.

Bird colonies relocated beyond the lower reaches of the Syr Darya. The gaming potential of bog lands was lost. In the 1950s hunting in this area yielded from 70000 to 230000 muskrat pelts. The 1968 brought 9000 pelts; this figure fell to 72 pelts in 1978 (at present muskrat is not hunted).

River water mineralization in the vicinity of the city of Kzlinsk reached 1.5-3.0 g L^{-1} as early as the 1970s. In recent years, there has been an increase in the river water pollution with salts and toxic chemical below Kzyl-Orda due to the growing discharge of drainage water from irrigated lands of the Kzyl-Orda region. As compared with 1982 the DDT content increased from 1 to 5 conversional MACs reaching 40 MACs (Zaletaev and Novikota, 1991).

The process of irrigation desertification is well expressed not only in the delta of the Syr Daria, but also in the valley of this river, in its low and middle flow.

On the cosmic photo (Photo 11.2) received from the station "cosmos" in 1985 (scale 1 : 1 000 000), the valley of the Syr Daria is well seen to the flow of Chardaria reservoir.

Against the dark background of the fields of cotton and rice patches flooded with water, the thick mosaic of white spots stands out. These spots correspond to the patches of salted lands that in previous times were irrigated fields, but nowadays they are desertified and abandoned.

The left lower corner of this photo shows us the picture of desertification on the extensive territory that occurred due to heavy salinization because of flow of saline water drainage from the fields. The most salinized and desertified are dune valleys of wedge-formed sand massif stretched from the north-east to the south-west.

2. The Amu Darya Delta

Reduction in the runoff, due to construction of the Takhia Tash and Tyuya Muyun water reservoirs in 1959 and by excessive water used to expand the area of irrigated lands led to the termination of floods and significant environmental changes in delta and flood-plain. The area of plant communities that can exist in the flood plain habitats, only on the condition of annual floods reduced from 35% to 8%, while the area taken up by communities characteristic of periodically watered habitats above the flood- plain reduced from 40% to 20%. At the same time the number of associations restricted to salinized and desertified areas of the flood plain and delta grew from 25% to 75% (Zaletaev and Novikova, 1991).

Vast reed growths used to cover over 300000 ha in the Amu Darya delta. They served as habitats for a rich complex of birds including 21 waterfowl species, 11 of which are nesting, Khiva peasant, 9 species of Ciconiiformes including *Platalea leucorodia,* several birds of prey species and a score of passer. Boars were abundant and jungle cats were common. Until the 1930s Central Asian tiger had habitated in the flooded areas. By the mid-1980s the reed areas reduced by about 5 times. Green mass productivity,

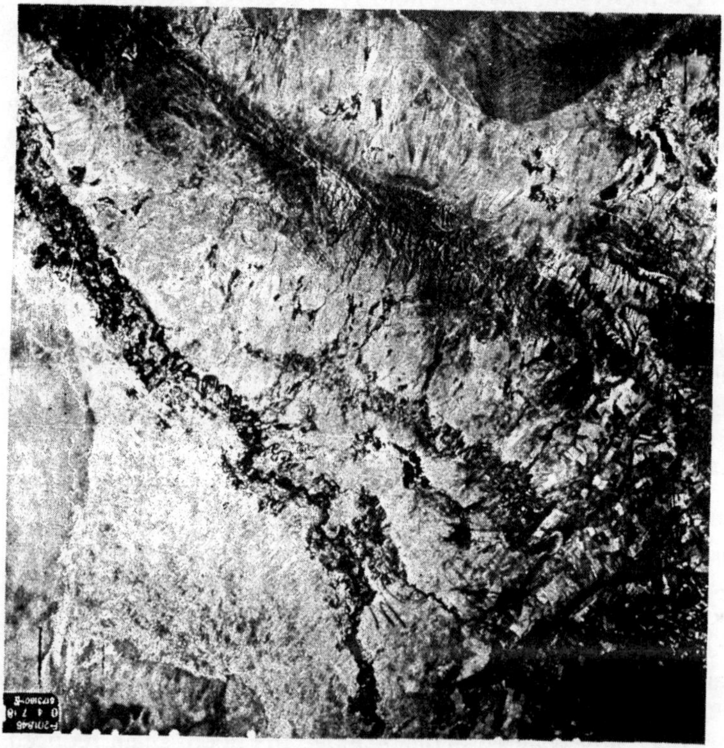

Photo 11.2. Cosmic photo made from the station "Cosmos" in 1985 (scale 1:1 000 000) showing the valley of the Syr Daria.

which in the 1960s reached 30000-40000 kg ha^{-1} in the 1980s decreased to 4000-12000 kg ha^{-1}; the animal population also decreased in numbers. For instance the abundance of boars had gone down 6 times by the 1980s.

The good idea of geoecological situation in the Amu Darya delta gives the map (Alekseeva et al., 1991) that shows the development of parallel processes of forming of agroirrigational landscapes, irrigational desertification, salinization of depressions with increase in level of groundwaters, drying of delta-alluvial plains, and development of postirrigational desertification (Fig.11.3).

Details of Fig. 11.3 are as follows :

Type fo change : IIA - natural - anthropogenic, A - anthropogenic. Character of change : I - formation of agroirrigative landscapes; II - irrigational desertification; III - pasture digression; IV - salinization and filling depressions with water; V - drying of lake bog complexes; VI - drying of delta alluvial plains; VII - postirrigational desertification.

Categories of landscapes : optimized : heavily + 3, weakly + 2;

compensated - 2; depressed : weakly - 3, heavily - 4; disturbed : weakly - 5 heavily - 6.

Desertification of delta leads to the replacement of the purely *Phragmites australis* of high productivity by mixed communities of low reeds with annual saltwort, often with the participation of rare tamarisk shrubs (*Tamarix ramosissima, T.laxa, T.hispida*) and, occasionally, *Karelinia caspica* and camel's thorn - *Alchagi persarum*. Productivity of such communities does not exceed 400-500 kg ha^{-1}.

By 1985 there had remained no more than 70000 ha of natural flooded *Phragmites* hay lands in the Amu DArya delta. Such economically valuable lands could be preserved only through flooding of dried or drying lakes and interbed depressions for no less than two months annually (Bakhiev et al., 1987).

Aridization of the Amu Darya delta dn flood-plain and poor natural conservation and forestry management became a threat to the existence of the unique tugai ecosystems. Whereas in the early 1930s tugai forests took up 300000 ha down stream, their area by

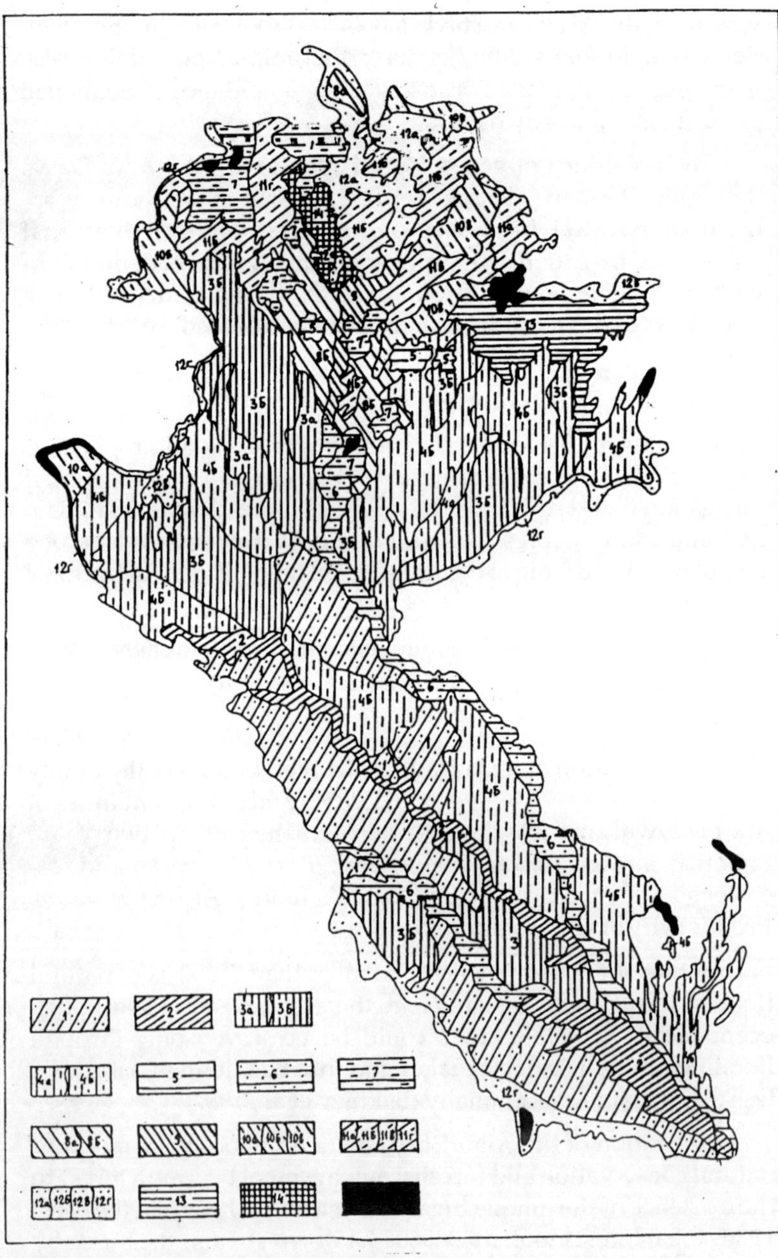

Fig. 11.3. The map of geoecological situation in the Amu Darya delta (scale 1:1 000 000). From Alekseeva et al. (1991).

1986 was no greater than 33000 ha. In the 50 years the area of tugai forests downstream and in the delta reduced by nearly 90%. Degradation and devastation of tugais was especially fast in the late 1970s and the early 1980s. Of 79000 ha of tugal forests recorded in 1978 only 33000 ha remained by 1986. In the course of 8 years the area of tugai forests, especially in the delta, reduced by twice. It follows that the rate of devastation of tugai ecosystems set the "record" of a kind and made up 5778 ha per year. To date, tugai forests constitute under 3% of the total areal of the state forest reserve in the Kara-Kalpak territory. Many remaining stands continue to dry and unauthorized cutting is still going on in some of them.

Drying and degradation of tugai ecosystems are the signs of the developing process of desertification involving the flood-lands and river deltas. The Amu Darya tugai forests are the richest among the tugai ecosystems of Asia in their floristic composition featuring 576 higher plant species, 29 species being endemics of Central Asia, with many relics of the Tertiary.

At present 54 plant species of tugais are vanishing. They represent 46 genera of 26 families, and disappear due to the deep deformation of the environment. By the late 1980s *Nymphaea alba, N.luteum, Aldrovanda, Elytrigia repens* and ferns were not encountered in the Amu Darya delta.

The number of birds, especially birds of prey, habitating in tugais also have gone down. Wild Khangul-Bukhara deer, the typical tugai inhabitant, can hardly be encountered nowadays. In 1969 there were 13 deers in Narazym - tugai (below the city of Kerka) and several individual in the Darganatin tugai. Today this deer can be found only in the lower reaches of the Amu Darya in the Badai-tugai reserve, where 10 individuals are kept semi-loose or in open cages.

Shrub communities appearing in oasis on irrigated lands along canals and irrigation ditches, despite their outward resemblance to tugais, cannot become genuine analogues to the latter because of the ecological differences in habitats, determining their inevitable floristic poverty and simplified structure. Floodings ensure a certain humid regime, fertile warp, and soil washing constitute the necessary conditions for the formation and resistivity of the typical biotic complex of the shrub in Amu Darya tugai ecosystems. This is why special emphasis should be given to preservation of tugai ecosystems in their natural state and a suitable programme of action should be developed. The existing regimes of passive protection of tugais in reserves are perfectly inadequate for the preservation of tugai and their recovery in the former boundaries of the Nazarkhan tugai, Nurumtubek tugai, Badai tugai, Seyid, and many other tugais located along the Amu Darya bed and dried delta arms.

Tugai plant communities may be preserved only given that the landscape-ecological conditions close to natural ones are recreated. To recover and maintain tugai complexes it is necessary to carry ameliorative engineering measures that would ensure drainage of flood water, flooding natural levees for the period of 10-12 days, and interbed depressions for no less than 30 days. According to expert appraisals, in order to preserve some tugais in the Amu Darya delta the flood water passed should amount for the Ak Darya to 1.5 m.km, for the Kipchak bed to 0.5 cu.km and for the Kazakh Darya to 0.5 cu.km. Besides, irrigation of natural meadows and hay lands in the coastal part of the delta over the area of 0.2-0.5 million ha will require up to 5 cu.km of water.

These meadows should be provided with a drainage system to prevent secondary salinisation.

3.The Postaquatic Land

The postaquatic land, i.e., the bottom of water bodies that emerged to the surface temporary or forever and still lacks the genuine soil cover but is already settled by plants and animals forming both short-lived communities and biotic complexes of the developing land ecosystems. The dried bottom of the Aral Sea (26000 km^2 in 1989) is a unique testing ground for the development of syngenesis of vegetation and settlement of animals and the formation of ecosystems against the background of desiccation and salinization of ecosystems against the background of desiccation of the new land, desertified as if from the inception. The dried bottom of Aral Sea is the area of "transmissional-irrigational desertification".

On the cosmic photograph (Photo 11.3) taken from the station. "Cosmos" in 1985 (scale 1:1 000 000) one can see well the development of process of desertification in the seaside part of the Amu Darya delta, provoked by unregulated irrigation, regulation of rivers flow by dams that caused the drying and desertification of Avandelta. On this photo we can also see the desertifying seashore of the Aral and patches of dried sea bottom turned into the desertified postaquatic land. This is the new desert originated owing to mistaken strategy of irrigation. It is the combination of extensive solonchaks (with crust and pudgy, from which the winds carry salt to the neighbouring oasis and desertified pastures) and comparatively small massifs of wandering sands without vegetation.

The dry belt of the Aral Sea bottom hosts 147 plant species, only 12 of which (4.2%) are not found on the native coast (Demeeva, 1990). The species spread on the coast thus also settle on the bottom. The genera, whose species are the most widely spread, include 15 species of *Calligonum,* 13 species of *Suaeda,* 12 species of *Atriplex,* 9 species of *Tamarix,* 9 species of *Astragalus,* 8 species of *Artemisia,* 6 species of *Salsola,* and 6 species of *Climacoptera.* The greater part of these species are haloxerophystes and halomesophytes. Syngenesis of plant communities on the dried bottom of the Aral is discussed in the rather copious literature (Vukhrer, 1979; Demeeva, 1990; Kurochkina, 1979. In the generalized form such processes are represented in scheme 1.

Scheme 1. Generalized scheme of plant communities succession on the dried bottom of the Aral Sea.

A. The Salicornia enropaea association. Sum of salts : 1.2-7.0%

Salinization type : Sodium chloride. Water mineralization : 10 $g\ L^{-1}$

B. The Aeluropus litoralis association. Sum of salts : 0.6-4.0%

Salinization type : sodium chloride. Water mineralization : 1-34 $g\ L^{-1}$.

(i) The Halostachys association. Sum of salts : 1.3-9.2%. Salinization type : sodium chloride. Water mineralization : 6-60 $g\ L^{-1}$

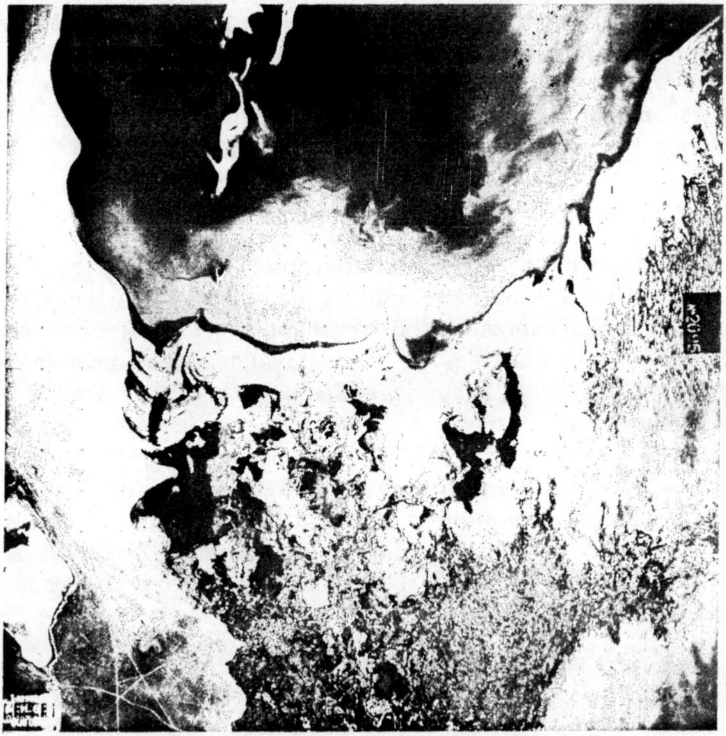

Photo 11.3. Cosmic photograph taken from the station "Cosmos" in 1985 (scale 1:1 000 000) showing the development of process of desertification in the seaside part of the Amu Darya.

(ii) The Halocnemum association. Sum of salts: 2-25%. Salinization type : sodium chloride. Water mineralization : 11-155 g L^{-1}

It should be noted that the coastal part of the delta and the areas of the drying sea bottom contacting with it evolve towards the formation of a solonchak desert. The expanding areas of Halocnemum with the salt-bearing pseudosandy substrates, which may become salt storm foci, are especially unfavourable.

The specifics of the biotic complexes formation on the postaquatic land of the Aral manifests themselves in fast intermittent successions of vegetation and sequential waves in settlement of small mammals on the drying bottom, which induced and reveals the mechanism of the "stepwise desertification" process characteristic of the territories where environmental changes occur at an increased rate (Zaletaev, 1989a).

The Aral's postaquatic land is settled with up to 25 mammal species, 15 bird species, about 10 reptile species, and a relatively small number of invertebrates, mostly of the halophilic species.

Fast changes in the groundwater level, the formation of the desiccation horizon and shifts in salinity of the postaquatic pseudosoil give rise to the "short-lived processes", i.e., fast exogenous successions in the plant cover or fast restructurings in the fauna complex, which interrupts and breaks down the developing cenogenesis into some autonomous states, i.e., "development steps". As a result, the whole process of the formation of an ecosystem, that has already started, does not reach completion. This peculiar class of natural processes occurring on the Aral Sea's postaquatic land underlies fast saltory desertification called "stepwise". Such desertification is defined by a number of autonomous stages of the formation and disruption of biogeocomplexes of the ecotonic character on the former sea bottom. As waters of the sea and firths retreat, the new land in the Amu Darya delta is populated by annual saltworts, by pest rats and house mice and, of birds, by Glareola nordmanni and Denauthe isabellina and lesser Alaudidae, Gryllotalpidae and Dermaptera are common among insects. At this initial stage the depth of groundwater does not exceed 20-40 cm and the surface of microsinks is wet. The phytomass productivity reaches 4000-6000 kg ha^{-1} in the air-dry form, there are living reed rhizomes, and rodents are provided with feed. At the second stage groundwater drops below 50-70 cm, there appears the surface desiccation horizon, annual saltworts die off, although Bassia is common; Atriplex dimorphostegias, Dipus sagitta and Ellobius begin to propagate on the postaquatic land and Nesoria indica in the driest areas leaves its colonies. Gerbillinae begin to populate the new land including Meriones mecidianus (on sandy areas with depositions of Zostera) and, later, M.libica and Rhombomys opimus; Calandrella ruffescens and C.cinerea begin to nest; Eurasian hamsters, Cricetulus migratorius and lesser jerboas Alactoga elater appear. Gerbils use abandoned colonies and passages of pest rats. At the third stage with groundwater dropping below 1- 1.2 m, the complexity of the desiccation horizon increasing, and there appearing crust and swell salinization, Atriplex dies off to give place to Salsola paulensii and the rarified cover of Halocucumm and Cl.aralas, Tamarix hispida as a halophyte replaces Tiromasissima and T. laxa. Due to a sharp depletion of the forage base gerbils leave the colonies, leaving behind a very low number of Cricetulus and Muscusculus that store seeds. At the fourth stage groundwater goes down below 2.5-5 m, sweating salinization stops and aeolian pulverization salinization is enhanced. The plant cover nearly completely degrades, the only survivors being single Climacoptera plants. Vertebrates disappear and only spiders feeding on air plankton remain. The outcome of the described stepwise process is the formation of the extreme forms of salt deserts that, ecologically, represent biotic badland. The total duration of all steps is 17-25 years.

The desertifying postaquatic land of the Aral has a negative economic potential since at present and in the years to come (until the year 2000 and beyond) it cannot be used as rangeland or for irrigated agriculture. Only small areas, mostly on the Ust Urt coast, are amenable to phytoamelioration whereas the emergence of the salt-dust deflation foci threatens the nearby rangelands and oasis lands with gradual salinization within the radius of up to 300-500 km.

Aridization of the territory and anthropogenic desertification are pivotal to the present day ecological processes and determine the dominant trend in the development of the natural environment in the Aral region as a whole. Until recently, however, prevailing processes were those of hydromorphization of the natural conditions. They took place at the time of the radical extension of irrigated areas and the creation of dense irrigation network and water reservoirs causing underflooding, rises in the water table, and hydromorphism of natural ecosystems across vast territories. These processes determined the ecologically fundamental readjustment of the structural-functional organization of natural systems.

The whole of the Aral region may probably be assessed as the area of the emergence of direct and intensive impacts of the "anthropogenic ecocide" type and as that of destruction of the human environment, as well as the area of modern and, in the foreseeable future, inescapable secondary unpremeditated environment changes developing as ecotonic rearrangements. At present the area affected by ecotonization of ecosystems in the Aral region is likely to exceed half of the total region's processes typical for the ecologically destabilized natural environment (Zaletaev, 1989a), which is regarded as a specific stage in state development, which essentially consists in the restructuring of the former, and the formation of a new system of the structural functional organization of the biogeocenotic and landscape cover.

VI. THE CONCEPTION OF ECOLOGICALLY DESTABILIZED ENVIRONMENT AS THE SCIENTIFIC BASE FOR INVESTIGATIONS AND PROGNOSIS OF ANTHROPOGENIC DESERTIFICATION

Ecologically destabilized environment (EDE) is the special stage in the development of environment that is characterized : a) by the disturbance of evolutionary formed scheme of organization of natural systems, b) by the emergence of new forms of ecosystem dynamics, c) by the increased rates of their rebuilding, and d) by removal of the natural rhythms.

During the evolution the biosphere repeatedly passed through the phase of EDE. In this phase natural systems loose the certain initial forms and number of components and simultaneously the new links and new forms of systems are originated. EDE is the phase of active rebuildings in the nature. In the epoch of scientific and technical progress the source of EDE is the large scale anthropogenic activity. In connection with this fact we must mark out the common fundamental laws of ecosystems rebuildings in the rapidly changing conditions, as well as those specific peculiarities of the modern phase of anthropogenic destabilized environment that are conditioned by the combination of natural, agrotechno- natural, irrigation, and technical systems and by phenomenon of active management of their creation and functioning. The problem of management of correlation of named categories of natural and artificial systems being very important has not been yet appraised properly, investigated, and prognosed. The disturbance of the integrity of ecosystems and their functioning ruins the zonal specific structures of ecosystems for time being or forever and changes the correlations of discreteness and continuity of environment. On the place of zonal ecosystems there appear the new ecosystems with ecotonal character, unstable or stable, which are notable for universal simplified structure and universalism of interecosystems links.

Biogeocenoses cover acquires the increased ecological contrastness together with common impoverishment of species composition of organisms and shortening of biological diversity. The patches of secondary life optimum for the limited number of species originate and disturbances in ecosystems take place. These changes are negative, but they could be useful for the development of new forms of agricultural use. The common picture of environmental transformation is complicated by disproportions of ecosystems destructions under the similar anthropogenic influences and by lag of rates of new formations from the destructive processes, which sharply lowers the quality and density of biotic cover and promotes the transport of substance: geochemical migration and the settling of organisms. On the arid lands, especially in connection with irrigation, the started process of ecotonisation of biogeocenoses cover becomes the biosphere process of global scale and significance. In arid zones it has involved more than one-third of the irrigated area, and together with the territory of desertification far in the desert and including the increase of the areas of technical systems and technogenic disturbances of environment soon it will involve more than one-half of the territories of arid lands. Increase of the carrying out of dust to the atmosphere more than 1.5 times and the change of albedo of underlying surface is connected with this circumstances and that is the essential transformation of input of arid territories into the global processes.

Investigations of the mechanisms of ecosystems reactions to the disturbance of the scheme of their organization, that leads finally to the desertifications of lands, the determination of the rates of ecosystems econtonisation, the scale and specificity of irrigational desertification in different regions of active anthropogenic influence, and the natural regions and the analysis of different stages of ecosystems desertification in relation to the effective application of meliorative influence for the regeneration of natural organization and productive characteristic of ecosystems must become the scientific base of the new strategy of nature management, which takes into account the peculiarities of the processes of ecologically destabilized environment of the irrigated regions. The tasks arise in connection with this management of the quality of EDE in connection with its role in the global change of the biosphere; the prognose of the rates of spontaneous development of EDE; the creation of the theory of irrigated ecology; definition of the place and amount of the area of "reserve regions" where the development of EDE must be strictly regulated. The special task is to study the transformation of energetic and bioproduction potential of EDE in its development and estimation of energetic expenditure for the compensative activity of the mankind.

REFERENCES

Alekseeva, N.N. and L.I. Kurakova. 1989. Ecological aspects of he hydromelioration. Chapter 1.3, "Asia". p. 56-96. Nature protection and development of nature resources. Vol. 23, Moscow, VINITI, 226 p.

Alekseeva, N.N., E.B. Glushko, T.J. Kondratieva, and A.V. Ptichnikov. 1991. Complex geoecological analysis of the zones of ecological disasters as per space photographs. "Reports of higher educational institutions. Geodesy and aerial photography". Moscow. Moscow Institute for Engineers of Geodesy, aerial photograph and carto- grashy. No. 1, p. 40-48.

Babaev, A.G. and N.S. Orolovsky. 1982. The processes of land desertifincation and their prevention in irrigated regions of USSR. "News of Vsesoyznogo Geographical Society". No.5, p. 369- 375.

Bachyev, A.G., N.M. Novikova, and M.E. Shenkareva. 1987. The change in the economical importance of plant resources while reducing the irrigation of delta towards Karakatnak. "Water Resources". No.2, p. 167-169.

Demeeva, L.A. 1990. Flora dn vegetation on sea shores and on Aral Sea dry bottom (Autoreferat of Desertification). Alma Ata. 27.p.

Dregne, H.E. 1978. Soil and Water Cons. 33: 11-14.

Glantz, M.H. and N.S. Orolovsky. 1983. Desertification : A review of the concept. Des. Con. Bull." December 1983.

Glushko, E.V. and I.N. Maslennikova. 1987. The study of formation of modern landscapes of lower Mesopotamia as per space photography. "Research of Earth Through Space". No 1, p. 51-58.

Glushko, E.V. and A.S. Shestakov. 1989. Th experience in studying the naturally occurring anthropogenic processes in Rajasthan (India) as per space photographs. "Geography and Natural Resources". No. 4, Novosibarsk.

Kharin, N.G. et al. 1986. Modern condition and forecasting of desertification in arid regions of USSR. "Problems of Desert Development".

Kovda, V.A. 1984. Problems of the desertification and irrigated soil's salinity prevention. Moscow : "Kolos" Publishaus, 304 p.

Kurochkina, L.J. 1979. "Problems of Desert Development". No. 3, p. 9-17.

Postel, Sandra. 1990. Saving water for agriculture. "State of world". p. 39-58. Red. by Lester R. Brown and Alan Darning and others. W.W. Norton and Company, New York-London,253 p.

Rozanov, B.G. 1984. Fundamentals of the study about environment. Publication of Moscow University.

Vukhrer, V.V. 1979. Primary overgrowth of drying shores of Aral- "Problems of Desert Development". No.2, p. 66-70.

Yadav, J.S.P. 1986. J. Indian Soc. Soil Sci. 34 : 660-674.

Zaletaev, V.S. 1989a. Ecologically destabilized environment (Ecosystems of arid zones under changing hydrological regime). Moscow, "Nauka". 150 p.

Zaletaev, V.S. 1989b. Irrigation and desertification: ecological contradictions. "Problems of Desert Development". No. 3, Allerton Press, Inc. p. 51-54.

Zaletaev, V.S. 1991. Peculiarities of the modern development and degradation of natural systems of deserts. "Problems of Desert Development". No.3-4, Ashkhabad, "Ylym". p. 56-64.

Zaletaev, V.S. and N.M. Novikova. 1991. Transformations of biota in Aral Sea region during 40 years. J. of Academy of Sciences, USSR, No. 10. Moscow. "Nauka".